PRAGMATIC MATHEMATICS
FOR SCIENTISTS AND ENGINEERS

PRAGMATIC MATHEMATICS
FOR SCIENTISTS AND ENGINEERS

Alexander Godunov
John A Adam

Old Dominion University, USA

World Scientific

NEW JERSEY · LONDON · SINGAPORE · BEIJING · SHANGHAI · HONG KONG · TAIPEI · CHENNAI · TOKYO

Published by

World Scientific Publishing Co. Pte. Ltd.
5 Toh Tuck Link, Singapore 596224
USA office: 27 Warren Street, Suite 401-402, Hackensack, NJ 07601
UK office: 57 Shelton Street, Covent Garden, London WC2H 9HE

Library of Congress Cataloging-in-Publication Data
Names: Godunov, Alexander, author. | Adam, John A, author.
Title: Pragmatic mathematics for scientists and engineers /
 Alexander Godunov, John A Adam.
Description: Hackensack, NJ : World Scientific, 2025. | Includes index.
Identifiers: LCCN 2024011365 | ISBN 9789811291333 (hardcover) |
 ISBN 9789811291340 (ebook) | ISBN 9789811291357 (ebook other)
Subjects: LCSH: Mathematics--Textbooks.
Classification: LCC QA39.3 .G63 2025 | DDC 510--dc23/eng20240720
LC record available at https://lccn.loc.gov/2024011365

British Library Cataloguing-in-Publication Data
A catalogue record for this book is available from the British Library.

Reprinted 2025 (in paperback edition)
ISBN 9789819815265 (pbk)

Copyright © 2025 by World Scientific Publishing Co. Pte. Ltd.

All rights reserved. This book, or parts thereof, may not be reproduced in any form or by any means, electronic or mechanical, including photocopying, recording or any information storage and retrieval system now known or to be invented, without written permission from the publisher.

For photocopying of material in this volume, please pay a copying fee through the Copyright Clearance Center, Inc., 222 Rosewood Drive, Danvers, MA 01923, USA. In this case permission to photocopy is not required from the publisher.

For any available supplementary material, please visit
https://www.worldscientific.com/worldscibooks/10.1142/13790#t=suppl

Desk Editors: Kannan Krishnan/Angeline Husni/Nijia Liu

Typeset by Stallion Press
Email: enquiries@stallionpress.com

Dedicated to my wife Claudia, daughter Dasha, son Leo, and my beautiful grandchildren
Alexander Godunov

Dedicated to my children and grandchildren: Rachel, Matthew and Lindsay, and John Mark, Caroline, Emma, Felix, Alice, Evie and Ethan
John A. Adam

Preface

> Mathematics is the language in which God has written the universe.
>
> — Galileo Galilei

The essence of this book aligns with its title, *Pragmatic Mathematics for Science and Engineering*. Pragmatic mathematics emphasizes the practical, applied use of mathematical concepts and methods. It prioritizes the utility of mathematics in addressing practical challenges rather than delving into purely theoretical or abstract considerations.

This is a textbook on basic to intermediate mathematics for undergraduate students majoring in the physical sciences and engineering. Many chapters, including the chapters on Green's functions, calculus of variations, and functions of a complex variable, are also well suited for graduate classes. This book can also be useful for researchers as a mathematical refresher for their professional work.

This book covers most of the common subjects in mathematics for physical sciences and engineering students, and it can be used for both one-semester and two-semester classes (see Contents).

This book provides readers with a fundamental understanding of underlying principles, using derivations based more on mathematical intuition rather than exposing them to multiple theorems, proofs, and lemmas. Simultaneously, for every chapter, this book offers highly relevant examples with detailed solutions and explanations. Therefore, this book is a practical one so that students/readers would be able to apply their knowledge to solving real problems in the physical sciences. This book has "end-of-chapter" exercises (with answers) that can easily be used by instructors for assignments. Most chapters also provide options for students to apply their freshly acquired knowledge for solving the "end-of-chapter" problems. Hence, if a student is capable of completing the reading and solving most problems in this textbook by the end of one or two semesters, then he or she will be proficient enough in using mathematics for other classes and indeed, for beginning professional research.

This book is not a replacement for advanced calculus textbooks but rather a guidebook to what mathematics is most relevant to the physical sciences and engineering. This book's size renders it easily "digestible" by students.

In conclusion, this book can be readily adapted for upper-level undergraduate and graduate classes, such as mathematical methods for students of physical sciences, applied mathematics, and engineering majors.

About the Authors

Alexander Godunov is an Associate Professor and University Professor of Physics at Old Dominion University, Norfolk, Virginia. Throughout his career, he has taught a wide range of courses, from introductory physics to mathematical methods in physics and computational physics at both undergraduate and graduate levels. Together with Dr. Andy Klein from Los Alamos National Laboratory, he co-authored the book *Introductory Computational Physics*.

John A. Adam is a University Professor Emeritus and Professor Emeritus of Mathematics & Statistics at Old Dominion University, Norfolk, Virginia. His books include *X and the City: Modeling Aspects of Urban Life, Guesstimation: Solving the World's Problems on the Back of a Cocktail Napkin, Mathematics in Nature: Modeling Patterns in the Natural World, A Mathematical Nature Walk*, and *Rays, Waves, and Scattering: Topics in Classical Mathematical Physics*.

Contents

Preface vii

About the Authors ix

1. Infinite Series 1
   - 1.1. Applications of series 1
   - 1.2. Basic concepts: Infinite sequences and infinite series 1
   - 1.3. Basic properties of infinite series 3
   - 1.4. Series with positive terms 6
   - 1.5. Alternating series 14
   - 1.6. Absolutely convergent series 16
   - 1.7. Series with complex numbers and series of vectors 17
   - 1.8. Series of functions 18
   - 1.9. Exercises and problems 28

2. Complex Numbers 33
   - 2.1. Complex numbers in science 33
   - 2.2. Definitions 33
   - 2.3. Complex numbers in $z = x + iy$ form 37
   - 2.4. Exponential form 40
   - 2.5. Hyperbolic functions 51
   - 2.6. Exercises 54

3. Vectors — 57

- 3.1. Introduction — 57
- 3.2. Vector algebra — 59
- 3.3. Base vectors and unit vectors — 62
- 3.4. Vector algebra with vector components — 68
- 3.5. Scalar and vector products — 69
- 3.6. Vector calculus — 79
- 3.7. Fields — 85
- 3.8. Exercises — 96

4. Matrices — 99

- 4.1. Introduction — 99
- 4.2. Matrices — 101
- 4.3. Determinants — 107
- 4.4. Linear systems of equations — 117
- 4.5. Eigenvalues and eigenvectors — 127
- 4.6. Exercises — 133

5. Partial Differentiation — 137

- 5.1. Complicated real world — 137
- 5.2. The partial derivatives — 138
- 5.3. The total differential and total derivative — 140
- 5.4. Second partial derivatives — 143
- 5.5. Taylor series for functions of two variables — 147
- 5.6. Stationary points of a function of two variables — 150
- 5.7. Stationary points with constrains — 156
- 5.8. Change of variables — 160
- 5.9. Implicit functions — 164
- 5.10. Exercises — 168

6. Line Integrals and Multiple Integrals — 173

- 6.1. Introduction — 173
- 6.2. Line integrals: First kind — 173
- 6.3. Line integrals: Second kind — 178
- 6.4. Application of line integrals — 189
- 6.5. Multiple integrals — 191
- 6.6. Change of variables in multiple integration — 208
- 6.7. Exercises — 214

7. Fourier Series and Transforms — 221

- 7.1. Function space and basis vectors 221
- 7.2. Generalized Fourier series 225
- 7.3. Trigonometric Fourier series 229
- 7.4. Fourier transform 244
- 7.5. Exercises 252

8. First-Order Ordinary Differential Equations — 255

- 8.1. A note on classification of ordinary differential equations 255
- 8.2. First-order ordinary differential equations: Basic concepts 258
- 8.3. Separable-variable equations 259
- 8.4. First-order linear ordinary differential equations 261
- 8.5. Exact equations: Integrating factors 265
- 8.6. Some first-order differential equations in physics 268
- 8.7. Exercises (Math) 275
- 8.8. Physics problems 276

9. Second-Order Linear Differential Equations — 279

- 9.1. General solution of the homogeneous equation . . . 280
- 9.2. General solution of the non-homogeneous equation 284
- 9.3 Homogeneous equations with constant coefficients 287
- 9.4. Non-homogeneous equations with constant coefficients 291
- 9.5. Series solutions 296
- 9.6. Periodic series solutions 305
- 9.7. Boundary value problem 307
- 9.8. WKB(J) approximation 310
- 9.9. Some physics examples 319
- 9.10. Mathematical problems 324
- 9.11. Physics problems 326

10. Green's Function Method — 331

- 10.1. Dirac's delta function 332
- 10.2. Green's function 336

	10.3.	Green's function solving for first-order ODE	339
	10.4.	Green's function for second-order ODE: Initial value problem	343
	10.5.	Green's function for second-order ODE: Boundary value problem	351
	10.6.	Sturm–Liouville problem and Green's function	373
	10.7.	Exercises	377
	10.8.	Problems	379
11.	**Calculus of Variations**	**381**
	11.1.	Introduction	381
	11.2.	From few degrees of freedom to many	383
	11.3.	A necessary condition for an extremum	387
	11.4.	Euler–Lagrange equation	389
	11.5.	Generalizations of the Euler–Lagrange equation	395
	11.6.	Variational problem with constraints	398
	11.7.	A variational problem with variable end points	402
	11.8.	Direct methods for solving variational problems	404
	11.9.	The principle of least action	405
	11.10.	Fermat's principle of least time	407
	11.11.	Exercises and problems	409
12.	**Functions of Complex Variables**	**413**
	12.1.	Introduction	413
	12.2.	Derivatives and the Cauchy–Riemann conditions	416
	12.3.	Integrals	422
	12.4.	Series	434
	12.5.	The residue theorem and its applications	446
	12.6.	Exercises and problems	461

Index   465

Chapter 1

Infinite Series

1.1. Applications of series

Infinite series have multiple applications in science and engineering. Very many problems cannot be solved exactly; however, using series we can either find a solution as an infinite series or find an approximate solution to a problem in hand by keeping as many terms as necessary. Both ordinary and partial differential equations can often be solved by using series. Power series are used to define functions of complex variables. Very many special functions can be represented as series. Also, Fourier series of various kinds are examples with a wide range of applications in physics and technology.

1.2. Basic concepts: Infinite sequences and infinite series

Suppose that an *infinite sequence* is given as a set of numbers

$$u_1, u_2, u_3, \ldots, u_n, \ldots.$$

A sequence may have a limit as $n \to \infty$. For example, the sequence

$$u_n = 1, \frac{1}{4}, \frac{1}{8}, \ldots, \frac{1}{2^n}, \ldots$$

has a limit since $u_n = 1/2^n \to 0$ as $n \to \infty$, but the sequence

$$u_n = 0, 1, 2, 3, 4, \ldots, n, \ldots$$

has no finite limit because $u_n = n \to \infty$ as $n \to \infty$.

Generally, elements of a sequence can be numbers, functions, or any objects, but the order of the elements is important. It is assumed that an algorithm for generating terms corresponding to each n is provided.

A sum of terms of an infinite sequence constitutes an *infinite series* as

$$u_1 + u_2 + \cdots + u_n + \cdots = \sum_{n=1}^{\infty} u_n. \tag{1.1}$$

By taking the sum of the first n terms

$$S_n = \sum_{k=1}^{n} u_k, \tag{1.2}$$

we can create a partial sum of an infinite series. An infinite series can be written as a limit of a partial sum

$$S = \lim_{n \to \infty} S_n. \tag{1.3}$$

If a series has a finite sum S, it is called *convergent*, otherwise if the sum is equal to $\pm\infty$ (or there is no sum at all), it is called *divergent*.

Example 1.1. A simple example of an infinite series is a geometric series, when the ratio of successive terms is a constant. A partial sum of a geometric series is written as

$$S_n = a + ar + ar^2 + \cdots + ar^{n-1} = \sum_{k=0}^{n-1} ar^k,$$

where a is a constant and r is the ratio of successive terms. The sum can be evaluated by considering S_n and rS_n:

$$S_n = a + ar + ar^2 + ar^3 + \cdots + ar^{n-1},$$
$$rS_n = ar + ar^2 + ar^3 + ar^4 + \cdots + ar^{n-1} + ar^n.$$

If we subtract the second equation from the first one, we have

$$(1 - r)S_n = a - ar^n,$$

and finally
$$S_n = \frac{a(1-r^n)}{1-r} \quad \text{(provided } r \text{ is not equal to 1)}.$$
For an infinite series with $abs(r) < 1$, we have the limit
$$S = \lim_{n\to\infty} \frac{a(1-r^n)}{1-r} = \frac{a}{1-r}. \tag{1.4}$$
However, for $|r| \geq 1$, a geometric series either diverges or oscillates (when $r = -1$). ∎

Example 1.2. In approximately 1500 AD, Indian mathematicians Somayaji and Madhava discovered an infinite series representation of π as
$$\frac{\pi}{4} = 1 - \frac{1}{3} + \frac{1}{5} - \frac{1}{7} + \frac{1}{9} + \cdots.$$
Later, it was independently rediscovered by European mathematicians James Gregory, Gottfried Wilhelm Leibniz, and Jakob Bernoulli. ∎

Example 1.3. Another fundamental number, Euler's number e, can also be represented as infinite series, namely
$$e = 1 + \frac{1}{1!} + \frac{1}{2!} + \frac{1}{3!} + \cdots \frac{1}{n!} + \cdots = 1 + \sum_{n=1}^{\infty} \frac{1}{n!}.$$
∎

1.3. Basic properties of infinite series

An infinite series can be written as a sum of two parts,
$$u_1 + u_2 + \cdots + u_n + \cdots = S_n + r_n, \tag{1.5}$$
where S_n is the partial sum (1.2) and r_n is the *reminder*
$$r_n = u_{n+1} + u_{n+2} + \cdots + u_{n+k} + \cdots = \sum_{k=1}^{\infty} u_{n+k}. \tag{1.6}$$
The remainder r_n is itself the sum of an infinite series, obtained from the given series by neglecting its first n terms.

Property 1. If series (1.5) converges, then any of its remainders (1.6) converges, and vice versa: the convergence of the remainder r_n implies the convergence of the original series. Indeed, let us consider a partial sum of terms from $n+1$ to $n+k$:

$$\tilde{S}_{n+k} = u_{n+1} + u_{n+2} + \cdots + u_{n+k}.$$

This partial sum can be be written as

$$\tilde{S}_{n+k} = S_{n+k} - S_n.$$

In the limit $k \to \infty$,

$$\lim_{k \to \infty} \tilde{S}_{n+k} \to r_n \quad \text{and} \quad \lim_{k \to \infty} (S_{n+k} - S_n) \to S - S_n,$$

thus,

$$r_n = S - S_n. \tag{1.7}$$

If S is a convergent, then r_n is a convergent as well since the partial sum S_n is just a number. And vice versa, if r_n is convergent, then S is a convergent too. It also follows that the rejection of a finite number of first terms of a series or an addition of several new terms at the beginning would not change series convergence or divergence.

We can only talk about a sum of an infinite series when it is convergent, in which case the sum S_n of the first n terms gives an approximate value for the sum S of the series. The error involved in this approximation is the reminder r_n. This remainder cannot be found accurately in the majority of the cases so that it is important to know the approximate error due to this remainder.

Property 2. From

$$S = \lim_{n \to \infty} S_n$$

and (1.7) follows that the reminder tends to zero as n increases, namely

$$\lim_{n \to \infty} r_n = 0. \tag{1.8}$$

Property 3. The necessary condition for the convergence. A common term u_n of a convergent series tends to zero on indefinite increase

of n:
$$\lim_{n\to\infty} u_n = 0. \tag{1.9}$$

Indeed, since $u_n = S_n - S_{n-1}$ and if the series converges and has the sum S as a limit for $n \to \infty$,
$$\lim_{n\to\infty} S_n = \lim_{n\to\infty} S_{n-1} = S$$
so that
$$\lim_{n\to\infty} u_n = \lim_{n\to\infty} S_n - \lim_{n\to\infty} S_{n-1} = S - S = 0.$$

Condition (1.9) is *a necessary condition for a convergence* of a series, but it is not sufficient: a series as a whole can diverge, while its general terms tends to zero. To illustrate this, we consider the *harmonic series*.

Example 1.4. The harmonic series is defined as
$$1 + \frac{1}{2} + \frac{1}{3} + \frac{1}{4} + \cdots + \frac{1}{n} + \cdots = \sum_{n=1}^{\infty} \frac{1}{n}.$$

We have that
$$u_n = \frac{1}{n} \to 0 \quad \text{for } n \to \infty.$$

However, we can show that the sum of the first n terms of the harmonic series increases indefinitely. We take the terms in groups of 1, 2, 4, 8, ... terms, starting with the second:
$$1 + \left(\frac{1}{2}\right) + \left(\frac{1}{3} + \frac{1}{4}\right) + \left(\frac{1}{5} + \cdots + \frac{1}{8}\right) + \left(\frac{1}{9} + \cdots + \frac{1}{16}\right) + \cdots$$

so that the kth group contains 2^{k-1} terms. If we replace all the terms of each group by the last, this being the smallest of the group, the resulting series
$$1 + \frac{1}{2} + \frac{1}{4} \cdot 2 + \frac{1}{8} \cdot 4 + \cdots = 1 + \frac{1}{2} + \frac{1}{2} + \cdots = 1 + \frac{1}{2}(n-1),$$
and one can see that the series diverges $S_n \to +\infty$. ∎

Property 4. If all terms of a convergent series are multiplied by the same multiplier c, then its convergence is not affected, and its sum S is only multiplied by c. In fact, let us multiply all terms of a partial sum S_n by c to create a new partial sum $\bar{S}_n$:

$$\bar{S}_n = cu_1 + cu_2 + \cdots + cu_n = c(u_1 + u_2 + \cdots + u_n) = cS_n.$$

In the limit $n \to \infty$,

$$\lim_{n \to \infty} \bar{S}_n = \lim_{n \to \infty} cS_n = c \lim_{n \to \infty} S_n = cS.$$

Property 5. Two converging series

$$A = a_1 + a_2 + \cdots + a_n + \cdots,$$
$$B = b_1 + b_2 + \cdots + b_n + \cdots$$

can be added (or subtracted) term by term so that the series

$$(a_1 \pm b_1) + (a_2 \pm b_2) + \cdots + (a_n \pm b_n) + \cdots$$

is also convergent, and its sum is $C = A \pm B$. Indeed, if A_n, B_n, and C_n are partial sums of the above series, then

$$C_n = (a_1 \pm b_1) + (a_2 \pm b_2) + \cdots + (a_n \pm b_n)$$
$$= (a_1 + a_2 + \cdots + a_n) \pm (b_1 + b_2 + \cdots + b_n) = A_n \pm B_n.$$

In the limit $n \to \infty$,

$$\lim_{n \to \infty} C_n = \lim_{n \to \infty} A_n \pm \lim_{n \to \infty} B_n \quad \text{or} \quad C = A \pm B.$$

1.4. Series with positive terms

There are two most important questions about infinite series, namely (1) convergence and (2) how to evaluate a sum of an infinite convergent series. The convergence question is easier to answer for series when *all* terms are non-negative $u_n \geq 0$. Such series are called *positive series*. For such series,

$$S_{n+1} = S_n + u_{n+1} \geq S_n.$$

The convergence or divergence of a series with positive terms can be determined by comparing it with another series that is known to converge or diverge. This comparison is based on the following *main comparison test*. Let there be two series with positive terms

$$u_1 + u_2 + \cdots + u_n + \cdots = \sum_{k=1}^{\infty} u_k \qquad (1.10)$$

and

$$v_1 + v_2 + \cdots + v_n + \cdots = \sum_{k=1}^{\infty} v_k. \qquad (1.11)$$

If, starting from some n, each term of the series (1.10) does not exceed the corresponding term of the series (1.11), i.e., $u_k \leq v_k$, then from convergence the series (1.11) follows the convergence of the series (1.10), or, if the series (1.10) is divergent, then the series (1.11) is divergent too.

From the main comparison test follows the *second comparison test*. If starting from some n,

$$\frac{u_{n+1}}{u_n} \leq \frac{v_{n+1}}{v_n},$$

then the convergence of the series v_k implies the convergence of the series u_k, or the divergence of the series u_k implies the divergence of the series v_k.

To apply the comparison tests, we need to have a second series that is known to be convergent.

1.4.1. *Cauchy's, d'Alembert's, and Raabe's tests*

It is very practical to compare a given series with a standard series, e.g., a geometrical series. There are two fundamental tests for convergence of series with positive terms based on comparisons with a geometrical series.

Cauchy's or the root test: If for a series with positive terms

$$\sqrt[n]{u_n} \leq r < 1, \qquad (1.12)$$

for all sufficiently large n, where r does not depend on n, then the series is convergent. The test (1.12) can be rewritten as $u_n \leq r^n < 1$.

Since r^n is the corresponding term of an infinite converging geometrical series, then $u_1 + u_2 + \cdots + u_n + \cdots$ series is convergent by the comparison test. If, however, $\sqrt[n]{u_n} > 1$, then the series must diverge. This test is particularly useful for analyzing properties of power series.

d'Alembert's or the ratio test: If the ratio of two successive terms of a series satisfies the inequality

$$\frac{u_{n+1}}{u_n} \leq r < 1 \tag{1.13}$$

for all sufficiently large n and r is independent of n, then the series $u_1 + u_2 + \cdots + u_n + \cdots$ is convergent. We can prove the test by writing (1.13) as

$$u_{n+1} \leq u_n r, \; u_n \leq u_{n-1} r, \ldots, \; u_2 \leq u_1 r,$$

then cross-multiplying term by term and canceling the common factors, we have

$$u_{n+1} \leq u_1 r^n,$$

i.e., the terms of the series are less than the terms of the converging geometrical series, and the original series is converged by the comparison test.

Example 1.5. Consider the series

$$1 + \frac{x}{1} + \frac{x^2}{1 \cdot 2} + \cdots + \frac{x^n}{1 \cdot 2 \cdot 3 \cdots n} + \cdots = \sum_{n=0}^{\infty} \frac{x^n}{n!}.$$

Applying d'Alembert's test we have

$$\frac{u_{n+1}}{u_n} = \frac{x}{n+1} \to 0 \quad \text{as } n \to \infty$$

so that the given series converges for all finite (positive) x. ∎

It is possible to show that, in general, Cauchy's test is stronger than d'Alembert's test. On the other hand, d'Alembert's test is easier to use.

Attention: In case of $r = 1$, both d'Alembert and Cauchy tests are inconclusive: some series with $r = 1$ converge and some diverge. In this case, we need to apply some other test.

Example 1.6. Consider this series

$$\sum_{n=1}^{\infty} \frac{1}{n^2} = 1 + \frac{1}{4} + \frac{1}{9} + \frac{1}{16} + \cdots.$$

For this series, we have

$$\frac{u_{n+1}}{u_n} = \frac{n^2}{(n+1)^2} = \left(\frac{n}{n+1}\right)^2 \to 1,$$

$$\sqrt[n]{u_n} = \sqrt[n]{\frac{1}{n^2}} = \left(\sqrt[n]{\frac{1}{n}}\right)^2 \to 1.$$

So both d'Alembert's and Cauchy's tests are inconclusive. But interestingly, this series converges to $\pi^2/6$. It is known as Euler's Basel problem. ∎

Raabe's test: D'Alembert and Cauchy tests are based on comparison with the geometric series that has a good speed of convergence. It would be very practical to have a test based on comparison with a series that has slower convergence comparing to the geometric series. Raabe's test employs a comparison with two harmonic series, namely a convergent one[1]

$$\sum_{n=1}^{\infty} \frac{1}{n^s} = 1 + \frac{1}{2^s} + \frac{1}{3^s} + \cdots + \frac{1}{n^s} + \cdots \quad (s > 1)$$

and a divergent one

$$\sum_{n=1}^{\infty} \frac{1}{n} = 1 + \frac{1}{2} + \frac{1}{3} + \cdots + \frac{1}{n} + \cdots.$$

---

[1]This is the Riemann zeta function defined as $\zeta(s) = \sum_{n=1}^{\infty} \frac{1}{n^s}$.

For a series of positive terms $u_n \geq 0$, the limit form of Raabe's test is

$$\rho = \lim_{n \to \infty} \left[n \left(\frac{u_n}{u_{n+1}} - 1 \right) \right], \qquad (1.14)$$

where the series converges if $\rho > 1$ and diverges for $\rho < 1$. For $\rho = 1$, the series may converge or diverge. Raabe's test is stronger comparing to d'Alembert's ratio test, i.e., it can provide an answer when d'Alembert's test is inconclusive, as demonstrated in Example 1.7.

Example 1.7. Consider the series

$$1 + \sum_{n=1}^{\infty} \frac{(2n-1)!!}{(2n)!!} \frac{1}{2n+1}.$$

The d'Alembert test is inconclusive since

$$\frac{(2n-1)^2}{2n(2n+1)} \to 1 \quad \text{as } n \to \infty.$$

Raabe's test gives

$$\rho = \lim_{n \to \infty} n \left(\frac{2n(2n+1)}{(2n-1)^2} - 1 \right) = \lim_{n \to \infty} \frac{(6n-1)n}{(2n-1)^2} = \frac{3}{2} > 1.$$

Thus, the above series is convergent. ∎

Of course, we may ask a question about the existence of a universal (extremely slowly convergent or divergent) series that could serve as a comparison test to make a conclusion about the convergence (or divergence) of any series with positive members. However, such universal series does not exist.

1.4.2. *Cauchy's integral test for convergence*

This test is different from all comparison tests. It is based on the idea of comparing a series with an integral. Indeed, definite integrals are defined as a limit of a Riemann sum. Thus, we can consider positive infinite series as a definite integral of a positive function.

Infinite Series

We assume that the terms of the series $u_1+u_2+u_3+\cdots+u_n+\cdots$ are positive and non-increasing, i.e.,

$$u_1 \geq u_2 \geq \cdots \geq u_n \geq u_{n+1} \ldots \geq 0.$$

We represent the terms graphically by plotting n as the independent variable, which only takes integral values, and $u_n(n)$ as a function of n. A continuous function $y = f(x)$ can always be found, such that it takes precisely the values u_n for integral values of $x = n$. With this graphical representation, the sum of the first n terms of the series is represented by the sum of the areas of the **exterior** rectangles so that we can write

$$S_n \geq \int_1^{n+1} f(x)dx \tag{1.15}$$

from $f(x)$ being monotonic decreasing. On the other hand, from Figure 1.1 follows that for the sum of the areas of **internal** rectangles

$$S_{n+1} - u_1 \leq \int_1^{n+1} f(x)dx. \tag{1.16}$$

These inequalities will lead us to Cauchy's integral test. First, for positive $f(x)$, we have

$$\int_1^{n+1} f(x)dx < \int_1^{\infty} f(x)dx. \tag{1.17}$$

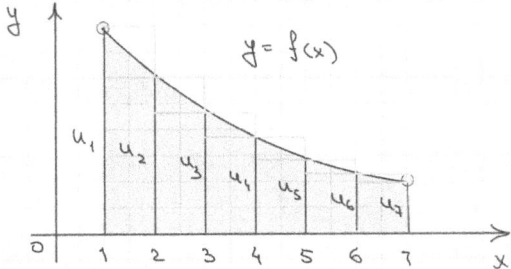

Fig. 1.1. Representing a set of terms as a function.

Second, using (1.16) and the fact that for positive u_i we have $S_n < S_{n+1}$, we can write

$$S_n < S_{n+1} \leq u_1 + \int_1^\infty f(x)dx.$$

Thus, we can see that the series $u_1 + u_2 + \cdots + u_n + \cdots$, $u_n = f(n)$ with positive terms that do not increase with increasing n, converges or strictly diverges, if the integral

$$I = \int_1^\infty f(x)dx \qquad (1.18)$$

converges or diverges. We recall that the $f(x)$ here must decrease with increasing x. Note that the integral I in Cauchy's test can be replaced by the integral

$$I = \int_a^\infty f(x)dx,$$

where a is any positive number, greater than one.

In summary, the Cauchy integral test states that if a_n is a sequence of positive terms, and $a_n = f(n)$ such that $f(n)$ is a continuous, positive, and decreasing function on the interval $[n, \infty)$ for some non-negative integer n, then the series $\sum_{n=1}^\infty a_n$ converges if and only if the improper integral

$$\int_n^\infty f(x)\,dx \qquad (1.19)$$

converges.

Example 1.8. Consider the harmonic series

$$1 + \frac{1}{2} + \frac{1}{3} + \frac{1}{4} + \cdots + \frac{1}{n} + \cdots = \sum_{n=1}^\infty \frac{1}{n}, \text{ and here we have } f(n) = \frac{1}{n}.$$

Then

$$I = \int_1^\infty \frac{1}{x}dx = \ln x\big|_1^\infty \to \infty,$$

thus the series diverges, as we have already seen in Example 1.4. ∎

Example 1.9. Consider the series from Example 1.6 (Euler's Basel problem):

$$\sum_{n=1}^{\infty} \frac{1}{n^2} = 1 + \frac{1}{4} + \frac{1}{9} + \frac{1}{16} + \cdots.$$

Then

$$I = \int_1^{\infty} \frac{1}{x^2} dx = 1,$$

thus the series converges. ∎

The two examples above are special cases of the Riemann zeta function $\zeta(s)$, which is defined for complex numbers s with a real part greater than 1 by the convergent series:

$$\zeta(s) = 1^s + 2^{-s} + 3^{-s} + 4^{-s} + \cdots = \sum_{k=1}^{\infty} \frac{1}{k^s}.$$

The Riemann zeta function has many applications in various branches of mathematics and physics, including harmonic analysis, and quantum field theory.

Note that the integral test can be used to estimate the sum of infinite series. In the same way as before, we can show that the reminder of series does not exceed the integral

$$\int_{n+1}^{\infty} f(x) dx.$$

Then,

$$S \approx S_n + \int_{n+1}^{\infty} f(x) dx,$$

or we calculate a sum of the first n terms and then we integrate the remainder from $n+1$ to ∞. For example, for the above series $1/n^2$, the sum of the first 10 terms is 1.5497. The integral from 11 to ∞ gives 0.0909, thus we have 1.6406 and the exact answer for the series is $\pi^2/6 \approx 1.6449$.

1.5. Alternating series

Series may have not only positive but also both positive and negative terms. First, we start with *alternating series*, in which the positive and negative terms alternate. It is convenient to write such series in the form

$$u_1 - u_2 + u_3 - u_4 \ldots \pm u_n \mp u_{n+1} \ldots,$$

where all numbers $u_1, u_2, \ldots, u_n, \ldots$ are considered to be positive. *Leibniz test* states that for an alternating series to converge, it is sufficient that the absolute values of its terms decrease $u_{n+1} < u_n$ and tend to zero as n increases, i.e., $\lim_{n \to \infty} u_n \to 0$. The remainder of such a series in absolute value does not exceed the absolute value of the first of the discarded term. Indeed, a partial sum of even order S_{2n} can be written as

$$S_{2n} = (u_1 - u_2) + (u_3 - u_4) + \cdots + (u_{2n-1} - u_{2n}).$$

Since every parenthesis is a non-negative number, then the partial sum above does not decrease as n increases. On the other hand, we can rewrite S_{2n} as

$$S_{2n} = u_1 - (u_2 - u_3) - (u_4 - u_5) - \cdots - (u_{2n-2} - u_{2n-1}) - u_{2n}.$$

Then, for any number n, we have $S_{2n} \leq u_1$. Thus, the partial sum S_{2n} is increasing but bounded from above. It means that as $n \to \infty$ the partial sum $\lim_{n \to \infty} S_{2n} = S$. From $S_{2n-1} = S_{2n} + u_{2m}$ and $\lim_{n \to \infty} u_{2n} = 0$ follows that both even and odd partial sums converge to the same number S.

Next, we estimate the reminder of the series

$$r_n = \pm u_{n+1} \mp u_{n+2} \pm u_{n+3} \mp u_{n+4} \pm \cdots$$
$$= \pm (u_{n+1} - u_{n+2} + u_{n+3} - u_{n+4} + \cdots).$$

Then, reasoning as before we can write

$$|r_n| = (u_{n+1} - u_{n+2}) + (u_{n+3} - u_{n+4}) + \cdots$$
$$= u_{n+1} - (u_{n+2} - u_{n+3}) - (u_{n+4} - u_{n+5}) - \cdots \leq u_{n+1}.$$

From
$$r_n = \pm[(u_{n+1} - u_{n+2}) + (u_{n+3} - u_{n+4}) + \cdots],$$
it follows that we have a set of non-negative terms enclosed in square brackets. This implies that the sign of the remainder of the alternating series coincides with the sign of the first discarded term, u_{n+1}.

Example 1.10. As an example we consider an alternating series
$$\sum_{n=1}^{\infty}(-1)^{n-1}\frac{1}{n} = 1 - \frac{1}{2} + \frac{1}{3} - \frac{1}{4} + \cdots + (-1)^{n-1}\frac{1}{n} + \cdots.$$
According to the Leibniz test, the series is convergent since $\frac{1}{n+1} < \frac{1}{n}$ and $\lim_{n\to\infty}\frac{1}{n} \to 0$. Indeed, the series has a sum converging to $\log(2)$. However, for the real calculation of $\log(2)$, this series is not suitable, since in order for its remainder r_n to be less than 0.001, one needs to take 1000 terms, since for the reminder
$$r_n < \frac{1}{n+1} \leq 1000, \quad n \geq 1000.$$
Although this series converges, it converges very slowly. In practical applications, slowly convergent series must be transformed into faster converging series. ∎

Improving the convergence of a series, often referred to as convergence acceleration, aims to achieve the series limit with increased efficiency, requiring fewer terms for a given level of accuracy.[2] Kummer's transformation is a popular method for convergence improvement. Let S be a series to evaluate:
$$S = \sum_{k=0}^{\infty} a_k.$$
And let
$$B = \sum_{k=0}^{\infty} b_k$$

---

[2]Here are several techniques commonly employed to enhance series convergence: acceleration methods, transformation techniques, series resummation, regrouping, and rearrangement.

be a convergent series with known B such that

$$\lim_{k\to\infty} \frac{a_k}{b_k} = \lambda \neq 0.$$

Then a series with more rapid convergence to the same value is given by

$$S = \lambda B + \sum_{k=0}^{\infty} \left(1 - \lambda \frac{b_k}{a_k}\right) a_k.$$

1.6. Absolutely convergent series

While there are many series with arbitrary terms, we will focus on series that are absolutely convergent ones.

A series

$$u_1 + u_2 + u_3 + \cdots + u_n + \cdots = \sum_{n=1}^{\infty} u_n \qquad (1.20)$$

is convergent if the series consisting of the *absolute values* of its terms is convergent, i.e., if the series

$$|u_1| + |u_2| + |u_3| + \cdots + |u_n| + \cdots = \sum_{n=1}^{\infty} |u_n| \qquad (1.21)$$

is convergent. A series of this sort is called *absolutely convergent*. Indeed,

$$|u_1 + u_2 + u_3 + \cdots + u_n + \cdots| \leq |u_1| + |u_2| + |u_3| + \cdots + |u_n| + \cdots.$$

It is possible that a series (1.20) is convergent, but the series (1.21) is not. In this case, the series (1.20) is called *non-absolutely convergent* (or conditionally convergent). For example, the series in Example 1.10 is a non-absolutely convergent series, since the series created from its absolute values is divergent (see Example 1.8).

Convergence tests used before for series with positive terms can be used for testing series on absolute convergence, provided that we replace u_n everywhere by $|u_n|$. In particular, in formulating Cauchy's and d'Alembert's tests, we have to replace

$$\sqrt[n]{u_n} \text{ and } \frac{u_{n+1}}{u_n} \text{ with } \sqrt[n]{|u_n|} \text{ and } \left|\frac{u_{n+1}}{u_n}\right|. \tag{1.22}$$

Example 1.11. The series

$$\sum_{n=1}^{\infty} \frac{x^n}{n!}$$

is absolutely convergent for all finite x, either positive or negative, since

$$\left|\frac{u_{n+1}}{u_n}\right| = \frac{|x|}{n+1} \to 0$$

for all finite x. ■

Example 1.12. The series

$$\sum_{n=1}^{\infty} \frac{x^n}{n}$$

is absolutely convergent for $|x| < 1$ and divergent for $|x| > 1$, since

$$\left|\frac{u_{n+1}}{u_n}\right| = \frac{n}{n+1}|x| \to |x|.$$ ■

1.7. Series with complex numbers and series of vectors

Series with terms as complex numbers

$$z_1 + z_2 + \cdots + z_n + \cdots, \qquad z_n = x_n + iy_n \quad (n = 1, 2, 3, \ldots \infty)$$

can be analyzed for convergence by analyzing series for real and imaginary terms

$$x_1 + x_2 + \cdots + x_n + \cdots, \qquad y_1 + y_2 + \cdots + y_n + \cdots.$$

If both series x_n and y_n are convergent and have the sums as x and y correspondingly, then the series with complex terms is convergent

to
$$z = x + iy.$$
If one of the series is divergent, then the original series is divergent too.

Series with vector terms can be studied by analyzing series of the corresponding vector components. For example, for a vector defined in the three-dimensional Cartesian coordinates as
$$\vec{u} = \vec{u}_1 + \vec{u}_2 + \cdots + \vec{u}_n + \cdots = \sum_{n=1}^{\infty} \vec{u}_n = \hat{i} \sum_{n=1}^{\infty} x_n + \hat{j} \sum_{n=1}^{\infty} y_n + \hat{k} \sum_{n=1}^{\infty} z_n,$$
we need to analyze series for every component. Then the vector series is convergent if all three series are convergent.

1.8. Series of functions

We consider a series
$$f_1(x) + f_2(x) + \cdots + f_n(x) + \cdots, \tag{1.23}$$
where terms of the series are functions defined on the same interval $a \leq x \leq b$. Similar to series with numbers, we say that a series of function converges if a partial sum of functions
$$S_n(x) = \sum_{k=1}^{n} f_k(x)$$
is convergent to a function $S(x)$ as $n \to \infty$, namely
$$S(x) = \lim_{n \to \infty} S_n(x).$$
If a series (1.23) uniformly[3] converges to $S(x)$ on $a \leq x \leq b$, then it converges for any c from this interval, and the sum is
$$f_1(c) + f_2(c) + \cdots + f_n(c) + \cdots = S(c).$$
There are two valuable properties of uniformly convergent series.

---

[3] A series of functions is said to be uniformly convergent to $S(x)$ if for $\varepsilon > 0$, there is an integer N such that $|S(x) - S_n(x)| < \varepsilon$ for $n > N$.

Property 1. A series that converges uniformly can be integrated term by term, namely

$$\int_{x_0}^{x} f_1(t)dt + \int_{x_0}^{x} f_2(t)dt + \cdots + \int_{x_0}^{x} f_n(t)dt + \cdots = \int_{x_0}^{x} S(t)dt,$$

and the new series converges uniformly on the interval $a \leq x \leq b$.

Property 2. A series of continuous functions that converges uniformly can be differentiated term-by-term, i.e.,

$$f_1'(x) + f_2'(x) + \cdots + f_n'(x) + \cdots = S'(x).$$

1.8.1. Power Series

A power series can be written as

$$P(x) = a_0 + a_1 x + a_2 x^2 + a_3 x^3 + \cdots, \tag{1.24}$$

where $a_0, a_1, a_2, a_3, \ldots$ are constants. Convergence of a power series $P(x)$ depends both on coefficients a_n and the value of x. The range of values of x for which $P(x)$ converges is also called the interval of convergence. The interval of convergence of a power series can be explored using the convergence tests explored before. Let us assume that there is a limit

$$\lim_{n \to \infty} \frac{|a_n|}{|a_{n+1}|} = R.$$

Using d'Alembert test we have for a power series

$$\lim_{n \to \infty} \frac{|a_{n+1} x^{n+1}|}{|a_n x^n|} = \lim_{n \to \infty} \frac{|x|}{|a_n/a_{n+1}|} = \frac{|x|}{|R|}.$$

Then for $|x| < R$, i.e.,

$$-R < x < R, \tag{1.25}$$

series (1.24) converges absolutely. The interval (1.25) is the interval of convergence, and R is the radius of convergence. For $-\infty < x < -R$ and $R < x < \infty$, the series is divergent. At $x = \pm R$, d'Alembert test is inconclusive.

Example 1.13. Let us apply d'Alembert's test to the series

$$\sum_{n=1}^{\infty} n! \left(\frac{x}{n}\right)^n.$$

$$\lim_{n\to\infty} \frac{u_{n+1}}{u_n} = \lim_{n\to\infty} \frac{(n+1)!\, x^{n+1}}{(n+1)^{n+1}} \frac{n^n}{n!\, x^n} = \lim_{n\to\infty} \frac{x}{\left(1+\frac{1}{n}\right)^n} = \frac{x}{e}.$$

Thus, for $x < e$, the series is convergent and the radius of convergence $R = e$, for $x > e$, the series is divergent, and for $x = e$, d'Alembert's test is inconclusive. ∎

The Cauchy–Hadamard theorem connects the radius of convergence R to coefficients of power series as

$$R = \frac{1}{\rho}, \quad \text{where } \rho = \lim_{n\to\infty} \sqrt[n]{|u_n|}.$$

For the example above, we have

$$\rho = \lim_{n\to\infty} \sqrt[n]{\frac{n!}{n^n}}.$$

Using Stirling's formula $n! = \sqrt{2\pi n}(n/e)^n(1 + O(1/n))$ for equation above gives

$$\rho = \lim_{n\to\infty} \sqrt[n]{\frac{n!}{n^n}} = \lim_{n\to\infty} (2\pi n)^{\frac{1}{2n}} \left(\frac{n}{e}\right)\left(\frac{1}{n}\right) = \frac{1}{e} \quad \text{or} \quad R = e$$

since

$$\lim_{n\to\infty} (2\pi n)^{\frac{1}{2n}} \to 1.$$

This theorem was discovered by Augustin-Louis Cauchy but it was forgotten until Jacques Hadamard found it again.

Power series can be easily differentiated and integrated. Besides, series resulting from differentiation or integration have the same radius of convergence as the original series. Indeed, differentiating

(1.24) gives

$$a_1 + \frac{a_2}{2}x + \frac{a_3}{3}x^2 + \cdots + \frac{a_n}{n}x^{n-1} + \frac{a_{n+1}}{n+1}x^n + \cdots.$$

The radius of convergence for this series is

$$\lim_{n\to\infty} \frac{|a_n|/n}{|a_{n+1}|/(n+1)} = \lim_{n\to\infty} \frac{n+1}{n} \frac{|a_n|}{|a_{n+1}|}$$
$$= \lim_{n\to\infty} \frac{n+1}{n} \lim_{n\to\infty} \frac{|a_n|}{|a_{n+1}|} = R,$$

that is, the same as that for the original series (1.24). However, the new series must be tested separately for the end points in order to determine whether the new series are convergent there.

Using term-by-term integration or differentiation, it is sometimes possible to reduce a given series to known series and thereby find its sum.

Example 1.14. We want to evaluate the sum of the series

$$S(x) = 2 + \frac{3}{1!}x + \frac{4}{2!}x^2 + \frac{5}{3!}x^3 + \cdots.$$

Applying d'Alembert test we can see that the radius of convergence is $R = \infty$. Now we multiply both sides by x and integrate from 0 to x

$$\int_0^x xS(x)dx = x^2\left(1 + \frac{x}{1!} + \frac{x^2}{2!} + \frac{x^3}{3!} + \cdots + \frac{x^n}{n!}\right) = x^2 e^x.$$

Differentiating both sides we get

$$xS(x) = (x^2 e^x)' = 2xe^x + x^2 e^x,$$

and finally

$$S(x) = (2+x)e^x. \qquad \blacksquare$$

1.8.2. Taylor series

Since power series can be easily differentiated and integrated, we wonder if it possible to represent a function $f(x)$ as a power series on $-R < x < R$, i.e.,

$$f(x) = a_0 + a_1 x + a_2 x^2 + a_3 x^3 + \cdots + a_n x^n + \cdots, \qquad (1.26)$$

where the coefficients a_n are to be determined. In fact, the coefficients of this series can be readily expressed in terms of derivatives of a function $f(x)$ leading to Taylor[4] series. Let us successively differentiate formula (1.26), and since the coefficients are independent of x, we can substitute $x = 0$ to find the coefficients a_n:

$$f(0) = a_0,$$
$$f'(x) = 1a_1 + 2a_2 x + 3a_3 x^2 + 4a_4 x^3 \ldots, \quad f'(0) = 1 \cdot a_1,$$
$$f''(x) = 2a_2 + 3!a_3 x + 3 \cdot 4a_4 x^2 + \ldots, \quad f''(0) = 1 \cdot 2a_2,$$
$$f'''(x) = 3!a_3 + 4!a_4 x + 5!a_5 x^2 + \ldots, \quad f'''(0) = 3!a_3,$$
$$f^{('''')}(x) = 4!a_4 + 5!a_5 x + 6!a_6 x^2 \ldots, \quad f^{('''')}(0) = 4!a_4,$$

$\ldots$ $\ldots$

---

[4] Brook Taylor (1685–1731) was an English mathematician and polymath known for his contributions to various fields of study, particularly mathematics and physics. Born in Edmonton, London, Taylor displayed an early aptitude for mathematics and quickly gained recognition for his intellectual prowess. He is most famous for formulating Taylor's theorem, a fundamental concept in calculus that allows for the approximation of functions using polynomial series. This theorem laid the groundwork for Taylor series expansions and provided a powerful tool for analyzing complex mathematical functions.

Beyond his work in calculus, Taylor made significant strides in other areas of mathematics as well. He contributed to the study of fluid dynamics, pioneering investigations into the motion of fluids and the behavior of solid objects within them. His insights into the mathematical principles governing fluid flow and resistance had a lasting impact on the field of physics. Additionally, Taylor was a respected figure in the scientific community of his time, maintaining correspondence with leading intellectuals and contributing to important discussions on various topics. His work demonstrated a deep understanding of mathematical principles and their applications, solidifying his place in the history of mathematics and science.

Then we have the power series representation as
$$f(x) = f(0) + \frac{f'(0)}{1!}x + \frac{f''(0)}{2!}x^2 + \frac{f'''(0)}{3!}x^3$$
$$+ \cdots + \frac{f^{(n)}(0)}{n!}x^n + \cdots. \qquad (1.27)$$

Three questions remain: What functions can be expanded into power series, what is accuracy if we keep n terms, and what is the radius of convergence? Taylor's theorem[5] gives answers to the first two questions. And the radius of convergence can be found by using tests for infinite series, e.g., d'Alembert test.

Taylor's theorem states that any function $f(x)$ having continuous derivative up to $n+1$ order in a vicinity of a point a can be expanded in series about the point $x = a$ as

$$f(x) = f(a) + \frac{f'(a)}{1!}(x-a) + \frac{f''(a)}{2!}(x-a)^2$$
$$+ \cdots + \frac{f^{(n)}(a)}{n!}(x-a)^n + R_{n+1}(x), \qquad (1.28)$$

where $R_{n+1}(x)$ is a remainder

$$R_{n+1}(x) = \frac{1}{n!}\int_a^x f^{(n+1)}(t)(x-t)^n dt. \qquad (1.29)$$

The remainder can be written in various forms, in particular

$$R_{n+1}(x) = \frac{(x-a)^{n+1}}{(n+1)!}f^{(n+1)}[a+\theta(x-a)], \quad \text{the Lagrange remainder,}$$

where $0 < \theta < 1$.

In numerical analysis, the Taylor series is written in a different form by using $a = x_0$ and $(x-a) = \Delta x$ such as

$$f(x_0 + \Delta x) = f(x_0) + \frac{f'(x_0)}{1!}\Delta x \frac{f''(x_0)}{2!}(\Delta x)^2$$
$$+ \cdots + \frac{f^{(n)}(x_0)}{n!}(\Delta x)^n. \qquad (1.30)$$

Such a form is the foundation for very many numerical methods from numerical differentiation and integration to solving ordinary and partial differential equations.

---

[5] Actually discovered first by Gregory nearly 40 years earlier.

The Taylor expansion about $a = 0$ is called *a Maclaurin series*

$$f(x) = f(0) + \frac{f'(0)}{1!}x + \frac{f''(0)}{2!}x^2 + \cdots + \frac{f^{(n)}(0)}{n!}x^n + R_{n+1}(x). \quad (1.31)$$

A Maclaurin series provides a representation of a function $f(x)$ in a vicinity of the point $x = 0$, where the Lagrange reminder is

$$R_{n+1}(x) = \frac{(x)^{n+1}}{(n+1)!} f^{(n+1)}(\theta x) \quad (0 < \theta < 1).$$

For a function to be expressible as a power series, we require it to be differentiable up to an order $n+1$, and the remainder term $R_{n+1}(x)$ to tends to zero as n tends to infinity. In this case, the infinite power series will represent the function within the interval of convergence of the series.

Example 1.15. Expansion of e^x, $\sin x$, and $\cos x$ in Maclaurin series.

A. $f(x) = e^x$. Since $f^{(n)}(x) = e^x$ and $f^{(n)}(0) = 1$, then

$$e^x = 1 + \frac{x}{1!} + \frac{x^2}{2!} + \cdots + \frac{x^n}{n!} + R_{n+1}(x),$$

where the Lagrange remainder is

$$R_{n+1}(x) = \frac{x^{n+1}}{(n+1)!} e^{\theta x} \quad (0 < \theta < 1).$$

For any $x > 0$, we have $e^{\theta x} < e^x$, and for $x < 0$, we have $e^{\theta x} < 1$, then estimation for the remainder is

$$|R_{n+1}(x)| < \frac{x^{n+1}}{(n+1)!} e^x.$$

B. $f(x) = \sin(x)$. Since

$$f(x) = \sin x, \ f'(x) = \cos x, \ f''(x) = -\sin x, \ldots$$

$$f^n(x) = \sin\left(x + \frac{n}{2}\pi\right),$$

and

$f(0) = 0$, $f'(0) = 1$, $f''(0) = 0$, $f'''(0) = -1$, $f^{2n} = 0$, $f^{2n+1} = (-1)^n$, then with the Lagrange remainder, we have

$$\sin(x) = \frac{x}{1!} - \frac{x^3}{3!} + \frac{x^5}{5!} + \cdots \frac{(-1)^n x^{2n+1}}{(2n+1)!}$$
$$+ \frac{x^{2n+3}}{(2n+3)!} \sin\left(\theta x + \frac{(2n+3)\pi}{2}\right). \qquad (1.32)$$

The estimation for the Lagrange remainder is

$$|R_{2n+3}(x)| \leq \frac{x^{2n+3}}{(2n+3)!}.$$

The expansion is valid for all x since the remainder tends to zero as $n \to \infty$ for all finite x (do you see why?).

We can show similarly that

$$\cos x = 1 - \frac{x^2}{2!} + \frac{x^4}{4!} - \frac{x^6}{6!} + \cdots + \frac{(-1)^n x^{2n}}{(2n!)} + \cdots \qquad (1.33)$$

is valid for all values of x. ∎

Series (1.32) and (1.33) are very practical for calculating the values of $\sin x$ and $\cos x$ for small values of x. They are alternating for all positive or negative x so that according to Leibniz test if we keep n terms, then the remainder does not exceed the absolute value of the $(n+1)$th term. The accuracy of the series approximation for $\sin x$ is presented in Figure 1.2. As one can see, the more terms that are kept in the approximation, the larger the interval in which the series representation is closer to the function.

Example 1.16. Binomial series for $f(x) = (1+x)^\alpha$, where α is a real number. Since

$$f^{(n)}(x) = \alpha(\alpha-1)\ldots(\alpha-n+1)(1+x)^{\alpha-1},$$
$$f^{(n)}(0) = \alpha(\alpha-1)\ldots(\alpha-n+1),$$

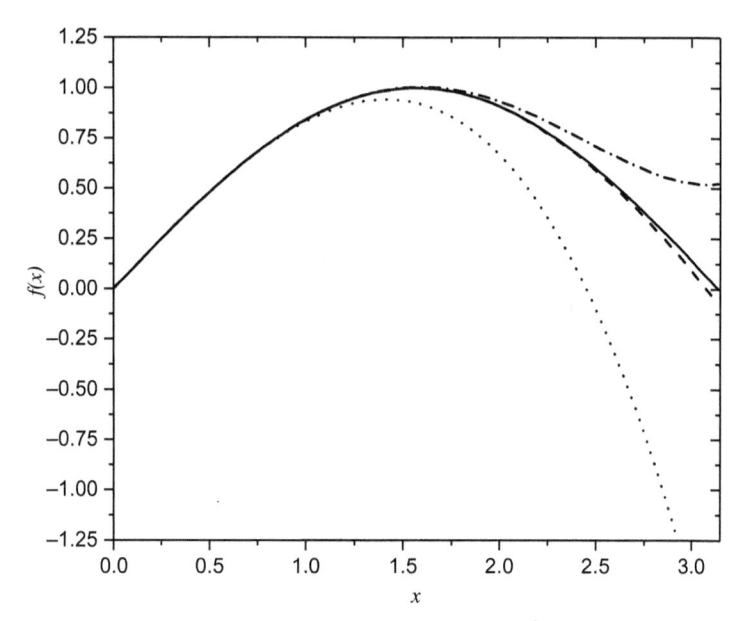

Fig. 1.2. Solid line is $\sin x$, dotted line is $f(x) = \frac{x}{1!} - \frac{x^3}{3!}$, dash-dot line is $f(x) = \frac{x}{1!} - \frac{x^3}{3!} + \frac{x^5}{5!}$, and dashed line is $f(x) = \frac{x}{1!} - \frac{x^3}{3!} + \frac{x^5}{5!} - \frac{x^7}{7!}$.

then using Maclaurin formula (1.31), we have the binomial series as

$$(1+x)^\alpha = 1 + \frac{\alpha}{1!}x + \frac{\alpha(\alpha-1)}{2!}x^2 + \cdots + \frac{\alpha(\alpha-1)\ldots(\alpha-n+1)}{n!}x^n + R_{n+1}(x), \qquad (1.34)$$

where the Lagrange remainder is

$$R_{n+1}(x) = \frac{\alpha(\alpha-1)\ldots(\alpha-n)}{(n+1)!}(1+\theta x)^{\alpha-(n+1)}x^{n+1} \qquad (0 < \theta < 1).$$

Using the d'Alembert test (1.13), we have

$$\left|\frac{n_{n+1}}{n_n}\right| = \left|\frac{m-n+1}{n}x\right| \to |x| \quad \text{as } n \to \infty.$$

Thus, the binomial series is absolutely convergent for $|x| < 1$ and divergent at $|x| > 1$. Convergence or divergence at the endpoints $x = \pm 1$ depends on the value of α.

Here are some useful special cases of the binomial expansion for $\alpha = -1$, $\alpha = 1/2$, and $\alpha = -1/2$:

$$\frac{1}{1 \pm x} = 1 \mp x + x^2 \mp x^3 + \cdots + x^n + \cdots,$$

$$\sqrt{1+x} = 1 + \frac{1}{2}x - \frac{1}{2\cdot 4}x^2 + \frac{1\cdot 3}{2\cdot 4\cdot 6}x^3 - \frac{1\cdot 3\cdot 5}{2\cdot 4\cdot 6\cdot 8}x^4 \cdots,$$

$$\frac{1}{\sqrt{1+x}} = 1 - \frac{1}{2}x + \frac{1\cdot 3}{2\cdot 4}x^2 - \frac{1\cdot 3\cdot 5}{2\cdot 4\cdot 6}x^3 + \frac{1\cdot 3\cdot 5\cdot 7}{2\cdot 4\cdot 6\cdot 8}x^4 - \cdots.$$

In a particular case, when $\alpha = n$, where n is an integer, $R_{n+1} = 0$, and then we have Newton's binomial theorem

$$(1+x)^n = 1 + \frac{n}{1!}x + \frac{n(n-1)}{2!}x^2 + \cdots x^n.$$

In case of $f(x) = (a+x)^n$, we can write it as

$$(a+x)^n = a^n \left(1 + \frac{x}{a}\right)^n$$

$$= a^n \left[1 + \frac{n}{1!}\left(\frac{x}{a}\right) + \frac{n(n-1)}{2!}\left(\frac{x}{a}\right)^2 + \cdots + \left(\frac{x}{a}\right)^n\right]$$

or in a more familiar form

$$(a+x)^n = a^n + \frac{n}{1!}a^{n-1}x + \frac{n(n-1)}{2!}a^{n-2}x^2 + \cdots + x^n. \quad \blacksquare$$

Example 1.17. In this example, we evaluate the following undetermined form:

$$\lim_{x \to 0} \frac{x - \sin x}{x^3}.$$

Such a limit can be evaluated using series expansions, namely

$$\lim_{x \to 0} \frac{x - \sin x}{x^3} = \lim_{x \to 0} \frac{x - x + \frac{x^3}{3!} - o(x^4)}{x^3} = \lim_{x \to 0}\left(\frac{1}{6} + o(x)\right) = \frac{1}{6}.$$

We can evaluate the same limit by using L'Hopital's rule that can be derived using series! $\blacksquare$

Series find numerous applications in mathematics and numerical analysis. Among the most common examples are the evaluation of definite integrals, the assessment of undetermined forms and limits, and finding approximate solutions for both ordinary and partial differential equations. Additionally, series are widely employed in various fields, such as physics, chemistry, engineering, computer science, economics, and more.

1.9. Exercises and problems

(1) Test the convergence of the following series by any appropriate technique:

(a) $\sum_{n=1}^{\infty} \dfrac{n}{2^n}$

(b) $\sum_{n=1}^{\infty} \dfrac{1}{\sqrt{n}}$

(c) $\sum_{n=1}^{\infty} \dfrac{n^2 - 1}{n^2 + 1}$

(d) $\sum_{n=2}^{\infty} \dfrac{1}{n^2 - n}$

(e) $\sum_{n=1}^{\infty} \dfrac{n^n}{n!}$

(f) $\sum_{n=2}^{\infty} \dfrac{\sqrt{n-1}}{(n+1)^2 - 1}$

(g) $\sum_{n=1}^{\infty} \dfrac{(-1)^{n-1}}{2n - 1}.$

(2) Find the interval of convergence of the following power series (make sure to check the end points):

(a) $\sum_{n=0}^{\infty} \dfrac{n}{n^2 + 1}(-x)^n$

(b) $-\dfrac{1}{2}\sum_{n=0}^{\infty} \dfrac{(x-3)^n}{(-2)^n}$

(c) $\sum_{n=1}^{\infty} nx^{n-1} \ (x > 0)$

(d) $\sum_{n=1}^{\infty} \dfrac{(nx)^n}{n!} \ (x > 0)$

(e) $\sum_{n=1}^{\infty} x^n \sin^2 nx \ (x > 0).$

(3) Find the first few terms of the Maclaurin series for the following functions:

(a) $e^x \cos x$
(b) $\sin^2 x \cos^2 x$
(c) $\sqrt{1+x^2}$
(d) $\dfrac{1}{1+x+x^2}$
(e) $\dfrac{\sin x}{x}$
(f) $\dfrac{2}{\sqrt{\pi}} \int_0^x e^{-u^2} du.$

(4) Use power series expansions to evaluate the following limits:

(a) $\lim\limits_{x \to 0} \left(\dfrac{1}{x} - \dfrac{1}{e^x - 1} \right)$

(b) $\lim\limits_{x \to 0} \dfrac{x^2 \ln(1+x^2)}{x^2 - \sin^2 x}$

(c) $\lim\limits_{x \to 0} \left(\dfrac{\ln(1+x)}{x^2} - \dfrac{1}{x} \right)$

(d) $\lim\limits_{x\to\infty} \left(1 - \dfrac{a^2}{x^2}\right)^{x^2}$

(e) $\lim\limits_{x\to 0} \dfrac{\cos(x) - e^{-x^2/2}}{x^3 \sin(x)}$.

(5) A ball falls on a horizontal plate (located a distance h below). Each time the ball hits the plate its vertical velocity component decreases. The ratio of the vertical component of the velocity after the impact to its value before the impact is constant and equal to β. Determine how much time does it take for the ball to stop.

Answers and solutions

(1) Test for convergence
  (a) convergent.
  (b) divergent.
  (c) divergent.
  (d) convergent.
  (e) divergent.
  (f) convergent.
  (g) non-absolutely convergent.

(2) Interval of convergence
  (a) convergent for $-1 < x \leq 1$.
  (b) convergent for $1 < x < 5$.
  (c) convergent for $x < 1$.
  (d) convergent for $x < 1/e$.
  (e) convergent for $x < 1$.

(3) Maclaurin series
  (a) $1 + x - \dfrac{x^3}{3} - \dfrac{x^4}{6} - \dfrac{x^5}{30} + \cdots$.
  (b) $x^2 - \dfrac{4}{3}x^4 + \dfrac{32}{45}x^6 - \dfrac{64}{315}x^8 + \cdots$.
  (c) $1 + \dfrac{x^2}{2} - \dfrac{x^4}{8} + \dfrac{x^6}{16} - \dfrac{5x^8}{128} + \cdots$.
  (d) $1 - x + x^3 - x^4 + \cdots$.

(e) $1 - \dfrac{x^2}{6} + \dfrac{x^4}{120} + \cdots$.

(f) $\dfrac{2}{\sqrt{\pi}}\left(x - \dfrac{x^3}{3} + \dfrac{x^5}{10} + \cdots\right)$.

(4) Limits (a) $\frac{1}{2}$, (b) 3, (c) $-\frac{1}{2}$, (d) $\exp(-a^2)$, (e) $-1/12$.

(5) Time for the ball to stop:

$$T = \sqrt{\dfrac{2h}{g}}\left(\dfrac{1+\beta}{1-\beta}\right).$$

Chapter 2

Complex Numbers

2.1. Complex numbers in science

Complex numbers should be viewed as an extension of real numbers. If we confine ourselves to real numbers, the operation of extracting a root is not always possible, a quadratic equation with real coefficients would not always possess real roots, and the root of a negative number would hold no meaning. By broadening the concept to allow for more complexity, it becomes much easier to solve advanced problems in science and engineering. In fact, the entirety of quantum mechanics is formulated in complex Hilbert spaces.[1] Many improper integrals can be easily evaluated using the residue theorem for functions of complex variables. Functions of complex variables are widely used in fluid dynamics, electrical engineering, signal processing, and many more areas.

However, complex numbers must only be defined in the case that all known basic arithmetical laws for real numbers remain valid.

2.2. Definitions

Complex numbers are defined as *ordered* pairs (x, y) of real numbers. We can interpret complex numbers as "two-dimensional" numbers, or as points in the complex plane, with rectangular coordinates

---

[1] A Hilbert space is like a generalization of an infinite-dimensional vector space.

x and y. Commonly, a complex number (x, y) is represented as z so that
$$z = (x, y). \tag{2.1}$$
The real numbers x and y are called the real and imaginary parts of z, respectively,
$$\text{Re}(z) = x, \quad \text{Im}(z) = y. \tag{2.2}$$
The complex numbers above include the real numbers as a subset when the imaginary part is equal to zero, i.e., a complex number $(x, 0)$ is just a real number x. Complex numbers of the form $(0, y)$ are called pure imaginary numbers. For the algebra of complex numbers, we need to define when two complex numbers are equal, and a set of linear operations, namely addition and multiplication of two complex numbers. And we want these operations defined in such a way that when the imaginary part $y = 0$, the addition and multiplication of complex numbers are simply reduced to addition and multiplication of real numbers.

2.2.1. Basic algebra of complex numbers

We say that two complex numbers $z_1 = (x_1, y_1)$ and $z_2 = (x_2, y_2)$ are equal whenever they have the same real parts and the same imaginary parts, or $z_1 = z_2$ when $x_1 = x_2$ and $y_1 = y_2$. It also means that z_1 and z_2 correspond to the same point in the complex plane.

The sum $z_1 + z_2$ of two complex numbers $z_1 = (x_1, y_1)$ and $z_2 = (x_2, y_2)$ is defined in a straightforward way as
$$z_1 + z_2 = (x_1, y_1) + (x_2, y_2) = (x_1 + x_2, y_1 + y_2). \tag{2.3}$$
The product $z_1 z_2$ is defined in a more complicated way as
$$z_1 z_2 = (x_1, y_1)(x_2, y_2) = (x_1 x_2 - y_1 y_2, x_1 y_2 + x_2 y_1). \tag{2.4}$$
While formulas (2.3) and (2.4) appear out of thin air, it becomes the usual operations of addition and multiplication when restricted to the real numbers, i.e., when the imaginary parts $y_1 = 0$ and $y_2 = 0$:
$$(x_1, 0) + (x_2, 0) = (x_1 + x_2, 0),$$
$$(x_1, 0)(x_2, 0) = (x_1 x_2, 0).$$
Therefore, complex numbers are extensions of the system of real numbers.

The various properties of addition and multiplication of complex numbers are the same as for real numbers. Thus, the commutative laws
$$z_1 + z_2 = z_2 + z_1, \quad z_1 z_2 = z_2 z_1$$
and the associative laws
$$(z_1 + z_2) + z_3 = z_1 + (z_2 + z_3), \quad (z_1 z_2) z_3 = z_1 (z_2 z_3)$$
follow easily from the definitions of addition (2.3) and multiplication (2.4) of complex numbers and the fact that real numbers obey these laws.

If we define zero complex number 0 as $(0,0)$, then we have additive identity
$$z + (0,0) = z$$
as for real numbers. However, defining the multiplicative identity for complex numbers requires a little more work. Assume that such a number is a complex number (a, b) where we need to find a and b, such as $(x, y) * (a, b) = (x, y)$. Using the definition (2.4) for the product of complex numbers, we get
$$x * a - y * b = x,$$
$$x * b + y * a = y.$$
Solving the above system of equations gives $a = 1$ and $b = 0$. Thus, the complex number corresponding to 1 for real numbers is $(1, 0)$ and
$$z * (1, 0) = z.$$
There is an additive inverse associated with each complex number $z = (x, y)$, namely
$$-z = (-x, -y),$$
satisfying the equation $z + (-z) = 0$. Moreover, there is only one additive inverse for any given z, since the equation $(x, y) + (a, b) = (0, 0)$ implies that $a = -x$ and $b = -y$. Additive inverses are used to define subtraction:
$$z_1 - z_2 = z_1 + (-z_2).$$
So, if $z_1 = (x_1, y_1)$ and $z_2 = (x_2, y_2)$, then the subtraction is written as
$$z_1 - z_2 = (x_1 - x_2, y_1 - y_2). \tag{2.5}$$

As for real numbers, for any non-zero complex number $z = (x, y)$, there is a number z^{-1} such that $zz^{-1} = (1, 0)$. This multiplicative inverse is less obvious than the additive one. To find it, we seek real numbers a and b, expressed in terms of x and y, such that

$$(x, y)(a, b) = (1, 0).$$

According to equation (2.4) which defines the product of two complex numbers, a and b must satisfy the system of linear equations:

$$xa - yb = 1,$$

$$ya + xb = 0.$$

The unknowns coefficients a and b can be easily found after simple computation. The unique solution is

$$a = \frac{x}{x^2 + y^2}, \quad b = -\frac{y}{x^2 + y^2}.$$

So, the multiplicative inverse of $z = (x, y)$ is

$$z^{-1} = \left(\frac{x}{x^2 + y^2}, -\frac{y}{x^2 + y^2} \right) \quad (z \neq 0). \tag{2.6}$$

Division by a non-zero complex number is defined as follows:

$$\frac{z_1}{z_2} = z_1 z_2^{-1} \quad (z_2 \neq 0).$$

Using the definition for the inverse z^{-1} we can easily get the equation for division of two complex numbers as

$$\frac{z_1}{z_2} = (x_1, y_1) \left(\frac{x}{x^2 + y^2}, -\frac{y}{x^2 + y^2} \right)$$

$$= \left(\frac{x_1 x_2 + y_1 y_2}{x_2^2 + y_2^2}, \frac{-x_1 y_2 + x_2 y_1}{x_2^2 + y_2^2} \right). \tag{2.7}$$

Soon we learn that there is a much more compact form for representing the division of two complex numbers.

Unlike real numbers, complex numbers do not have ordering. Therefore, there are no complex-valued inequalities. As we noted before, complex numbers are viewed as being elements in the complex plane, and points in a plane lack a natural ordering.

2.3. Complex numbers in $z = x + iy$ form

Any complex number $z = (x, y)$ can be written $z = (x, 0) + (0, y)$. We are interested in representing the pure imaginary part $(0, y)$ as a real number $(y, 0)$ multiplied by some complex number, or as

$$(0, y) = (y, 0) * (a, b).$$

Using the definition for the product of two complex number, we can find that $a = 0$ and $b = 1$, then $(0, y) = (y, 0) * (0, 1) = (0, 1) * (y, 0)$. Hence,

$$z = (x, y) = (x, 0) + (0, 1)(y, 0),$$

and, if we think of a real number as either x or $(x, 0)$ and let i denote the imaginary number $(0, 1)$, then we can write

$$z = x + iy. \tag{2.8}$$

Since $z^2 = zz$, then $(0, 1)(0, 1) = i^2 = (-1, 0)$ or

$$i^2 = -1, \quad i = \sqrt{-1}. \tag{2.9}$$

Clearly, no real numbers – either positive or negative – possess such a square.

In view of expression (2.8), definitions (2.3) and (2.4) for the sum and product become

$$z_1 + z_2 = (x_1 + iy_1) + (x_2 + iy_2) = (x_1 + x_2) + i(y_1 + y_2) \tag{2.10}$$

and

$$z_1 z_2 = (x_1 + iy_1)(x_2 + iy_2) = (x_1 x_2 - y_1 y_2) + i(x_1 y_2 + x_2 y_1). \tag{2.11}$$

Observe that the right-hand sides of these equations can be obtained by formally manipulating the terms on the left as if they involved only real numbers and by replacing i^2 by -1 when it occurs.

The subtraction (2.5) and division (2.7) are then written as

$$z_1 - z_2 = (x_1 - x_2) + i(y_1 - y_2) \tag{2.12}$$

and

$$\frac{z_1}{z_2} = \frac{x_1 + iy_1}{x_2 + iy_2} = \frac{(x_1 x_2 + y_1 y_2) + i(-x_1 y_2 + x_2 y_1)}{x_2^2 + y_2^2}. \tag{2.13}$$

2.3.1. *Rectangular and polar representations*

Just as a real number x may be represented by a point on a line, so may a complex number $z = (x, y)$ be represented by a point in the plane.

Each complex number corresponds to one and only one point. The x and y axes are referred to as the real axis and the imaginary axis, respectively, while the xy plane is called the complex plane or the z plane.

There is yet another interpretation of the complex numbers. Each point (x, y) of the complex plane determines a two-dimensional vector from $(0, 0)$ the initial point, to (x, y), the terminal point, as shown in Figure 2.1. The magnitude of the vector (x, y) is called the modulus or absolute value of the complex number z and denoted by $|z|$; its value is $|z| = r = \sqrt{x^2 + y^2}$.

A measurement of the angle θ that the vector $z(\neq 0)$ makes with the positive real axis is called an argument of z. Thus, we may express the point $z = (x, y)$ in the polar coordinate representation form

$$z = (x, y) = (r\cos\theta, r\sin\theta),$$

where

$$r = |z| = \sqrt{x^2 + y^2} \quad \text{and} \quad \tan\theta = \frac{y}{x}.$$

The real numbers r and θ, like x and y, uniquely determine the complex number z. Unfortunately, the converse is not completely true. While z uniquely determines the x and y, hence r, the value of θ is determined up to a multiple of 2π. There are infinitely many

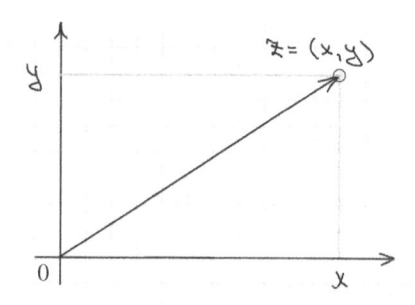

Fig. 2.1. Cartesian representation of a complex number z in (x, y) plane.

distinct arguments for a given complex number z, and the symbol arg(z) is used to indicate any one of them. Thus, the arguments of the complex number $(2,2)$ are

$$\frac{\pi}{4} + 2k\pi \quad (k = 0, \pm 1, \pm 2, \ldots).$$

This inconvenience can sometimes (although not always) be ignored by distinguishing (arbitrarily) one particular value of arg(z). We use the symbol Arg(z) to stand for the unique determination of θ for which $-\pi < \text{arg}(z) \leq \pi$. This θ is called the principal value of the argument. Note that some textbooks may define $0 \leq \theta < 2\pi$ as the principal value range. Alternately, it may be advantageous to consider this range of arg(z) instead of the principal value range.

Note that a number of inequalities for the modulus of complex numbers can be derived using the vector form of the representation, like the triangle inequality, which provides an upper bound for the modulus of the sum of two complex numbers, e.g., $|z_1 + z_2| \leq |z_1| + |z_2|$.

2.3.2. Complex conjugate

The complex conjugate, or simply the conjugate, of a complex number $z = x + iy$ is defined as the complex number $x - iy$ and is denoted by two equivalent notations, namely as $\bar{z}$ or as z^*, that is,

$$\bar{z} = z^* = x - iy. \tag{2.14}$$

We mostly use the latter.

If we view the real axis as a two-way mirror, then z^* is the mirror image of z.

There are several useful properties involving complex conjugates. The sum $z + z^*$ of a complex number $z = x + iy$ and its conjugate $z^* = x - iy$ is the real number $2x$ and the difference $z - z^*$ is the pure imaginary number $2iy$. Hence,

$$\text{Re } z = \frac{z + z^*}{2}, \quad \text{Im } z = \frac{z - z^*}{2i}.$$

An important identity relating the conjugate of a complex number $z = x + iy$ to its modulus is

$$zz^* = |z|^2, \tag{2.15}$$

where each side is equal to $x^2 + y^2$.

If $z_1 = x_1 + iy_1$ and $z_2 = x_2 + iy_2$, then
$$(z_1 + z_2)^* = (x_1 + x_2) - i(y_1 + y_2) = (x_1 - iy_1) + (x_2 - iy_2).$$
So, the conjugate of the sum is the sum of the conjugates:
$$(z_1 + z_2)^* = z_1^* + z_2^*.$$
In like manner, it is easy to show that
$$(z_1 - z_2)^* = z_1^* - z_2^*, \quad (z_1 z_2)^* = z_1^* z_2^*, \quad \left(\frac{z_1}{z_2}\right)^* = \frac{z_1^*}{z_2^*}.$$
Complex conjugates are widely used in physics.

Example 2.1. In quantum mechanics, the complex conjugate often appears when dealing with wavefunctions. The probability density $P(x)$ of finding a particle in a region $[a, b]$ is given by
$$P(x) = \int_a^b |\psi(x)|^2 \, dx,$$
where $\psi(x)$ is the wavefunction. By using (2.15), the probability density $P(x)$ is written as
$$P(x) = \int_a^b |\psi(x)|^2 \, dx = \int_a^b \psi(x)\psi^*(x) \, dx. \quad \blacksquare$$

2.4. Exponential form

Quite often the exponential form for complex numbers is introduced by simply using the Euler's formula[2]
$$e^{ix} = \cos(x) + i\sin(x).$$
However, a more careful approach includes a brief introduction to complex elementary functions. Let's consider one of the elementary

---
[2]Leonhard Euler (1707–1783) was a pioneering Swiss mathematician and physicist who made immense contributions to a wide array of fields, earning him a place among the most prolific and influential mathematicians in history. Born in Basel, Switzerland, Euler demonstrated an early aptitude for mathematics and quickly rose to prominence with his groundbreaking work in areas such as number theory, graph theory, and calculus. His extraordinary output included over 800

functions of a real variable x, namely e^x. Wishing to extend the definition of the exponential function to complex numbers, we cannot start from the expression $e^x = \exp(x)$ because we have not yet considered powers with complex exponents. The best we can do is start from the well-known expansion of e^x into powers of x

$$\exp(x) = 1 + \frac{x}{1!} + \frac{x^2}{2!} + \frac{x^3}{3!} + \cdots \qquad (2.16)$$

and replace x by z. Thus, we define the exponential function for complex values of the variable by the series

$$\exp(z) = 1 + \frac{z}{1!} + \frac{z^2}{2!} + \frac{z^3}{3!} + \cdots. \qquad (2.17)$$

Since we already defined the product of two complex numbers, then we have a recipe to calculate z^n. Note that in this chapter we do not concentrate on considerations of convergence. If x and x' are real numbers, then

$$e^x \cdot e^{x'} = e^{x+x'} \qquad (2.18)$$

by the ordinary rule for the multiplication of powers with the same base. This formula leads us to examine the product $e^z \cdot e^{z'}$, where z and z' denote arbitrary complex numbers. Applying definition (2.17) and analyzing the product of the two infinite series we can get

$$\exp(z)\exp(z') = \exp(z+z')$$

for arbitrary complex numbers z and z'. Thus, we identify e^z with $\exp(z)$ and write henceforward

$$1 + \frac{z}{1!} + \frac{z^2}{2!} + \frac{z^3}{3!} + \cdots = e^z. \qquad (2.19)$$

---

books and papers on diverse mathematical topics, making him one of the most productive scholars of his time.

Euler's contributions extended far beyond theoretical mathematics. He played a vital role in developing the modern notation and terminology used in mathematics, which continues to be the backbone of the discipline today. Euler's formula, known as "Euler's identity," is celebrated for its elegance and deep mathematical significance, linking five of the most important numbers in mathematics in a single equation. Beyond his work in pure mathematics, Euler made valuable contributions to physics, mechanics, astronomy, and even music theory. His lasting impact on the scientific world is a testament to his brilliant mind and relentless dedication to advancing human knowledge.

This definition is reasonable because it implies the generalization of (2.18) that is

$$e^z e^{z'} = e^{z+z'}. \tag{2.20}$$

We now apply the method that has succeeded so well in the case of the exponential function to the trigonometric functions sine and cosine. We define these functions for arbitrary complex z by the expansions

$$\cos(z) = 1 - \frac{z^2}{2!} + \frac{z^4}{4!} - \frac{z^6}{6!} + \cdots, \tag{2.21}$$

$$\sin(z) = z - \frac{z^3}{3!} + \frac{z^5}{5!} - \frac{z^7}{7!} + \cdots, \tag{2.22}$$

$$\cos(-z) = \cos(z) \quad \sin(-z) = -\sin(z)$$

since the series (2.21) contains only even powers of z and (2.22) only odd powers. The series (2.21) and (2.22) appear to be related to the exponential series, (2.19). We may observe that the signs in the last series for the trigonometric functions recur periodically with the period 4. We also observe that the powers of i have the same sort of periodicity:

$$i = i, \quad i^2 = -1, \quad i^3 = -i, \quad i^4 = 1, \quad i^5 = i, \ldots,$$

and at this point, we come very near to the discovery of Euler, who, upon introducing the number i, found a remarkable relation between the functions $\cos(z), \sin(z)$, and e^z. The foregoing remarks suggest the idea of substituting iz for z in the exponential series. Disposing the terms appropriately, we write

$$e^{iz} = 1 - \frac{z^2}{2!} + \frac{z^4}{4!} - \frac{z^6}{6!} + \cdots + i\left(\frac{z}{1!} - \frac{z^3}{3!} + \frac{z^5}{5!} - \cdots\right).$$

Using (2.21) and (2.22), we obtain a very important formula

$$e^{iz} = \cos(z) + i\sin(z). \tag{2.23}$$

This is one of many theorems by Euler! It has many and various applications; its discovery was, in fact, one of the principal incentives to further investigation of complex numbers.

Supposing as usual $z = x + iy$, where x and y are real, then
$$e^z = e^{x+iy} = e^x \cdot e^{iy} = e^x(\cos y + i \sin y) \qquad (2.24)$$
and
$$e^{iy} = \cos y + i \sin y. \qquad (2.25)$$
For $x = 0$ and $y = \pi$, we have
$$e^{i\pi} = \cos(\pi) + i\sin(\pi) = -1$$
or
$$e^{i\pi} + 1 = 0.$$
This formula is considered by many as the "most beautiful" formula since it combines the five "most important" numbers $0, 1, i, \pi$, and e.

Let us make one more step and relate our results to a polar coordinate representation. Since in the polar form $x = r\cos\theta$ and $y = r\sin\theta$, then
$$z = x + iy = r\cos\theta + ir\sin\theta = r(\cos\theta + i\sin\theta),$$
and from (2.24) follows the exponential representation of complex numbers
$$z = x + iy = re^{i\theta}, \qquad (2.26)$$
where
$$r = \sqrt{x^2 + y^2}, \quad \theta = \tan^{-1}\left(\frac{y}{x}\right). \qquad (2.27)$$
The exponential form (2.26) of representation of complex numbers is widely used in physics.

Example 2.2. Let's consider the example of a free particle in one dimension. In this case, the particle is not subject to any potential energy, and its behavior is described by the time-dependent Schrödinger equation:
$$i\hbar \frac{\partial \psi(x,t)}{\partial t} = -\frac{\hbar^2}{2m} \frac{\partial^2 \psi(x,t)}{\partial x^2}.$$
The solution in this case takes the form
$$\psi(x,t) = Ae^{i(kx-\omega t)}.$$
Here, A is a normalization constant, k is the wavenumber, and ω is the angular frequency. ∎

2.4.1. Products and quotients in exponential form

It is often convenient to use Euler's formula when we want to multiply or divide complex numbers. From (2.24), we easily obtain

$$e^{i\theta_1} \cdot e^{i\theta_2} = e^{i(\theta_1+\theta_2)},$$
$$e^{i\theta_1} \div e^{i\theta_2} = e^{i(\theta_1-\theta_2)},$$

then

$$z_1 z_2 = r_1 e^{i\theta_1} r_2 e^{i\theta_2} = r_1 r_2 e^{i(\theta_1+\theta_2)},$$
$$\frac{z_1}{z_2} = \frac{r_1 e^{i\theta_1}}{r_2 e^{i\theta_2}} = \frac{r_1}{r_2} e^{i(\theta_1-\theta_2)}.$$

Since $1 = 1e^{i0}$, it follows from the expression above that the inverse of any non-zero complex number $z = re^{i\theta}$ is

$$z^{-1} = \frac{1}{z} = \frac{1}{r} e^{-i\theta}.$$

Example 2.3. We wish to derive the well-known formulas for $\cos(\alpha + \beta)$ and $\sin(\alpha + \beta)$ for real values of α and β. We proceed as follows:

$$\begin{aligned}\cos(\alpha + \beta) + i\sin(\alpha + \beta) &= e^{i(\alpha+\beta)} = e^{i\alpha} e^{i\beta} \\ &= (\cos\alpha + i\sin\alpha)(\cos\beta + i\sin\beta) \\ &= \cos\alpha\cos\beta - \sin\alpha\sin\beta \\ &\quad + i(\sin(\alpha)\cos\beta + \cos\alpha\sin\beta).\end{aligned}$$

Comparing the first line with the last line, and separating the real part from the imaginary part, we obtain

$$\cos(\alpha + \beta) = \cos\alpha\cos\beta - \sin\alpha\sin\beta,$$
$$\sin(\alpha + \beta) = \sin(\alpha)\cos\beta + \cos\alpha\sin\beta.$$

This derivation supposes that the arguments involved are real but, nevertheless, the result remains valid even if the values involved are complex. ∎

2.4.2. Power and roots of complex numbers

Using the rules for multiplication and division of complex numbers in the exponential form we can write

$$z^n = (re^{i\theta})^n = r^n e^{in\theta}, \qquad (2.28)$$

for any integral n. The case $r = 1$ is of particular interest. Then, (2.28) becomes

$$e^{in\theta} = (\cos(\theta) + i\sin(\theta))^n$$
$$= \cos(n\theta) + i\sin(n\theta) \quad (n = 0, \pm 1, \pm 2, \ldots), \qquad (2.29)$$

this is known as *de Moivre's formula*. This equation can be easily used to find the formulas for $\sin(2\theta), \cos(2\theta), \sin(3\theta)$, etc.

Example 2.4. Express $\cos(2\theta)$ and $\sin(2\theta)$ in terms of $\cos(\theta)$ and $\sin(\theta)$.

Using de Moivre's formula (2.29), we can write

$$\cos 2\theta + i\sin 2\theta = (\cos\theta + i\sin\theta)^2 = \cos^2\theta + 2i\cos\theta\sin\theta - \sin^2\theta.$$

Equating the imaginary parts of the first and last terms, we have

$$\cos 2\theta = \cos^2\theta - \sin^2\theta,$$
$$\sin 2\theta = 2\cos\theta\sin\theta.$$

In the same way, we may consider expressions of multiple-angle functions in terms of polynomials in the single-angle functions. ∎

Consider now a point $z = re^{i\theta}$ lying on a circle centered at the origin with radius r. As θ is increased, z moves around the circle in the counterclockwise direction. In particular, when θ is increased by 2π, we arrive at the original point, and the same is true, θ is decreased by 2π. It is, therefore, evident from Figure 2.2 that two non-zero complex numbers

$$z_1 = r_1 e^{i\theta_1} \quad \text{and} \quad z_2 = r_2 e^{i\theta_2}$$

are equal if and only if

$$r_1 = r_2 \quad \text{and} \quad \theta_1 = \theta_2 + 2k\pi,$$

where k is some integer ($k = 0, \pm 1, \pm 2, \ldots$).

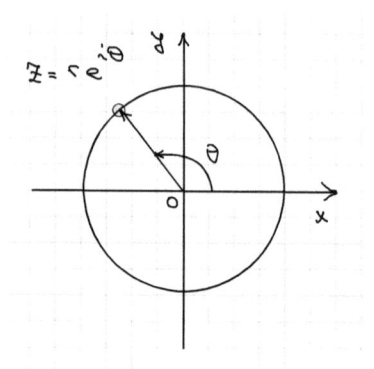

Fig. 2.2. Exponential representation of a complex number z in (x,y) plane.

This observation, together with the expression $z^n = r^n e^{in\theta}$ for integral powers of complex numbers, is useful in finding the nth roots of any non-zero complex number $z_0 = r_0 e^{i\theta_0}$, where n has one of the values $n = 2, 3, \ldots$. The method starts with the fact that an nth root of z_0 is a non-zero number $z = re^{i\theta}$ such that $z^n = z_0$, or

$$r^n e^{in\theta} = r_0 e^{i\theta_0},$$

then

$$r^n = r_0 \quad \text{and} \quad n\theta = \theta_0 + 2k\pi \quad (k = 0, \pm 1, \pm 2, \ldots).$$

Consequently, the complex numbers

$$z = \sqrt[n]{z_0} = \sqrt[n]{r_0 e^{i\theta_0}} = \sqrt[n]{r_0} \exp\left[i\left(\frac{\theta_0}{n} + \frac{2k\pi}{n}\right)\right]$$

are the nth roots of z_0. We are able to see immediately from this exponential form of the roots that they all lie on the circle $|z| = \sqrt[n]{z_0}$ about the origin and are equally spaced every $2\pi/n$ radians, starting with argument θ_0/n. All of the distinct roots are obtained when $k = 0, \pm 1, \pm 2, \ldots, n-1$, and no further roots arise with other values of k. We let z_k denote these distinct roots and write

$$z_k = \sqrt[n]{r_0} \exp\left[i\left(\frac{\theta_0}{n} + \frac{2k\pi}{n}\right)\right] \quad (k = 0, 1, 2, \ldots, n-1). \quad (2.30)$$

The number $\sqrt[n]{r_0}$ is the length of each of the radius vectors representing the n roots. The first root z_0 has argument $\frac{\theta_0}{n}$ and the two roots when $n = 2$ lie at the opposite ends of a diameter of the circle $|z| = \sqrt[n]{r_0}$, the second root being $-z_0$. When $n \geq 3$, the roots lie at the vertices of a regular polygon of n sides inscribed in that circle.

Example 2.5. Find the value of the root $\sqrt[4]{16}$. We can write 16 as a complex number in the exponential form $16 = 16e^{i0}$. Then, according to (2.30),

$$\sqrt[4]{16} = \sqrt[4]{16} \exp\left[i\left(\frac{0}{n} + \frac{2k\pi}{4}\right)\right] \quad (k = 0, 1, 2, 3)$$

and

$$z_0 = 2, \quad z_1 = 2e^{i\pi/2} = 2i, \quad z_2 = 2e^{i\pi} = -2, \quad z_3 = 2e^{i3\pi/2} = -2i.$$

∎

2.4.3. Complex logarithm and complex powers

We now extend to the complex domain the concept of natural logarithms, which we shortly call logarithms. We say that w is the logarithm of z and write

$$w = \ln(z) \tag{2.31}$$

if and only if

$$z = e^w. \tag{2.32}$$

Thus, we have defined the logarithmic function for complex values of the variable. By our definition, the two relations above are fully equivalent; both have exactly the same meaning. We have discussed the exponential function before in Section 2.4 defining it by a series. Knowing how to pass from w to z by (2.32), we are going to study how to pass in the inverse direction, from z to w, by (2.33). In mathematical terminology, we are going to study the logarithmic function as the inverse function of the exponential function.

We write

$$z = re^{i\theta}, \quad w = u + iv,$$

with real θ, u, v and positive r. Then, we may write (2.32) in the form

$$re^{i\theta} = e^{u+iv}.$$

The modulus of the left-hand side is r and its argument θ. Now, if two complex numbers are equal, their absolute values are equal, and the difference in their arguments is an integral multiple of 2π. Therefore,

$$e^u = r \quad v = \theta + 2\pi n,$$

where n is an integer ($n = 0, \pm 1, \pm 2, \ldots$). But r is a positive number and u a real number. Therefore, the connection between r and u is well known; it is the connection between a positive number and its natural logarithm, considered in the theory of real functions. With this meaning of the term "logarithm," we rewrite the last relations in the form $u = \ln r$. Then, we can write

$$\ln z = \ln r + i(\theta + 2n\pi). \tag{2.33}$$

We use the concept of logarithm that we have just acquired to extend the concept of power to complex numbers. Being given a complex constant a, we define z^a by the equation

$$z^a = e^{a \ln z}. \tag{2.34}$$

If we use polar coordinates r and θ in expressing z, and also (2.33), we may rewrite the above equation in the form

$$z^a = \exp\{a[\ln r + i(\theta + 2n\pi)]\}. \tag{2.35}$$

The last factor contains the indeterminate integer n, and so the function z^a of z may be multi-valued. Whether it actually is depends on the nature of the number a.

For $a = u + iv$, we can rewrite the past equation to be more specific as

$$z^{u+iv} = \exp\{u[\ln r + i(\theta + 2n\pi)]\} + \exp\{iv[\ln r + i(\theta + 2n\pi)]\}$$
$$= \exp[(u \ln r) - v(\theta + 2n\pi)] \cdot \exp[i(u\theta + 2nu\pi + v \ln r)]. \tag{2.36}$$

Example 2.6. Let's consider some values for a

If $a = 1/2$, then by (2.35),

$$z^{1/2} = \exp\left[\frac{1}{2}(\ln r + i\theta) + in\pi\right] = (-1)^n r^{1/2} e^{i(\theta/2)} = \pm r^{1/2} e^{i(\theta/2)}.$$

If $a = i$, then then by (2.35),

$$z^i = \exp[i(\ln r + i\theta + i2n\pi)].$$

We are interested in a specific case when $z = i$. In the exponential form, $z = i$ can be written as

$$z = i = e^{i\left(\frac{\pi}{2} + 2k\pi\right)}.$$

Then,

$$i^i = e^{-\left(\frac{\pi}{2} + 2k\pi\right)}.$$

For $k = 0$, we obtain the principal value as

$$i^i = e^{-\frac{\pi}{2}}. \qquad \blacksquare$$

2.4.4. The exponential and trigonometric functions

As we have seen, there is a relationship between complex exponentials and trigonometric functions of real angles (see equation (2.25)). Let's write this relationship for θ and $-\theta$ using properties of trigonometric functions, namely $\cos(-\theta) = \cos(\theta)$ and $\sin(-\theta) = -\sin(\theta)$. Then, we have

$$e^{i\theta} = \cos\theta + i\sin\theta,$$
$$e^{-i\theta} = \cos\theta - i\sin\theta.$$

These two equations can be solved for $\cos\theta$ and $\sin\theta$ as

$$\cos\theta = \frac{e^{i\theta} + e^{-i\theta}}{2}, \qquad (2.37)$$

$$\sin\theta = \frac{e^{i\theta} - e^{-i\theta}}{2i}. \qquad (2.38)$$

These formulas are useful in evaluating integrals since the products of exponentials are easier to integrate than the products of sines and cosines. Many trigonometric identities can also be derived with these formulas.

Example 2.7. Evaluate the integral

$$I = \int_0^\pi \sin(3x)\cos(4x)dx.$$

Let us use the exponential representation for the two trigonometric functions above, then

$$\sin(3x) = \frac{1}{2i}(e^{i3x} - e^{-i3x})$$

and

$$\cos(4x) = \frac{1}{2}(e^{i4x} + e^{-i4x}).$$

Then, the original integral can be written as

$$\int_0^\pi \sin(3x)\cos(4x)dx = \int_0^\pi \frac{1}{2i}(e^{i3x} - e^{-i3x})\frac{1}{2}(e^{i4x} + e^{-i4x})dx$$

$$= \frac{1}{4i}\int_0^\pi (e^{i7x} + e^{-ix} - e^{ix} - e^{-i7x})dx.$$

Now, we can use

$$\int e^{ax}dx = \frac{1}{a}e^{ax},$$

and then

$$I = \frac{1}{4i}\left[\frac{1}{7i}e^{i7x} + \frac{1}{-i}e^{-ix} - \frac{1}{i}e^{ix} - \frac{1}{-7i}e^{-i7x}\right]_0^\pi$$

$$= \frac{1}{4i}\left[\frac{1}{7i}(e^{i7x} + e^{-i7x}) - \frac{1}{i}(e^{ix} + e^{-ix})\right]_0^\pi.$$

Using the above equations for $e^{i\theta}$ and $e^{-i\theta}$, we can rewrite the integral as

$$I = \frac{1}{4i}\left[\frac{1}{7i}2\cos(7x) - \frac{1}{i}2\cos(x)\right]_0^\pi = \left[-\frac{1}{14}\cos 7x + \frac{1}{2}\cos x\right]_0^\pi.$$

Since $\cos(7\pi) = \cos(\pi) = -1$ and $\cos 0 = 1$, then

$$I = \frac{1}{14} - \frac{1}{2} + \frac{1}{14} - \frac{1}{2} = \frac{1}{7} - 1 = -\frac{6}{7}. \blacksquare$$

Complex Numbers

Example 2.8. Evaluate the integral

$$I = \int e^{ax} \cos(bx)dx.$$

There are two ways to evaluate the integral, either using the exponential form for $\cos bx$ or replacing $\cos bx$ on $\cos bx + i \sin bx$ with subsequent $\cos bx + i \sin bx = e^{ibx}$ and keeping real only terms at the end. Here, we are going to use the second approach and consider the following integral:

$$\int e^{ax}(\cos bx + i \sin bx)dx = \int e^{ax} e^{ibx} dx$$

$$= \int e^{(a+ib)x} dx = \frac{1}{a+ib} e^{(a+ib)x}.$$

Now, using Euler's formula, we get

$$\frac{1}{a+ib} e^{(a+ib)x} = e^{ax} \frac{1}{a+ib}(\cos bx + i \sin bx)$$

$$= e^{ax} \frac{a-ib}{a^2+b^2}(\cos bx + i \sin bx).$$

Equating the real parts of the integral above and last terms, we have

$$\int e^{ax} \cos bx\, dx = \frac{e^{ax}}{a^2+b^2}(a \cos bx + b \sin bx).$$

By the way, for the imaginary parts, we get

$$\int e^{ax} \sin bx\, dx = \frac{e^{ax}}{a^2+b^2}(a \sin bx - b \cos bx). \blacksquare$$

2.5. Hyperbolic functions

The hyperbolic functions sinh, cosh, and tanh (hyperbolic sine, hyperbolic cosine, and hyperbolic tangent) are defined by removing i appearing in the complex exponentials. For example, for $\cos x$

and $\sin x$,

$$\cos x = \frac{e^{ix} + e^{-ix}}{2}, \quad \sin x = \frac{e^{ix} - e^{-ix}}{2i}.$$

Then, the two fundamental hyperbolic functions are defined as combinations of the two exponential functions:

$$\cosh x = \frac{e^x + e^{-x}}{2}, \tag{2.39}$$

$$\sinh x = \frac{e^x - e^{-x}}{2}, \tag{2.40}$$

as shown in Figure 2.3 where

$$\cosh(x) + \sinh(x) = e^x.$$

The hyperbolic functions of x are connected to the trigonometric functions of ix as

$$\cos ix = \cosh x, \quad \sin ix = i \sinh x,$$
$$\cosh ix = \cos x, \quad \sinh ix = i \sin x.$$

As always, it is useful to plot new functions to better know their behaviors.

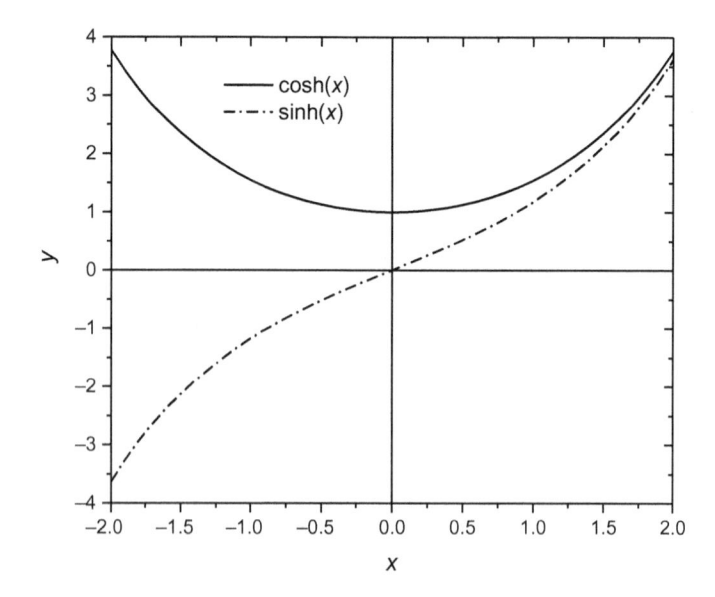

Fig. 2.3. Hyperbolic functions $\cosh x$ and $\sinh x$.

As a combination of the exponential functions, the hyperbolic functions simplify many mathematical expressions.

There is the following reasonable question: Why do we consider completely real hyperbolic functions inside this chapter on complex numbers? The answer is related to the similarity between the definitions for the basic hyperbolic functions and equations (2.37). By analogy with the trigonometric functions, more hyperbolic functions can be defined as

$$\tanh = \frac{\sinh x}{\cosh x}, \quad \coth x = \frac{1}{\tanh x}.$$

Hyperbolic functions have many properties similar to trigonometric functions. Here are a couple of examples:

$$\cosh^2(x) - \sinh^2(x) = 1,$$
$$\sinh(2x) = 2\cosh(x)\sinh(x).$$
$$\frac{d}{dx}\cosh(x) = \sinh(x),$$
$$\frac{d}{dx}\sinh(x) = \cosh(x).$$

All properties for the hyperbolic functions (identities, derivatives, and integrals) are easily derived from the definitions of the hyperbolic functions.

Hyperbolic functions are widely used in physics, most often as solutions of various ordinary and partial differential equations. Probably one of the most famous problems is the catenary problem, or the shape of a hanging chain under its own weight. The solution is given as

$$y(x) = a \cosh\left(\frac{x}{a}\right),$$

where a is a constant determined by initial conditions. The problem has a long history and has been of interest to mathematicians, physicists, and engineers. The problem was studied by Galileo Galilei, Robert Hooke, James and Johann Bernoulli, Gottfried Leibniz, and Leonhard Euler.

2.6. Exercises

(1) Two complex numbers z_1 and z_2 are given by $z_1 = 2 + 3i$ and $z_2 = 4 - 5i$. Find
  (a) $z_1 + z_2$
  (b) $z_1 - z_2$
  (c) $z_1 z_2$
  (d) z_1/z_2.

(2) Write the following numbers in the exponential form $z = re^{i\theta}$:
  (a) -1
  (b) $1 + i$
  (c) $1 - i\sqrt{3}$
  (d) $(i + \sqrt{3})^2$.

(3) Express the following complex numbers in the $x + iy$ form:
  (a) $e^{-i\pi/4}$
  (b) $4e^{2(1-i\pi)}$
  (c) $(1 + i)^{16}$
  (d) $\left(\dfrac{1}{2} + i\dfrac{\sqrt{3}}{2}\right)^9$.

(4) Find all solutions of the following equations:
  (a) $2ix + 3 = y - i$
  (b) $e^z = 1 + i$
  (c) $\sin z = i$
  (d) $\cos z = 2$.

(5) Find all the values of the following powers and roots:
  (a) $(-1)^i$
  (b) $\sqrt[5]{32}$
  (c) $\sqrt[3]{-8i}$
  (d) $\sqrt{2 + 2i\sqrt{3}}$.

(6) Using Euler formula or de Moivre's theorem prove that
  (a) $\cos(\alpha + \beta) = \cos\alpha\cos\beta - \sin\alpha\sin\beta$
  (b) $\cos^2\theta + \sin^2\theta = 1$
  (c) $\sin 3\theta = 3\cos^2\theta\sin\theta - \sin^3\theta$.

(7) Using the sines and cosines in exponential forms, evaluate the integrals:
   (a) $\int_{-\pi}^{\pi} \cos^2(3x)\,dx$
   (b) $\int_0^{\pi} \sin 2x \cos 3x\,dx$
   (c) $\int_0^{2\pi} \sin^2(4x)\,dx$.

(8) Demonstrate the orthogonality of the Fourier set of functions $\sin nx$ and $\cos mx$ on the interval $[-\pi, \pi]$:
   (a) $\int_{-\pi}^{\pi} \sin(nx)\cos(mx)\,dx = 0$
   (b) $\int_{-\pi}^{\pi} \sin(nx)\sin(mx)\,dx = \pi\delta_{n,m}$
   (c) $\int_{-\pi}^{\pi} \cos(nx)\cos(mx)\,dx = \pi\delta_{n,m}$,
   where $\delta_{n,m}$ is the Kronecker delta.

(9) Has the following series a limit; if it has one, what is the limit?
   (a) $1 + i + \dfrac{i^2}{2!} + \dfrac{i^3}{3!} + \dfrac{i^4}{4!} + \cdots$
   (b) $1 + \dfrac{i}{2} - \dfrac{1}{3} - \dfrac{i}{4} + \dfrac{1}{5} + \dfrac{i}{6} - \dfrac{1}{7} - \dfrac{i}{8} + \dfrac{1}{9} + \cdots$
   (c) $\dfrac{i+1}{2} + \left(\dfrac{i+1}{2}\right)^2 + \left(\dfrac{i+1}{2}\right)^3 + \cdots + \left(\dfrac{i+1}{2}\right)^n + \cdots$.

(10) Compute $(2+i)(3+i)$ and prove that $\dfrac{\pi}{4} = \arctan\dfrac{1}{2} + \arctan\dfrac{1}{3}$.

(11) Establish the identity
$$1 + z + z^2 + \cdots + z^n = \dfrac{1 - z^{n+1}}{1 - z}$$
and then use it to derive Lagrange's trigonometric identity
$$1 + \cos\theta + \cos 2\theta + \cdots + \cos n\theta$$
$$= \dfrac{1}{2} + \dfrac{\sin[(2n+1)\theta/2]}{2\sin\theta/2} \quad (0 < \theta < 2\pi).$$

(12) Verify each of the following:
   (a) $\cos(z) = \cos(x)\cosh(y) - i\sin(x)\sinh(y)$
   (b) $\cosh^2(z) - \sinh^2(z) = 1$
   (c) $\cosh(x) - \cosh(y) = 2\sinh\left(\dfrac{x+y}{2}\right)\sinh\left(\dfrac{x-y}{2}\right)$.

Answers and solutions

(1) (a) $6-2i$, (b) $-2+8i$, (c) $23+2i$, (d) $(-7+22i)/41 = -0.1707 + 0.5366i$.

(2) (a) $e^{i\pi}$, (b) $\sqrt{2}e^{i\pi/4}$, (c) $2e^{-i\pi/3}$, (d) $4e^{i\pi/3}$.

(3) (a) $\dfrac{\sqrt{2}}{2}(1-i)$, (b) $4e^2 + i0$, (c) $2^8 + i0$, (d) $-1 + i0$.

(4) (a) $-\dfrac{1}{2} + 3i$, (b) $\dfrac{1}{2}\ln 2 + i\dfrac{\pi}{4}$, (c) $n\pi + i\ln(1+\sqrt{2})$ for even n and $n\pi + i\ln(-1+\sqrt{2})$ for odd n, (d) $i\ln(2\pm\sqrt{3})$.

(5) (a) $e^{-\pi}$, (b) $2, 2e^{\pm i\frac{2}{5}\pi}, 2e^{\pm i\frac{4}{5}\pi}$, (c) $2i, \pm\sqrt{3}-i$, (d) $\sqrt{3}+i, -\sqrt{3}-i$.

(7) (a) π, (b) -0.8, (c) π.

(9) (a) e^i (b) $\dfrac{\pi}{4} + \dfrac{\ln(2)}{2}i$, (c) i.

Chapter 3

Vectors

3.1. Introduction

Our daily lives are ruled by quantities specified by their magnitudes, i.e., the object and the units in which it is measured. Whether it's how many minutes your commute takes, what temperature it is outside, or the density of the ice cubes in your water, these are all examples of *scalars*, otherwise known as physical quantities that can be solely defined by a number.

However, there is a second type of quantity that requires both a *magnitude* (that is always positive) and a *direction* — these are called *vectors*. The most familiar example of a vector is force, i.e., a quantity that has a magnitude, in this case, strength, and a direction of application. Vectors are key to measuring physical quantities, such as velocity, acceleration, displacement, momentum, electric fields, and many more. A vector is usually indicated by either an arrow over a letter representing a physical quantity (e.g., $\vec{a}$) or by a boldface letter (e.g., **a**).

A vector is best described as a *directional segment*, or as a line with a direction and magnitude, and conveniently represented as an arrow in space. If a vector $\vec{a}$ originates from a point A, then the vector $\vec{a}$ is applied to point A, as shown in Figure 3.1. The length of the arrow representing a vector $\vec{a}$ is called the length or the magnitude of a (written as $|a|$ or just a, where $a \geq 0$). Note the use of a to mean the magnitude of $\vec{a}$; for this reason, it is important to make it clear whether you mean a vector or its magnitude (which is a scalar). The magnitude together with the angle provides a complete

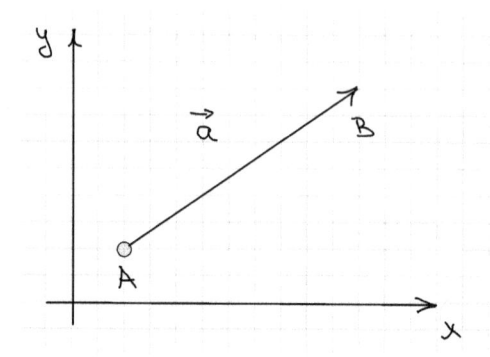

Fig. 3.1. A displacement from point A to point B is a vector $\vec{a}$ that originates from point A and ends at point B.

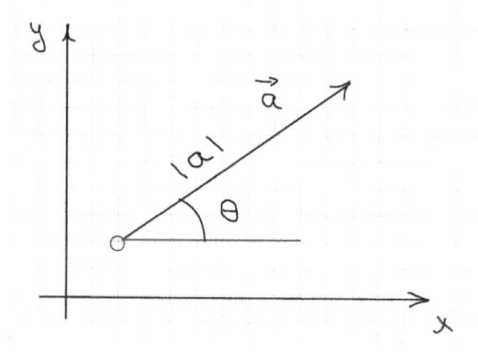

Fig. 3.2. A vector $\vec{a}$ in the xy-plane with a magnitude $|a|$ and direction θ.

description of a vector. For example, in a two-dimensional case a set of the two numbers a and θ uniquely describe vector $\vec{a}$, as illustrated in Figure 3.2. Nevertheless, we also need to define how vectors are added, multiplied by scalars and other vectors, etc. This is known as vector algebra and is discussed in the following.

One interesting mathematical fact is that scalars and vectors do not change their basic properties even if the coordinate system used to describe them is rotated — this is fundamentally their most important feature. The laws of physics written in terms of scalars and vectors do not change simply because we choose to change the orientation of our coordinate systems.

3.2. Vector algebra

For vector algebra, we need to define (a) zero or "null" vector, (b) when two vectors are equal, and (c) a set of linear operations, namely addition, subtraction, multiplication of vectors by scalars, and multiplication of two vectors.

Zero vector: A zero, or null, vector is a vector for which the beginning and end points are the same. A null vector does not have a direction and its magnitude is equal to zero, i.e., $|a| = 0$. Thus, a null vector corresponds to zero in real numbers.

Equal vectors: Two vectors $\vec{a}$ and $\vec{b}$ are equal $\vec{a} = \vec{b}$ when they have the same magnitude $|a| = |b|$ **and** are parallel or located on the same line. Note that vectors in Figure 3.3 are equal even though they have different starting points. This property allows us to move a vector to a position parallel to itself in a diagram without affecting the vector.

A comment for a curious student: Note that strictly speaking in physics we use three kinds of vectors: (a) *free vectors* — they are equal if they have the same magnitude and direction, (b) *sliding vectors* — they are not only equal but also located on the same line, and (c) *tied vectors* — they are not only equal but also originate from the same point (example — a position vector counted from an origin). The good news is that the vector algebra is the same for all three kinds of vectors.

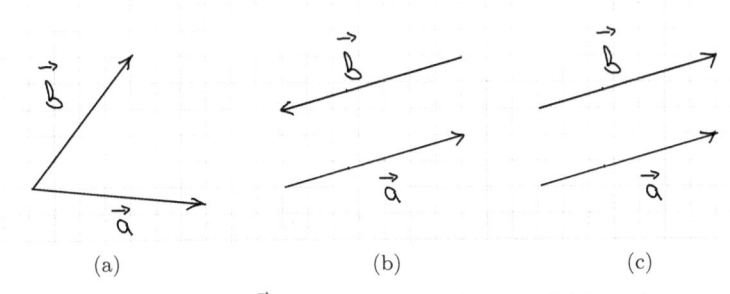

Fig. 3.3. Two vectors $\vec{a}$ and $\vec{b}$ are not equal on (a) and (b) but they are equal on (c).

3.2.1. Addition of vectors

The rules for adding vectors are conveniently described by geometric methods. To add vector $\vec{b}$ to vector $\vec{a}$, first draw vector $\vec{a}$, with its magnitude represented by a convenient scale and then draw vector $\vec{b}$ to the same scale with its tail starting from the tip of $\vec{a}$, as shown in Figure 3.4. The resultant vector $\vec{c} = \vec{a} + \vec{b}$ is the vector drawn from the tail of $\vec{a}$ to the tip of $\vec{b}$. This method is called a triangle method of addition. An alternative graphical procedure for adding two vectors is called the parallelogram rule of addition. In this construction, the tails of the two vectors $\vec{a}$ and $\vec{b}$ are joined together and the resultant vector $\vec{c}$ is the diagonal of a parallelogram formed with $\vec{a}$ and $\vec{b}$ as two of its four sides.

Note that the addition of vectors has the same basic properties as the addition of real numbers, namely

(a) $\vec{a} + \vec{b} = \vec{b} + \vec{a}$ the commutative law of addition,
(b) $(\vec{a} + \vec{b}) + \vec{c} = \vec{a} + (\vec{b} + \vec{c})$ associative law for addition,
(c) there is a zero vector $\vec{0}$ such as $\vec{a} + \vec{0} = \vec{a}$,
(d) for any vector $\vec{a}$, there is an opposite vector $\vec{a}'$ such that $\vec{a} + \vec{a}' = 0$.

The above properties allow us to operate with a sum of vectors in the same way as with a sum of real numbers. And with these properties, we can define the subtraction of vectors.

Note that addition of a vector to a scalar is meaningless because it is not defined. Apples and oranges are not the same types of fruits!

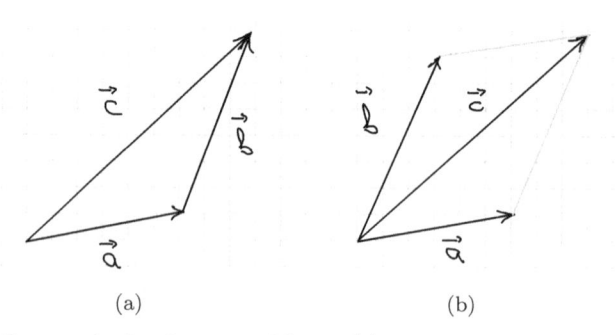

Fig. 3.4. Two methods of vector addition (a) the triangle method and (b) the parallelogram method of addition.

3.2.2. *Subtraction of vectors*

Using the property of addition of vectors, we can define the subtraction of vectors $\vec{a} - \vec{b}$. As we noted in property (d) above, for any vector $\vec{a}$, there is such a vector $\vec{a}'$ that $\vec{a} + \vec{a}' = 0$. From the definition of the sum of two vectors follows that the vector $\vec{a}'$ has the same magnitude as $\vec{a}$ but points in the opposite direction. Thus, for a negative vector $\vec{a}'$, we use a notation $-\vec{a}$ and then $\vec{a} + (-\vec{a}) = 0$.

The operation of vector subtraction makes use of the definition of the negative of a vector. We define the operation $\vec{a} - \vec{b}$ as vector $-\vec{b}$ added to vector $\vec{a}$:

$$\vec{c} = \vec{a} - \vec{b} = \vec{a} + (-\vec{b}). \tag{3.1}$$

Or rearranging equation (3.1), we can write $\vec{c} + \vec{b} = \vec{a}$. This rule is illustrated in Figure 3.5.

3.2.3. *Multiplying a vector by a scalar*

Multiplication of a vector $\vec{a}$ by a scalar α produces a new vector $\vec{b} = \alpha \vec{a}$ that has a magnitude $|\alpha| \cdot |a|$ and pointing in the same direction is $\alpha > 0$ and in the opposite direction if $\alpha < 0$. For $\alpha = 0$, the product gives a null vector. Geometrically, the linear operation of multiplication by a scalar can be viewed as stretching (for $\alpha > 0$) or compressing (for $\alpha < 0$) a vector in α times.

It is important to note that the multiplication of a vector by a scalar should not to be confused with the "scalar product" of two vectors, which is to be discussed in Section 3.5.

Unlike real numbers, there are no positive or negative vectors. Also, for vectors, there is no comparison (larger or smaller).

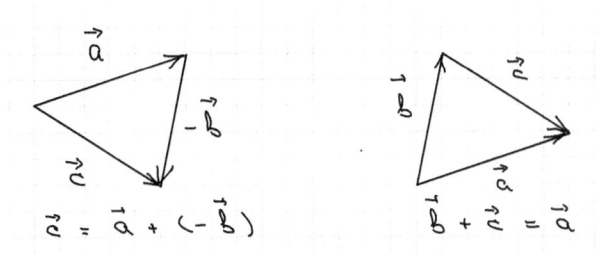

Fig. 3.5. Subtraction of two vectors.

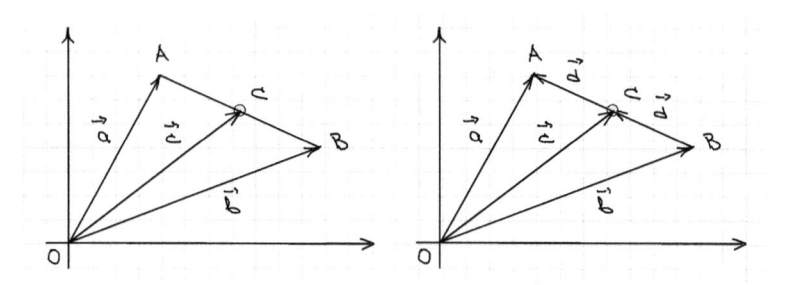

Fig. 3.6. Illustrations for Example 3.1.

Example 3.1. Let $\vec{a}$ and $\vec{b}$ be two vectors originating from the point O. We need to find such a vector $\vec{c}$ in terms of vectors $\vec{a}$ and $\vec{b}$, where the point C is in the middle of the line AB (see Figure 3.6).

First, we draw two vectors: one from C to A and the second from B to C (see the right part of the figure). These two vectors are equal (do you see why), and we name it a vector $\vec{d}$. Then, by using the rule of addition of two vectors, we obtain

$$\vec{c} + \vec{d} = \vec{a},$$
$$\vec{b} + \vec{d} = \vec{c}.$$

Solving this system for $\vec{c}$ gives

$$\vec{c} = \frac{\vec{a} + \vec{b}}{2}.$$
∎

3.3. Base vectors and unit vectors

While geometric methods for adding and subtracting vectors are straightforward, they are unfortunately impractical in applications. Instead, using vector components is much more accurate and leaves less room for mistakes.

3.3.1. *Linear independence*

A linear combination of n vectors is defined as a sum of products of vectors with arbitrary scalars, i.e.,

$$\alpha_1 \vec{a}_1 + \alpha_2 \vec{a}_2 + \cdots \alpha_n \vec{a}_n, \qquad (3.2)$$

where $\alpha_1, \alpha_2, \ldots \alpha_n$ are real numbers.

We call vectors $\vec{a}_1, \vec{a}_2, \ldots, \vec{a}_n$ as *linearly dependent* if there are real numbers $\alpha_1, \alpha_2, \ldots \alpha_n$, such that at least one of them is not equal to zero, the linear combination of the vectors is equal to zero, or

$$\alpha_1 \vec{a}_1 + \alpha_2 \vec{a}_2 + \cdots \alpha_n \vec{a}_n = 0.$$

If at least one of the vectors $\vec{a}_1, \vec{a}_2, \ldots \vec{a}_n$ is zero, then these vectors are also linearly dependent since in this case a coefficient corresponding to a zero vector can be a non-zero number.

We call vectors *linear independent* if the linear combination is equal to zero when *all* the scalar coefficients are equal to zero. We can say that vectors that are not dependent are independent. Note that if among the n vectors there are any $n-1$ vectors that are linearly dependent, then all n vectors are linearly dependent.

Linear combination of two vectors: Two vectors are dependent if and only if they are parallel. Indeed, if vectors $\vec{a}$ and $\vec{b}$ are dependent, then $\alpha \vec{a} + \beta \vec{b} = 0$. In this case, we can express vector $\vec{b}$ as $\vec{b} = -(\alpha/\beta)\vec{a}$ or we have vector $\vec{b}$ as vector $\vec{a}$ multiplied by a scalar. By definition, such vectors are parallel.

Linear combination of three vectors: Three vectors are dependent if they are located on the same plane or on parallel planes. In fact, we assume that three vectors are linear dependent (necessary condition), i.e.,

$$\alpha \vec{a} + \beta \vec{b} + \gamma \vec{c} = 0,$$

when at least one of coefficients is not equal to zero. For our example, let $\gamma \neq 0$, then we can write

$$\vec{c} = -\frac{\alpha}{\gamma} \vec{a} - \frac{\beta}{\gamma} \vec{b}.$$

Introducing notations $\mu = -\alpha/\gamma$ and $\nu = -\beta/\gamma$ we can write for vector $\vec{c}$

$$\vec{c} = \mu \vec{a} + \nu \vec{b}.$$

That is a sum of two vectors, and by the definition of the sum, vector $\vec{c}$ is located in the same plane as vectors $\vec{a}$ and $\vec{b}$. Now, we assume that three vectors are located in the same plane (sufficient condition). First, we exclude the case when any pair of the three vectors is parallel, otherwise that pair is dependent, and then all three

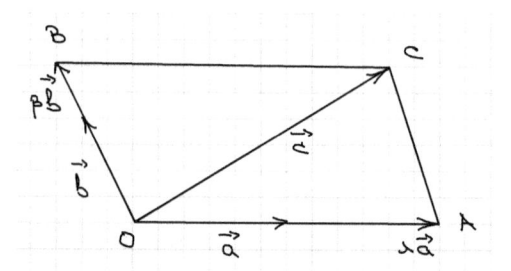

Fig. 3.7. Three vectors in the same plane.

vectors are dependent. Next we move the vectors $\vec{a}, \vec{b}$, and $\vec{c}$ onto the same plane and bring them to the same origin, as in Figure 3.7. Using the definition of the sum of two vectors, we see that vector $\vec{c}$ can be written as

$$\vec{c} = \vec{OA} + \vec{OB}.$$

Since vector $\vec{OA}$ can be represented as $\vec{OA} = \lambda \vec{a}$, and similarly $\vec{OB} = \beta \vec{b}$, then we have

$$\vec{c} = \lambda \vec{a} + \beta \vec{b}.$$

Thus, three vectors are dependent if all three are located in the same plane (a necessary and sufficient condition for the dependence of three vectors).

Linear combination of four vectors: Any four vectors in a three-dimensional space are dependent.[1] Thus, for any three independent vectors $\vec{a}, \vec{b}$ and $\vec{c}$, there are such numbers α, β, γ that any vector $\vec{d}$ can be represented as

$$\vec{d} = \alpha \vec{a} + \beta \vec{b} + \gamma \vec{c}.$$

3.3.2. *Basis vectors*

Three linearly independent vectors $\vec{a}, \vec{b}$, and $\vec{c}$ form a basis in three-dimensional space if any vector $\vec{d}$ can be represented as some linear

---

[1]The proof is similar to the one for linear combination of three vectors.

combination vectors $\vec{a}$, $\vec{b}$, and $\vec{c}$, i.e., if for any vector $\vec{d}$ there are such real numbers α, β, γ that

$$\vec{d} = \alpha\vec{a} + \beta\vec{b} + \gamma\vec{c}. \tag{3.3}$$

There are two fundamental statements: (1) any three vectors not belonging to the same plane (or parallel planes) form a basis in this space and (2) any pair of non-parallel vectors in a plane form a basis. The main advantage of having a basis is that then the linear operations on vectors become the standard linear operations on numbers, i.e., operations on the coordinates of these vectors. For example, when adding two vectors $\vec{a}$ and $\vec{b}$, their coordinates (with respect to any basis α, β, γ) add up. When $\vec{a}$ is multiplied by any number μ, its coordinates are multiplied by that number. Any three linear independent vectors together with an origin point form an *affine* coordinate system. Each vector $\vec{d}$ can be expanded in terms of the basis vectors so that the vector is uniquely described by three numbers, namely α, β, γ corresponding to this vector.

The Cartesian rectangular coordinate system is a special case of an affine coordinate system corresponding to a triple of mutually orthogonal and unit basis vectors. They are defined in the following.

3.3.3. *The Cartesian coordinates and unit vectors*

We define a unit vector as a vector with a magnitude precisely equal to *one*, pointing in a specific direction. It lacks dimension, such as meters, because its primary purpose is to specify a direction In the case of a Cartesian rectangular system, we indicate the basis vectors not by the letters $\vec{a}$, $\vec{b}$, and $\vec{c}$ but also by the letters $\hat{i}$, $\hat{j}$, and $\hat{k}$. The three vectors are mutually orthogonal and with the directions coinciding with the directions of the x, y and z axes respectively. For example, Figure 3.8 shows unit vectors $\hat{i}$ and $\hat{j}$ in the x,y-plane.

Thus, using unit vectors in the Cartesian coordinate system, we can write any vector in the form

$$\vec{d} = \alpha\hat{i} + \beta\hat{j} + \gamma\hat{k}.$$

The numbers α, β, and γ are called Cartesian rectangular coordinates of the vector $\vec{d}$. For a vector $\vec{a}$ in an x,y plane, we write

$$\vec{a} = \alpha\hat{i} + \beta\hat{j}.$$

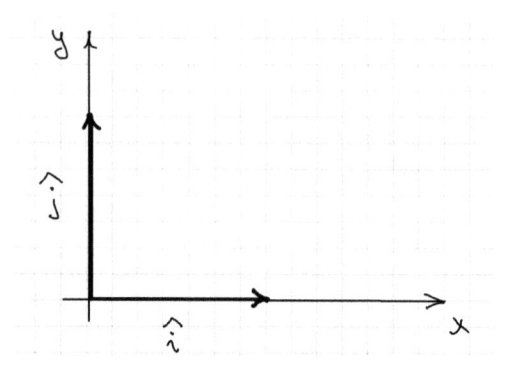

Fig. 3.8. Unit vectors $\hat{i}$ and $\hat{j}$ in the x,y-plane.

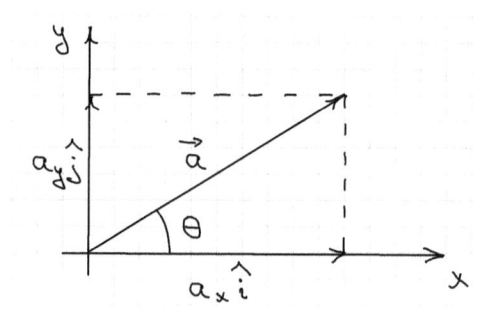

Fig. 3.9. A vector $\vec{a}$ written by using vector components $\vec{a} = a_x\hat{i} + a_y\hat{j}$.

Let us be specific by considering a vector $\vec{a}$ lying in the xy-plane and making an arbitrary angle θ with the positive x-axis (see Figure 3.9). Then, using trigonometry, we can easily see that in equation (3.4) $\alpha = a_x$ and $\beta = a_y$ thus giving

$$\vec{a} = a_x\hat{i} + a_y\hat{j}, \qquad (3.4)$$

where

$$a_x = |a|\cos\theta, \quad a_y = |a|\sin\theta. \qquad (3.5)$$

Note that these components can be positive or negative, and the signs of the components a_x and a_y depend on the angle θ. When solving problems, you can specify a vector $\vec{a}$ in a plane either with its components a_x and a_y or with its magnitude $|a|$ and direction θ, where

$$|a| = \sqrt{a_x^2 + a_y^2}, \quad \theta = \arctan\left(\frac{a_y}{a_x}\right). \tag{3.6}$$

Therefore, a vector in two-dimensional space requires two components to fully describe both its direction and its magnitude. For example, a displacement in space may be thought of as the sum of displacements along the x and y directions.

Example 3.2. The initial point of vector $\vec{a}$ is located at a point $x_1 = 3, y_1 = 1$, and the terminal point is at this point $x_2 = -2$, $y_2 = 4$, as shown in Figure 3.10. Find the vector components a_x and a_y. While the vector does not start from the origin, it can easily be written as $a_x = x_2 - x_1 = -5$ and $a_y = y_2 - y_1 = 3$, or

$$\vec{a} = -5\hat{i} + 3\hat{j}.$$ ∎

It is important to note that in this chapter we are primarily working with two-dimensional coordinates (x, y). However, a generalization for a three-dimensional (x, y, z) system is straightforward. For example, a position of a particle in three-dimensional space can be written in a unique way as

$$\vec{r} = x\hat{i} + y\hat{j} + z\hat{k}, \tag{3.7}$$

where x, y, z are vector components in the Cartesian coordinate system.

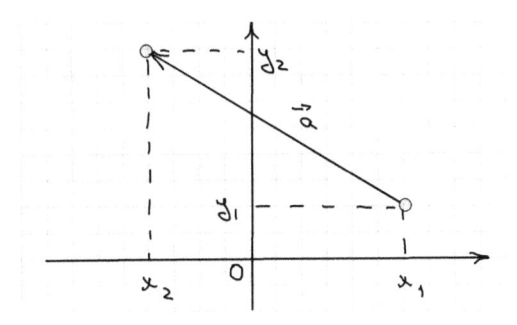

Fig. 3.10. A vector originating from a point different from the origin.

Example 3.3. Let $\vec{a} = 4\hat{i} + 2\hat{j} - 4\hat{k}$. Find a unit vector in the same direction as the vector $\vec{a}$. The norm of vector $\vec{a}$ is

$$|a| = \sqrt{a_x^2 + a_y^2 + a_z^2} = \sqrt{4^2 + 2^2 + 4^4} = 6.$$

Then, the unit vector in the same direction as vector $\vec{a}$ is

$$\hat{u} = \frac{1}{6}\vec{a} = \frac{2}{3}\hat{i} + \frac{1}{3}\hat{j} - \frac{2}{3}\hat{k}. \qquad \blacksquare$$

3.4. Vector algebra with vector components

Now, we can consider the addition and subtraction of vectors in terms of their components. The sum of two vectors $\vec{a}$ and $\vec{b}$ is found by simply adding their components, i.e.,

$$\vec{c} = \vec{a} + \vec{b} = a_x\hat{i} + a_y\hat{j} + b_x\hat{i} + b_y\hat{j}$$
$$= (a_x + b_x)\hat{i} + (a_y + b_y)\hat{j} = c_x\hat{i} + c_y\hat{j}.$$

The components of the resultant vector $\vec{c}$ are

$$\begin{aligned} c_x &= a_x + b_x, \\ c_y &= a_y + b_y. \end{aligned} \qquad (3.8)$$

And the difference between the two vectors can be written by subtracting their components

$$\vec{c} = \vec{a} - \vec{b} = a_x\hat{i} + a_y\hat{j} - b_x\hat{i} - b_y\hat{j}$$
$$= (a_x - b_x)\hat{i} + (a_y - b_y)\hat{j} = c_x\hat{i} + c_y\hat{j}$$

with

$$\begin{aligned} c_x &= a_x - b_x, \\ c_y &= a_y - b_y. \end{aligned} \qquad (3.9)$$

We obtain the magnitude of $\vec{c}$ and the angle it makes with the x-axis from its components by using the relationships

$$c = \sqrt{c_x^2 + c_y^2} = \sqrt{(a_x + b_x)^2 + (a_y + b_y)^2},$$

$$\tan\theta = \frac{c_y}{c_x} = \frac{a_y + b_y}{a_x + b_x}.$$

Multiplication of a vector by a scalar λ is written as

$$\vec{c} = \lambda\vec{a} = \lambda a_x \hat{i} + \lambda a_y \hat{j}, \qquad (3.10)$$

with $c_x = \lambda a_x$ and $c_y = \lambda a_y$.

Example 3.4. Find vectors $\vec{a}$ and $\vec{b}$, if $\vec{a} + \vec{b} = 2\hat{i} + 3\hat{j}$ and $\vec{a} - \vec{b} = 4\hat{i} + 5\hat{j}$.

Using the component form for addition (3.8) and subtraction (3.9) of two vectors, we can write

$$a_x + b_x = 2,$$
$$a_y + b_y = 3,$$
$$a_x - b_x = 4,$$
$$a_y - b_y = 5.$$

By solving the above linear system of equations, we obtain $a_x = 3$, $a_y = 4$, $b_x = -1$, and $b_y = -1$, or $\vec{a} = 3\hat{i} + 4\hat{j}$ and $\vec{b} = -1\hat{i} - 1\hat{j}$. ∎

3.5. Scalar and vector products

There are two kinds of products of vectors, but neither can be made using common algebraic multiplication. The first kind is called the scalar (or dot) product. It produces a result that is *a scalar* quantity (just a number). The second kind is called the vector (or cross) product. It yields *a new vector*.

3.5.1. *Scalar product of two vectors*

A scalar product of two multiplied vectors is a number equal to the product of the magnitudes (lengths) of these vectors by the cosine of the angle between them. Mathematically, the scalar product of the vectors $\vec{a}$ and $\vec{b}$ is written as $\vec{a} \cdot \vec{b}$ and defined as

$$\vec{a} \cdot \vec{b} = ab\cos\theta, \qquad (3.11)$$

where a is the magnitude of the vector $\vec{a}$, b is the magnitude of $\vec{b}$, and θ is the angle between the directions of $\vec{a}$ and $\vec{b}$. Note that there are actually two such angles: θ and $2\pi-\theta$. Either can be used because their cosines are the same $\cos(\theta) = \cos(2\pi - \theta)$.

The scalar product (3.11) can be rewritten as

$$\vec{a} \cdot \vec{b} = (a\cos\theta)b = a(b\cos\theta).$$

Then, from the equation above and Figure 3.11, we can see that $a\cos\theta$ is a projection of vector $\vec{a}$ on axis defined by the direction of vector $\vec{b}$. Then, we have the second definition for the scalar product, namely a scalar product of two vectors is a product of a magnitude of one of the vectors by a projection of the second vector on the direction determined by the first vector.

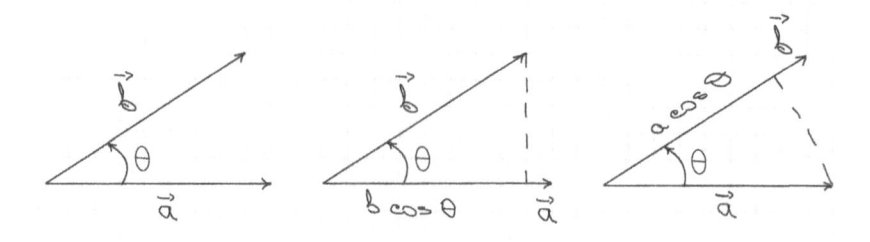

The scalar product of two vectors has the following properties:

(a) $\vec{a} \cdot \vec{b} = \vec{b} \cdot \vec{a}$,
(b) $(\vec{a} + \vec{b}) \cdot \vec{c} = \vec{a} \cdot \vec{c} + \vec{b} \cdot \vec{c}$,
(c) $\vec{a} \cdot \vec{a} \geq 0$.

From equation (3.11) follows a geometric property of the scalar product, namely if the magnitudes of vectors $\vec{a}$ and $\vec{b}$ are not equal to zero but their scalar product is zero, then the *two vectors are orthogonal* because $\cos\theta = 0$.

When two vectors are defined by their Cartesian coordinates, then the scalar product can be written as

$$\vec{a} \cdot \vec{b} = a_x b_x + a_y b_y + a_z b_z. \qquad (3.12)$$

Equation (3.12) can be easily proved using properties of unit vectors $\hat{i}$, $\hat{j}$, and $\hat{k}$ since they are all orthogonal to each other. For our purpose, we are going to work with vectors located in the

Fig. 3.11. The scalar product of two vectors.

x, y-plane. From the definition for the dot product and the property that $\cos 0 = 1$ and $\cos(\pi/2) = 0$, we have

$$\hat{i} \cdot \hat{i} = (1)(1)\cos 0 = 1, \quad \hat{i} \cdot \hat{j} = \hat{j} \cdot \hat{i} = (1)(1)\cos \pi/2 = 0,$$
$$\hat{j} \cdot \hat{j} = (1)(1)\cos 0 = 1.$$

Taking into account that $\vec{a} = a_x\hat{i} + a_y\hat{j}$ and $\vec{b} = b_x\hat{i} + b_y\hat{j}$, we obtain after multiplying all terms

$$\vec{a} \cdot \vec{b} = a_x b_x \hat{i} \cdot \hat{i} + a_x b_y \hat{i} \cdot \hat{j} + a_y b_x \hat{j} \cdot \hat{i} + a_y b_y \hat{j} \cdot \hat{j} = a_x b_x + a_y b_y.$$

Thus, the scalar product of two vectors can be written in two forms:

$$\begin{aligned} \vec{a} \cdot \vec{b} &= a \cdot b \cos \theta, \\ \vec{a} \cdot \vec{b} &= a_x b_x + a_y b_y. \end{aligned} \quad (3.13)$$

An important example of a scalar product application is the work done by a *constant* force $\vec{F}$ acting on an object over a displacement $\vec{d}$. If the force is directed along the displacement, then the work is equal to a product of the two magnitudes, i.e., $W = Fd$. If the direction of the force does not coincide with the direction of displacement, then the work is equal

$$W = Fd \cos \theta,$$

where θ is the angle between the direction of the force and the displacement.[2] Mathematically, this can be expressed as a scalar product of the two vectors:

$$W = \vec{F} \cdot \vec{d}. \quad (3.14)$$

If the force is *not constant*, or the direction of the displacement changes with time, then for a small displacement $d\vec{r}$, we can write

$$dW = \vec{F} \cdot d\vec{r}. \quad (3.15)$$

---

[2]The work done by a force can be positive, negative, or zero, depending on the orientation of the force and the displacement vectors. Positive work is done when the force and the displacement are in the same direction, negative work is done when they are in opposite directions, and zero work is done when the force is perpendicular to the displacement.

The total work can be evaluated by integrating the scalar product over the path of the object

$$W = \int_L \vec{F} \cdot d\vec{r}, \tag{3.16}$$

where $d\vec{r}$ is an infinitesimal displacement vector along the path of integration. This is a line integral of second kind considered in Chapter 6, Section 6.3.

Note the following useful property for a length or magnitude of a vector:

$$\vec{a} \cdot \vec{a} = a_x a_x + a_y a_y = a_x^2 + a_y^2 = a^2. \tag{3.17}$$

In some cases, we may want to know the angle between two vectors. Using the definitions for the scalar product equation (3.13), we write

$$\cos\theta = \frac{\vec{a} \cdot \vec{b}}{ab}.$$

Together with equation for a magnitude of a vector (3.17), it gives

$$\cos\theta = \frac{a_x b_x + a_y b_y}{\left(\sqrt{a_x^2 + a_y^2}\right)\left(\sqrt{b_x^2 + b_y^2}\right)}. \tag{3.18}$$

From the definition of the scalar product follows that two vectors $\vec{a}$ and $\vec{b}$ are parallel if the angle between them is equal to zero. This condition is equivalent to $\vec{b} = \lambda \vec{a}$. Then, we can show that two vectors are parallel if

$$\frac{a_x}{b_x} = \frac{a_y}{b_y} = \frac{a_z}{b_z}.$$

Two vectors are perpendicular if the angle between them is $\pi/2$. But this means that the scalar product is zero since $\cos(\pi/2) = 0$. Using vector components, we can write the condition of orthogonality as

$$a_x b_x + a_y b_y + a_z b_z = 0. \tag{3.19}$$

There is an interesting property of the scalar product, namely the scalar product is invariant under all possible rotational transformations. This is a part of more general observation that laws of physics

Vectors

are invariant under rotations of the coordinate systems. Rotational symmetry, in particular, implies conservation of angular momentum.

Example 3.5. In this example, we demonstrate that the scalar product remains invariant under rotational transformations. First, let's explore the relationship between the components of a vector in two coordinate systems that are rotated relative to each other. Let x and y be components of vector $\vec{a}$ in the original coordinate system, and x' and y' are components in the system rotated by angle θ (see Figure 3.12). Let r and φ be the polar coordinates of the vector $\vec{a}$ in the rotated system. Since the magnitude of the vector $\vec{a}$ does not change under rotation, we can write

$$x = r\cos(\varphi + \theta) = r\cos\varphi\cos\theta - r\sin\varphi\sin\theta = x'\cos\theta - y'\sin\theta,$$
$$y = r\sin(\varphi + \theta) = r\cos\varphi\sin\theta + r\sin\varphi\cos\theta = x'\sin\theta + y'\cos\theta.$$
(3.20)

By assuming that we know x and y, we can solve the system of two linear equations above to determine x' and y' as follows:

$$\begin{aligned} x' &= x\cos\theta + y\sin\theta, \\ y' &= -x\sin\theta + y\cos\theta. \end{aligned}$$
(3.21)

Next, we consider a scalar product of two vectors $\vec{a}$ and $\vec{b}$ by using the components (3.13). Then, in the original coordinates, the scalar product is

$$\vec{a} \cdot \vec{b} = a_x b_x + a_y b_y.$$

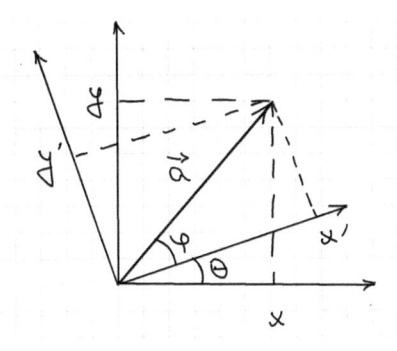

Fig. 3.12. Vector components in a rotated coordinate system.

By using transformation (3.21) between rotated coordinate systems, we obtain in the rotated coordinate systems

$$a'_x = a_x \cos\theta + a_y \sin\theta,$$
$$a'_y = -a_x \sin\theta + a_y \cos\theta,$$
$$b'_x = b_x \cos\theta + b_y \sin\theta,$$
$$b'_y = -b_x \sin\theta + b_y \cos\theta.$$

By substituting it into

$$\vec{a} \cdot \vec{b} = a'_x b'_x + a'_y b'_y,$$

we obtain that the scalar product is invariant under rotational transformation, i.e.,

$$a'_x b'_x + a'_y b'_y = a_x b_x + a_y b_y.$$ ∎

3.5.2. Vector product of two vectors

The vector product of two vectors $\vec{a}$ and $\vec{b}$ is a vector $\vec{c}$ that is represented as

$$\vec{c} = \vec{a} \times \vec{b}. \tag{3.22}$$

The vector product satisfies to the three following conditions:

(1) The magnitude (or length) of vector $\vec{c}$ is a product

$$c = ab \sin\varphi, \tag{3.23}$$

where φ is the angle between vectors $\vec{a}$ and $\vec{b}$, and we take it to be the smallest of two possible angles, so φ ranges from 0° to 180° (because the magnitude c is always positive).

(2) Vector $\vec{c}$ is orthogonal to both vectors $\vec{a}$ and $\vec{b}$, i.e., the plane formed by the two vectors, as shown in Figure 3.13.

(3) There are two directions perpendicular to a given plane. The direction of the vector product follows the right-hand rule. Envision rotating vector $\vec{a}$ around the perpendicular line until it is aligned with $\vec{b}$, selecting the smaller of the two possible angles between $\vec{a}$ and $\vec{b}$. To execute this, curl the fingers of your right hand around the perpendicular line so that the fingertips point in the direction of rotation; your thumb will then indicate the direction of $\vec{c}$. This right-hand rule is also referred to as the

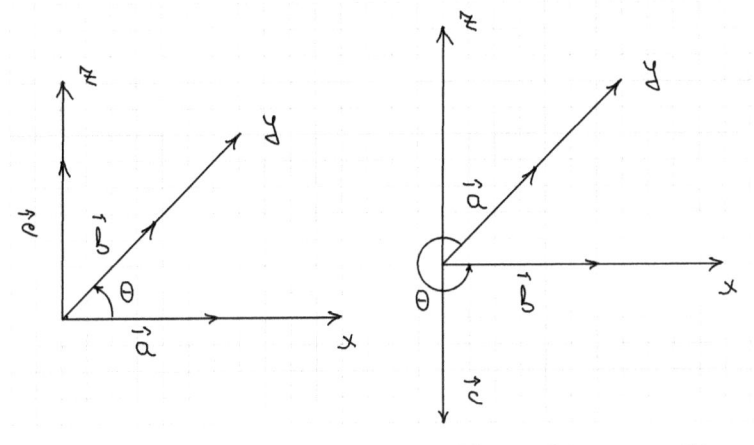

Fig. 3.13. The vector product of two vectors. The resulting vector $\vec{c}$ is perpendicular to the plane formed by $\vec{a}$ and $\vec{b}$ vectors.

right-handed screw rule, which involves the following steps:

(1) position a screw perpendicular to the plane created by vectors $\vec{a}$ and $\vec{b}$,
(2) rotate the screw in the direction from $\vec{a}$ to $\vec{b}$, and
(3) the direction of the screw's movement indicates the direction of the resulting vector $\vec{c} = \vec{a} \times \vec{b}$.

The algebra of the vector product of two vectors has the following properties:

(a) $\vec{a} \times \vec{b} = -(\vec{b} \times \vec{a})$,
(b) $\vec{a} \times (\vec{b} + \vec{c}) = \vec{a} \times \vec{b} + \vec{a} \times \vec{c}$,
(c) $\vec{a} \times \vec{a} = 0$ for any vector $\vec{a}$.

If we know the components of $\vec{a}$ and $\vec{b}$, we can calculate the components of the vector product $\vec{c} = \vec{a} \times \vec{b}$ using the definition of the vector product applied to unit vectors:

$$\hat{i} \times \hat{i} = \hat{j} \times \hat{j} = \hat{k} \times \hat{k} = 0,$$
$$\hat{i} \times \hat{j} = -\hat{j} \times \hat{i} = \hat{k},$$
$$\hat{j} \times \hat{k} = -\hat{k} \times \hat{j} = \hat{i},$$
$$\hat{k} \times \hat{i} = -\hat{i} \times \hat{k} = \hat{j}.$$

If you are feeling curious, you can check the above identities by using the definition for the vector product.

Now, we can write

$$\vec{c} = \vec{a} \times \vec{b} = (a_x\hat{i} + a_y\hat{j} + a_z\hat{k}) \times (b_x\hat{i} + b_y\hat{j} + b_z\hat{k}).$$

Or by multiplying and regrouping terms we obtain

$$\vec{c} = (a_yb_z - a_zb_y)\hat{i} + (a_zb_x - a_xb_z)\hat{j} + (a_xb_y - a_yb_x)\hat{k}.$$

Thus, the components of the resulting vector are given by

$$\begin{aligned} c_x &= a_yb_z - a_zb_y, \\ c_y &= a_zb_x - a_xb_z, \\ c_z &= a_xb_y - a_yb_x. \end{aligned} \quad (3.24)$$

In most problems and applications in this book, vectors $\vec{a}$ and $\vec{b}$ are located in the x, y-plane. Therefore, only one component of vector $\vec{c}$ is not equal to zero, namely

$$c_z = a_xb_y - a_yb_x. \quad (3.25)$$

Note that the sign of c_z (positive or negative) is determined by the values of the Cartesian coordinates of vectors $\vec{a}$ and $\vec{b}$. By using vector components, we do not need to worry about using the right-hand rule.

The module of a vector product $\vec{a} \times \vec{b}$ has a useful property, namely it is equal to the area of a parallelogram with sides $\vec{a}$ and $\vec{b}$ started from the same origin.

Here we bring examples on applications of the vector product (or also called as a cross product). The magnetic force acting on a moving charged particle in a magnetic field is a classic example of the application of the cross product. When a charged particle with charge q moves with a velocity $\vec{v}$ in a magnetic field $\vec{B}$, the force $\vec{F}$ acting on the particle, according to the Lorentz force law, is given by the cross product of the velocity and the magnetic field:

$$\vec{F} = q\vec{v} \times \vec{F}. \quad (3.26)$$

The resulting force is perpendicular to both the velocity of the particle and the magnetic field, and its direction is determined by the right-hand rule.

In physics, torque is a measure of the tendency of a force to rotate an object about an axis. When a force is applied to a rigid body and causes it to rotate, the torque $\vec{\tau}$ is calculated as the cross product of the position vector $\vec{r}$ and the force vector $\vec{F}$:

$$\vec{\tau} = \vec{r} \times \vec{F}, \qquad (3.27)$$

where $\vec{r}$ is the position vector pointing from the axis of rotation to the point where the force $\vec{F}$ is applied. Torque is an important concept in various areas of physics, particularly in mechanics and dynamics. It is essential in understanding rotational motion, the equilibrium of objects, and the behavior of rotating systems.

Angular momentum is one more example for applications of the cross product. In classical mechanics, the angular momentum of a particle is defined as the cross product of the position vector $\vec{r}$ and the linear momentum $\vec{p}$ of the particle:

$$\vec{l} = \vec{r} \times \vec{p}, \qquad (3.28)$$

where $\vec{r}$ is the position vector of the particle relative to the axis of rotation.

Example 3.6. Show that vectors $\vec{a} = 3\hat{i} + 2\hat{j} + 1\hat{k}$ and $\vec{b} = 1\hat{i} - 2\hat{j} + 1\hat{k}$ are orthogonal, and find a vector $\vec{c}$ that is perpendicular to both $\vec{a}$ and $\vec{b}$.

By definitions of the orthogonality in the component form (3.19), we have

$$\vec{a} \cdot \vec{b} = a_x b_x + a_y b_y + a_z b_z = 3 - 4 + 1 = 0.$$

A vector $\vec{c}$ that is a vector product

$$\vec{c} = \vec{a} \times \vec{b}$$

is perpendicular to both $\vec{a}$ and $\vec{b}$. Then, using the component form of the vector product (3.24), we obtain $c_x = a_y b_z - a_z b_y = 4$, $c_y =_z b_x - a_x b_z = -2$ and $c_z = a_x b_y - a_y b_x = -8$, or $\vec{c} = 4\hat{i} - 2\hat{j} - 8\hat{k}$. ∎

3.5.3. *Scalar and vector triple products*

The *scalar triple product* of three vectors $\vec{a}$, $\vec{b}$, and $\vec{c}$ is defined as

$$[\vec{a} \times \vec{b}] \cdot \vec{c}. \qquad (3.29)$$

The triple product has a number of a useful property, namely

$$[\vec{a} \times \vec{b}] \cdot \vec{c} = \vec{a} \cdot [\vec{b} \times \vec{c}].$$

It allows us to write the scalar triple product in the form $[\vec{a}, \vec{b}, \vec{c}]$, without specifying which two vectors (the first two or the last two) are multiplied as a vector product. Using vectors' components $\vec{a} = a_x \hat{i} + a_y \hat{j} + a_z \hat{k}$, $\vec{b} = b_x \hat{i} + b_y \hat{j} + b_z \hat{k}$, and $\vec{c} = c_x \hat{i} + c_y \hat{j} + c_z \hat{k}$, we can write

$$[\vec{a}, \vec{b}, \vec{c}] = \begin{vmatrix} a_x & a_y & a_z \\ b_x & b_y & b_z \\ c_x & c_y & c_z \end{vmatrix}. \tag{3.30}$$

In a geometric sense, the scalar triple product is equal to the volume of the parallelepiped constructed from the vectors $\vec{a}$, $\vec{b}$, and $\vec{c}$ transformed to a common origin. If vector $\vec{c}$ is itself a vector product, e.g., $\vec{c} \times \vec{d}$, then Lagrange identity states that

$$(\vec{a} \times \vec{b}) \cdot (\vec{c} \times \vec{d}) = (\vec{a} \cdot \vec{c})(\vec{b} \cdot \vec{d}) - (\vec{a} \cdot \vec{d})(\vec{b} \cdot \vec{c}).$$

The *vector triple product* is defined as $\vec{a} \times (\vec{b} \times \vec{c})$, and it can be written as

$$\vec{a} \times (\vec{b} \times \vec{c}) = \vec{b}(\vec{a} \cdot \vec{c}) - \vec{c}(\vec{a} \cdot \vec{b}). \tag{3.31}$$

Example 3.7. Find the value of

$$(\vec{a} \times \vec{b})^2 + (\vec{a} \cdot \vec{b})^2.$$

There are three ways to find the value: (1) by using the component forms for the vector and scalar products, (2) by using the Lagrange identity, and (3) by using definitions (3.11) for the scalar and (3.23) for the vector products. We proceed with the latest, as the shortest. Then, we have

$$\vec{a} \times \vec{b} = ab \sin \theta, \quad \vec{a} \cdot \vec{b} = ab \cos \theta,$$

$$(\vec{a} \times \vec{b})^2 + (\vec{a} \cdot \vec{b})^2 = a^2 b^2 \sin^2 \theta + a^2 b^2 \cos^2 \theta = a^2 b^2 (\sin^2 + \cos^2) = a^2 b^2.$$

∎

3.6. Vector calculus

Vector calculus deals with vector differentiation and integration as well as scalar and vector fields. These concepts are indispensable in many areas of physics and engineering, including electromagnetism, fluid dynamics, computer graphics and computer vision, electrical circuit analysis, and many more.

3.6.1. *Differentiation of vectors*

Let vector $\vec{a}(t)$ be a function of a scalar argument t. It means that for every value of t we associate by a certain rule a vector.[3] For example, a position vector $\vec{r}(t)$ and a velocity vector $\vec{v}(t)$ are functions of time t.

We define a derivative of the vector similar to a derivative of a function $f(x)$ as

$$\frac{d\vec{a}(t)}{dt} = \lim_{\Delta t \to 0} \frac{\vec{a}(t + \Delta t) - \vec{a}(t)}{\Delta t}, \qquad (3.32)$$

assuming that the limit exists. Note that the derivative of a vector $d\vec{a}(t)/dt$ is also a vector, which in general, is not parallel to the original vector $\vec{a}(t)$, as illustrated in Figure 3.14.

Equation (3.32) can be rewritten as

$$\vec{a}(t+dt) = \vec{a}(t) + \frac{d\vec{a}(t)}{dt} dt + o(dt). \qquad (3.33)$$

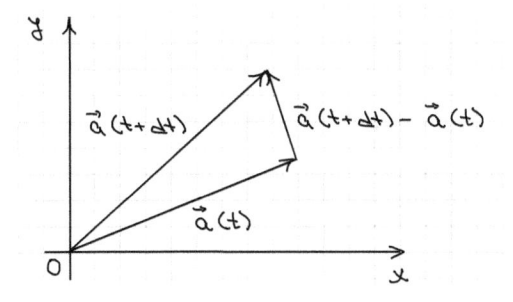

Fig. 3.14.   A derivative of a vector.

---

[3] Note that such a vector originates from the same point, i.e., for $\vec{a}(x, y, t)$, variables x and y do not change as the parameter t changes.

As in case of scalar functions, we can write a Taylor series for vectors as

$$\vec{a}(t) = \vec{a}(t_0) + \frac{d\vec{a}(t_0)}{dt}(t-t_0) + \frac{1}{2}\frac{d^2\vec{a}(t_0)}{dt^2}(t-t_0)^2 + \cdots. \quad (3.34)$$

Using the definition of derivative (3.32), we can derive two useful properties:

(1) For a linear combination of two vectors, we have

$$\frac{d}{dt}(c_1\vec{a}(t) + c_2\vec{b}(t)) = c_1\frac{d\vec{a}(t)}{dt} + c_2\frac{d\vec{b}(t)}{dt},$$

where c_1 and c_2 are constants.

(2) And for a product of a scalar function on a vector function,

$$\frac{d}{dt}[f(t)\,\vec{a}(t)] = \frac{df(t)}{dt}\vec{a}(t) + f(t)\frac{d\vec{a}(t)}{dt}.$$

In particular, if $\vec{a}$ is a constant, then

$$\frac{d}{dt}[f(t)\,\vec{a}] = \frac{df(t)}{dt}\vec{a}.$$

Thus, we see that the derivative of a constant vector multiplied by a scalar function is a vector parallel to the vector itself.

Let us find the derivative of a scalar product of two vectors $\vec{a}(t)$ and $\vec{b}(t)$. From the definition of derivative (3.32), we have

$$\frac{d}{dt}(\vec{a}(t) \cdot \vec{b}(t)) = \frac{\vec{a}(t+dt) \cdot \vec{b}(t+dt) - \vec{a}(t) \cdot \vec{b}(t)}{dt}.$$

By applying (3.33), we obtain

$$\frac{d}{dt}(\vec{a}(t) \cdot \vec{b}(t)) = \frac{\left(\vec{a}(t) + \frac{d\vec{a}(t)}{dt}dt\right) \cdot \left(\vec{b}(t) + \frac{d\vec{b}(t)}{dt}dt\right) - \vec{a}(t) \cdot \vec{b}(t)}{dt}$$

or

$$\frac{d}{dt}(\vec{a}(t) \cdot \vec{b}(t))$$
$$= \frac{\vec{a}(t) \cdot \vec{b}(t) + \vec{a}(t) \cdot \frac{d\vec{b}(t)}{dt}dt + \frac{d\vec{a}(t)}{dt} \cdot \vec{b}(t)dt + \frac{d\vec{a}(t)}{dt} \cdot \frac{d\vec{b}(t)}{dt}(dt)^2 - \vec{a}(t) \cdot \vec{b}(t)}{dt}.$$

By neglecting the term with $(dt)^2$, we finally obtain for a scalar product of two vectors

$$\frac{d}{dt}(\vec{a}\cdot\vec{b}) = \frac{d\vec{a}}{dt}\cdot\vec{b} + \vec{a}\cdot\frac{d\vec{b}}{dt}. \quad (3.35)$$

In the same manner, we can obtain for a vector product

$$\frac{d}{dt}(\vec{a}\times\vec{b}) = \frac{d\vec{a}}{dt}\times\vec{b} + \vec{a}\times\frac{d\vec{b}}{dt}. \quad (3.36)$$

It is interesting to consider a velocity vector as a product of its magnitude on the unit vector in the direction of the velocity vector, namely

$$\vec{v} = |v(t)|\hat{v}(t).$$

Then, the acceleration is

$$\vec{a}(t) = \frac{\vec{v}(t)}{dt} = \frac{d|v(t)|}{dt}\hat{v}(t) + |v(t)|\frac{\hat{v}(t)}{dt}.$$

We can now infer that the acceleration can be non-zero even if the magnitude of velocity $|v(t)|$ is constant, but the direction of the velocity changes. This is what we define as uniform circular motion.

From the scalar product of two vectors $\vec{b}\cdot\vec{c} = b\,c\cos(\theta)$ (where θ is the angle between the two vectors) follows that $\vec{b}\cdot\vec{b} = b^2$. Assume that vector $\vec{b}$ is a function of time. Applying the rule of differentiating a product to both side of the last equation we get

$$\frac{d}{dt}(\vec{b}\cdot\vec{b}) = \frac{d\vec{b}}{dt}\cdot\vec{b} + \vec{b}\cdot\frac{d\vec{b}}{dt} = 2\vec{b}\cdot\frac{d\vec{b}}{dt}$$

and

$$\frac{d}{dt}b^2 = 2b\frac{db}{dt}.$$

Then,

$$2\vec{b}\cdot\frac{d\vec{b}}{dt} = 2b\frac{db}{dt}.$$

We can rewrite the last equation as

$$\frac{db}{dt} = \frac{\vec{b}\cdot\frac{d\vec{b}}{dt}}{b}.$$

If the magnitude of vector $\vec{b}$ is constant, then $db/dt = 0$ and from the above equation follows

$$\vec{b} \cdot \frac{d\vec{b}}{dt} = 0,$$

or in this case, vector $\vec{b}$ is orthogonal to its derivative $d\vec{b}/dt$. From this property, we can see that for circular motion with constant radius $r = \text{const}$, the velocity $\vec{v} = d\vec{r}/dt$ is orthogonal to the radius $\vec{v} \perp \vec{r}$, and for uniform circular motion with constant speed $v = \text{const}$, we have that the acceleration vector $\vec{a} = d\vec{v}/dt$ is orthogonal to the velocity $\vec{a} \perp \vec{v}$.

Let us consider the derivative of the angular momentum

$$\vec{L} = \vec{r} \times m\vec{v}.$$

The derivative then can be written as

$$\frac{d\vec{L}}{dt} = \frac{d}{dt}(\vec{r} \times m\vec{v}) = m\frac{d\vec{r}}{dt} \times \vec{v} + m\vec{r} \times \frac{d\vec{v}}{dt} = m\vec{v} \times \vec{v} + \vec{r} \times m\vec{a}.$$

Since the first term is equal to zero (do you see why?) and $\vec{r} \times m\vec{a} = \vec{r} \times \vec{F} = \vec{\tau}$, that is, torque, it follows that the derivative of the angular momentum is equal to torque (otherwise known as Newton's second law for rotational motion):

$$\frac{d\vec{L}}{dt} = \vec{r} \times \vec{F} = \vec{\tau}.$$

Differentiation by using vector components: It is helpful to use vector components for differentiation. For example, in Cartesian coordinates, a vector $\vec{a}$ is represented as

$$\vec{a}(x, y, t) = a_x(x, y, t)\hat{i} + a_y(x, y, t)\hat{j}.$$

Then,

$$\frac{d\vec{a}(x, y, t)}{dt}$$
$$= \lim_{\Delta t \to 0} \frac{a_x(x, y, t+\Delta t)\hat{i} + a_y(x, y, t+\Delta t)\hat{j} - a_x(x, y, t)\hat{i} + a_y(x, y, t)\hat{j}}{\Delta t}$$

or

$$\frac{d\vec{a}(x,y,t)}{dt} = \lim_{\Delta t \to 0} \frac{(a_x(x,y,t+\Delta t) - a_x(x,y,t))\hat{i} + (a_y(x,y,t+\Delta t) - a_y(x,y,t))\hat{j}}{\Delta t}.$$

Since both x and y are independent of the parameter t, we obtain

$$\frac{d\vec{a}(t)}{dt} = \frac{da_x}{dt}\hat{i} + \frac{da_y}{dt}\hat{j}. \tag{3.37}$$

Such vector component differentiation is common in physics and engineering. For example, for a position vector

$$\vec{r}(t) = x(t)\hat{i} + y(t)\hat{j} + z(t)\hat{k},$$

we have

$$\frac{d\vec{r}(t)}{dt} = \vec{v}(t) = \frac{dx}{dt}\hat{i} + \frac{dy}{dt}\hat{j} + \frac{dz}{dt}\hat{k} = v_x\hat{i} + v_y\hat{j} + v_z\hat{k}. \tag{3.38}$$

In the same way, for acceleration, we can write

$$\frac{d\vec{v}(t)}{dt} = \frac{dv_x}{dt}\hat{i} + \frac{dv_y}{dt}\hat{j} + \frac{dv_z}{dt}\hat{k} = \frac{d^2x}{dt^2}\hat{i} + \frac{d^2y}{dt^2}\hat{j} + \frac{d^2z}{dt^2}\hat{k}. \tag{3.39}$$

Thus, by using vector components, we can use the same techniques we use for differentiation of functions of a single variable.

Example 3.8. A particle moves in the (x, y) plane on a circle of radius R with a constant angular speed ω. The position vector of the moving particle is

$$\vec{r}(t) = R(\cos(\omega t)\hat{i} + \sin(\omega t))\hat{j}.$$

Find the velocity and speed of the particle as a function of time. Show that the acceleration of the particle is always directed along the radius toward the center of the circle.

The velocity is $\vec{v}(t) = R\omega(-\sin(\omega t)\hat{i} + \cos(\omega t)\hat{j})$ and the speed is $R\omega$, i.e., the particle moves with a constant speed. The acceleration is $\vec{a}(t) = R\omega^2(-\cos(\omega t)\hat{i} - \sin(\omega t)\hat{j})$. We can see that the direction of the acceleration is in the direction opposite to the radius vector of the particle! ∎

3.6.2. *Integration*

First, we consider a vector that is a function of a scalar parameter; let it be $\vec{u}(t)$. Then, using vector components, we can readily write

$$\int_a^b \vec{u}(t)dt = \hat{i}\int_a^b u_x(t)dt + \hat{j}\int_a^b u_y(t)dt + \hat{k}\int_a^b u_z(t)dt. \quad (3.40)$$

This integration process is similar to the integration of regular functions, but it is applied separately to each component of the vector.

However, integration for $\vec{u}(x,y)$ is more complicated if we integrate along one of the space variables, such as x or y. In this case, we normally integrate along a path L in (x,y) space. There are two types of such integrals.

The first type is

$$\int_L f(x,y)d\vec{r}. \quad (3.41)$$

Since $d\vec{r} = x\hat{i} + y\hat{j}$, integral (3.41) can be represented as

$$\int_L f(x,y)d\vec{r} = \hat{i}\int_L f(x,y)dx + \hat{j}\int_L f(x,y)dy. \quad (3.42)$$

The second type is

$$\int_L \vec{u}(x,y) \cdot d\vec{r}. \quad (3.43)$$

By using vector components in Cartesian coordinates

$$\int_L \vec{u}(x,y) \cdot d\vec{r} = \int_L (u_x(x,y)\hat{i} + u_y(x,y)\hat{j}) \cdot (\hat{i}dx + \hat{j}dy),$$

we finally obtain

$$\int_L \vec{u}(x,y) \cdot d\vec{r} = \int_L a_x dx + \int_L a_y dy. \quad (3.44)$$

Both types of the integrals above, i.e., (3.42) and (3.44), are the second kind of line integral. Such integrals are considered in Chapter 6, Section 6.3.

3.7. Fields

A field is a concept that assigns a value to every point in space. It can be either scalar or vector depending on its intrinsic properties. For instance, a temperature distribution within a space is a scalar field, denoted as $T(x, y, z)$, where each point in the field is associated with a single numerical value. Conversely, an electric field, determined by the force acting on a test charge, is an example of a vector field $\vec{E}(x, y, z)$. Formally, a stationary field can be expressed as a function of three variables, namely x, y, z. It is often emphasized that a scalar or vector field is defined as a function of the spatial coordinates, as the determination of the point in space completely specifies the corresponding value of the field at that location. Furthermore, a field can be classified as either stationary, indicating no temporal variation at each point in space, or non-stationary, implying that it changes with time.

3.7.1. Gradient of a scalar field

Let us examine a two-dimensional scalar field, such as a temperature distribution on a surface denoted as $T(x, y)$. If the temperature remains constant over time, we might aim to assess the rate of change of temperature with respect to the *distance* from a point (x_0, y_0) to another point (x, y), treated as a directional derivative (Figure 3.15). For this purpose, we introduce a unit vector $\hat{u} = u_x\hat{i} + u_y\hat{j}$ that originates from (x_0, y_0) and terminates at (x, y). We can write $x - x_0 = lu_x$ and $y - y_0 = lu_y$ by using a scaling factor l that is the distance between the points. Hence, the temperature at (x, y) can be expressed as $T(x, y) = T(x_0 + lu_x, y_0 + lu_y)$, as a function of the scaling parameter l. Utilizing the rules for partial differentiation (5.5), we can

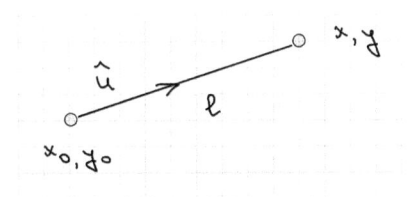

Fig. 3.15. A directional derivative.

calculate the partial derivatives of T with respect to x and y as follows:

$$\frac{dT}{dl} = \frac{\partial T}{\partial x}\frac{dx}{dl} + \frac{\partial T}{\partial y}\frac{dy}{dl} = \frac{\partial T}{\partial x}u_x + \frac{\partial T}{\partial y}u_y.$$

This result can be represented as a scalar product of two vectors:

$$\frac{dT}{dl} = \left(\frac{\partial T}{\partial x}\hat{i} + \frac{\partial T}{\partial y}\hat{j}\right) \cdot (u_x\hat{i} + u_y\hat{j}).$$

The first vector is called a gradient of a scalar field $T(x, y)$:

$$\text{grad } T = \vec{\nabla}T = \frac{\partial T}{\partial x}\hat{i} + \frac{\partial T}{\partial y}\hat{j}. \tag{3.45}$$

There are two common notations for the vector gradient operator, namely $\vec{\nabla}$ and grad. Throughout the text, we have employed both notations interchangeably to represent the same mathematical concept.

The gradient of a scalar field $T(x, y)$ at a point (x, y) is a *vector* that indicates the direction of the steepest ascent of the scalar field and whose magnitude signifies the rate of change in that direction.

Let two differentiable scalar fields φ and μ be given in the same domain. From the definition of gradient (3.45), the following relations follow:

$$\begin{aligned}\text{grad }(\varphi + \mu) &= \text{grad } \varphi + \text{grad } \mu, \\ \text{grad }(\varphi\mu) &= \varphi \text{ grad } \mu + \mu \text{ grad } \varphi, \\ \text{grad }\frac{\varphi}{\mu} &= \frac{\mu \text{ grad } \varphi - \varphi \text{ grad } \mu}{\mu^2}.\end{aligned} \tag{3.46}$$

Now, we can write the change in the temperature in the direction of $\hat{u}$:

$$\frac{dT}{dl} = \vec{\nabla}T \cdot \hat{u}. \tag{3.47}$$

That is called a directional derivative in the direction of $\hat{u} = \vec{u}/|u|$. The first term in the right hand side of the above equation for a given field T depends only on a choice of a point (x, y), and the second term depends only on the direction of the vector $\vec{u}$, as shown in Figure 3.16. The next objective is to determine when the directional derivative

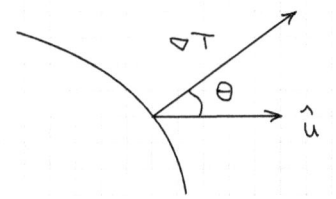

Fig. 3.16. Change in temperature in the $\hat{u}$ direction.

(3.47) is maximized for a given field T at a particular point (x, y). The directional derivative can be written as

$$\frac{dT}{dl} = |\vec{\nabla}T| \cos \theta,$$

where θ is the angle between the vectors $\vec{\nabla}T$ and $\hat{u}$. Hence, the gradient vector $\vec{\nabla}T$ at the point (x, y) indicates the direction of the most rapid increase of the field. Additionally, this maximum rate per unit length equates to $|\vec{\nabla}T|$. If the scalar field T represents a temperature distribution generated by a fireplace, then $\vec{\nabla}T$ points toward the fireplace, and $-\vec{\nabla}T$ points away from the fireplace.

Let's consider a line where the temperature remains constant, denoted as $T(x, y) = T_0$. Such a line is termed an isothermal line. Consequently, the directional derivative between two points on this line is zero, as there is no variation in temperature. Since the vector $\vec{u}$ between two adjacent points on the isothermal line is tangential to the line, we can deduce that from

$$\frac{dT}{dl} = 0 = \vec{\nabla}T \cdot \hat{u}$$

follows that the gradient vector $\vec{\nabla}T$ is perpendicular to the isothermal line $T(x, y) = T_0$. In the case of three dimensions, the gradient is perpendicular to the surface of constant temperature.

The physical interpretation of the gradient reveals that the gradient vector $\vec{\nabla}T$ (a) physically represents the direction and magnitude of the steepest ascent of a scalar field, (b) is perpendicular to the isothermal surface (or to an isothermal line in the two-dimensional case) of the constant scalar field T, and (c) remains unaffected (invariant) by replacing Cartesian coordinates with some other coordinate system. Thus, the gradient of the scalar field constitutes a vector field.

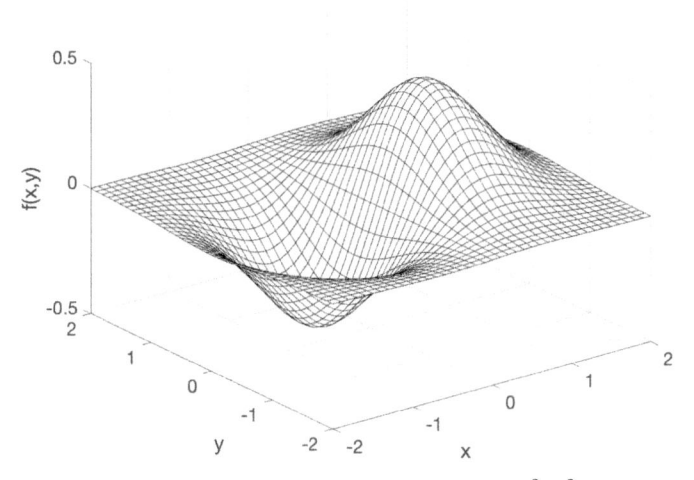

Fig. 3.17. The function $T(x,y) = ye^{-x^2-y^2}$.

Example 3.9. Find the gradient of the function $T(x,y) = ye^{-x^2-y^2}$ shown in Figure 3.17.

By using (3.45), we obtain

$$\frac{\partial T}{\partial x} = -2xye^{-x^2-y^2},$$

$$\frac{\partial T}{\partial y} = (1-2y^2)e^{-x^2-y^2},$$

and

$$\vec{\nabla} T = \left(-2xye^{-x^2-y^2}\right)\hat{i} + \left((1-2y^2)e^{-x^2-y^2}\right)\hat{j}.$$

Figure 3.18 shows both the contour map of the scalar field $T(x,y) = ye^{-x^2-y^2}$ and its gradient, pictured as arrows. In the context of our contour map, the gradient vector at a point will point uphill, perpendicular to the contour lines, and its magnitude will be proportional to the steepness of the slope at that point. ∎

Example 3.10. Find the directional derivative of $\varphi(x,y,z) = xy - z^2$ at point $(2,1,-3)$ in the direction of vector $\vec{a} = \hat{i} + 3\hat{k}$.

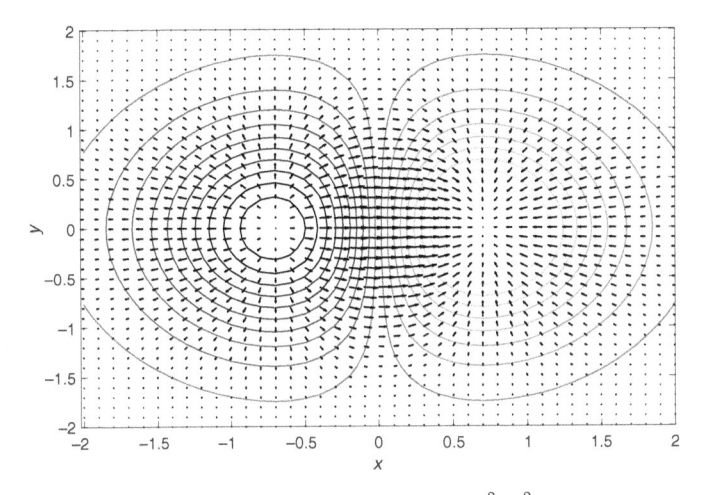

Fig. 3.18. The contour map of $T(x,y) = ye^{-x^2-y^2}$ and its gradient.

The gradient of the function $\varphi(x, y, z)$ is

$$\text{grad}\varphi = \vec{\nabla}\varphi = \frac{\partial \varphi}{\partial x}\hat{i} + \frac{\partial \varphi}{\partial y}\hat{j} + \frac{\partial \varphi}{\partial z}\hat{k} = y\hat{i} + x\hat{j} - 2z\hat{k}.$$

Then, the gradient at point $(2, 1, -3)$ is

$$\text{grad}\varphi = \hat{i} + 2\hat{j} + 6\hat{k}.$$

The unit vector $\vec{u}$ is obtained as

$$\vec{u} = \frac{\vec{a}}{|a|} = \frac{\hat{i} + 0\hat{j} + 3\hat{k}}{\sqrt{10}}.$$

Then,

$$\text{grad}\varphi \cdot \vec{u} = (\hat{i} + 2\hat{j} + 6\hat{k}) \cdot (\hat{i} + 0\hat{j} + 3\hat{k})\frac{1}{\sqrt{10}} = \frac{19}{\sqrt{10}}. \quad \blacksquare$$

Example 3.11. Suppose a height of a hill can be described in meters as

$$H(x, y) = 2xy - 3x^2 - 4y^2 - 18x + 28y + 12.$$

Figure 3.19 shows the function as a contour map.

(a) Where is the top of the hill located?
(b) How high is the hill?
(c) If we start at the point $(3, 2)$ and in the direction $\hat{i} + \hat{j}$, are we going uphill or downhill, and how fast?

The first question can be answered using partial derivatives, namely by setting both partial derivatives to zero (see Chapter 5 on partial derivative, Section 5.6),

$$\frac{\partial H}{\partial x} = 2y - 6x - 18 = 0,$$

$$\frac{\partial H}{\partial y} = 2x - 8y + 28 = 0,$$

with the solution $x = -2$, $y = 3$, and then $H(-2, 3) = 72$ meters. The gradient of the hill is

$$\vec{\nabla} H = (-6x + 2y - 18)\hat{i} + (2x - 8y + 28)\hat{j}.$$

At the point $(3, 2)$, the gradient is equal to $\vec{\nabla} H(3, 2) = -32\hat{i} + 18\hat{j}$. The unit vector in the direction of $\hat{i} + \hat{j}$ is

$$\hat{u} = \frac{1}{\sqrt{2}}\hat{i} + \frac{1}{\sqrt{2}}\hat{j}.$$

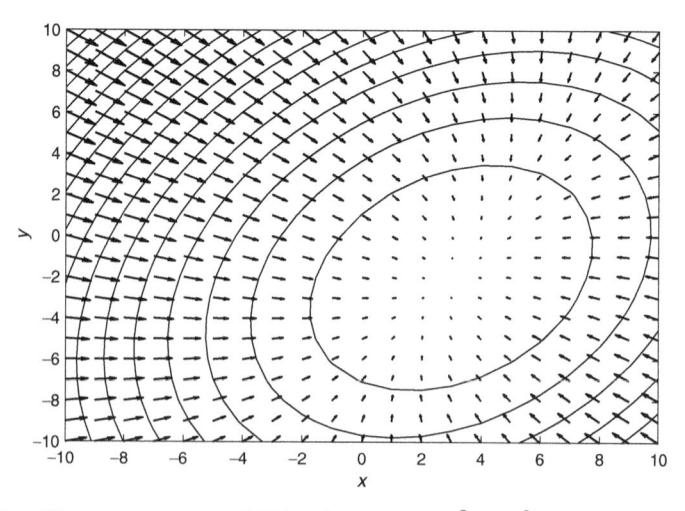

Fig. 3.19. The contour map of $H(x, y) = 2xy - 3x^2 - 4y^2 - 18x + 28y + 12$ and its gradient.

Then, the directional gradient at the point of interest is
$$\vec{\nabla}H \cdot \hat{u} = -\frac{32}{\sqrt{2}} + \frac{18}{\sqrt{2}} = -7\sqrt{2}.$$
It means that we are going downhill with a rate of $7\sqrt{2}$. ∎

The gradient of a scalar field has numerous applications. Here is a known example from general physics courses, i.e., a connection between potential energy $U(x, y, z)$ and corresponding conservative force $\vec{F}$,
$$\vec{F} = -\nabla U. \tag{3.48}$$
Or by using vector components, we obtain
$$\vec{F} = F_x\hat{i} + F_y\hat{j} + F_z\hat{k} = -\frac{\partial U}{\partial x}\hat{i} - \frac{\partial U}{\partial y}\hat{j} - \frac{\partial U}{\partial z}\hat{k}.$$
Some other examples include heat transfer, electric potential and field analysis, fluid flow and pressure analysis, and optimization and machine learning.

3.7.2. *Divergence and curl of a vector field*

The gradient definition (3.45) can be rewritten as
$$\vec{\nabla}\varphi = \frac{\partial \varphi}{\partial x}\hat{i} + \frac{\partial \varphi}{\partial y}\hat{j} + \frac{\partial \varphi}{\partial z}\hat{k} = \left(\frac{\partial}{\partial x}\hat{i} + \frac{\partial}{\partial y}\hat{j} + \frac{\partial}{\partial z}\hat{k},\right)\varphi,$$
where $\vec{\nabla}$ is a vector operator
$$\vec{\nabla} = \frac{\partial}{\partial x}\hat{i} + \frac{\partial}{\partial y}\hat{j} + \frac{\partial}{\partial z}\hat{k}. \tag{3.49}$$
There are three basic multiplication operations with vectors: multiplication by a scalar, and then scalar (dot) and vector (cross) products of two vectors. Since $\vec{\nabla}$ is an operator, it must always go first. We must not leave hungry operators! Now, the multiplication by a scalar, i.e., $\vec{\nabla}\varphi$, is the gradient of a scalar field!

There are three fundamental operations for multiplying vectors: multiplication of a vector by a scalar, multiplication of two vectors as the scalar (dot) product, and multiplication of two vectors as the vector (cross) product. Since $\vec{\nabla}$ is an operator, it must always come first; we must not leave hungry operators unfed! Feeding the operator $\vec{\nabla}$ with a scalar field yields the gradient of a scalar field.

Let $\vec{A}(x, y, z)$ describe a vector field, when for a point in space x, y, z we have a corresponding vector $\vec{A}(x, y, z)$ that has both magnitude and direction. The vector field can be represented in the component form (in Cartesian coordinates) as

$$\vec{A}(x, y, z) = A_x(x, y, z)\hat{i} + A_y(x, y, z)\hat{j} + A_z(x, y, z)\hat{k}. \quad (3.50)$$

Let us form for a scalar product of the vector operator (3.49) on vector $\vec{A}(x, y, z)$

$$\vec{\nabla} \cdot \vec{A}(x, y, z) = \left(\frac{\partial}{\partial x}\hat{i} + \frac{\partial}{\partial y}\hat{j} + \frac{\partial}{\partial z}\hat{k},\right) \cdot (A_x\hat{i} + A_y\hat{j} + A_z\hat{k}).$$

After the component by component multiplication (3.13), we obtain the *divergence* of a vector field:

$$\text{div}\vec{A} = \vec{\nabla} \cdot \vec{A} = \frac{\partial A_x(x, y, z)}{\partial x} + \frac{\partial A_y(x, y, z)}{\partial y} + \frac{\partial A_z(x, y, z)}{\partial z}. \quad (3.51)$$

The divergence of a vector field is a scalar function that measures the rate at which the vector field's points are moving away from or toward a given point. Figure 3.20 shows a vector field $\vec{A} = x\hat{i} + y\hat{j}$ with a positive divergence $\text{div}\vec{A} = 2$. Such a field represents a *positive source*.

The divergence of a vector field has several practical applications across different fields of science and engineering. Some notable examples include fluid flow analysis, electric charge analysis, heat transfer and thermal conduction, and understanding source and sink behavior.

Next, we explore a vector product of $\vec{\nabla} \times \vec{A}(x, y, z)$. Following the rule of vector product for components (3.24), we obtain the curl of a vector field:

$$\text{curl}\vec{A} = \vec{\nabla} \times \vec{A} = \left(\frac{\partial A_z}{\partial y} - \frac{\partial A_y}{\partial z}\right)\hat{i} + \left(\frac{\partial A_x}{\partial z} - \frac{\partial A_z}{\partial x}\right)\hat{j}$$

$$+ \left(\frac{\partial A_y}{\partial x} - \frac{\partial A_x}{\partial y}\right)\hat{k}. \quad (3.52)$$

The curl of a vector field is another vector field that represents the rotation or "circulation" of the original vector field at a given point in

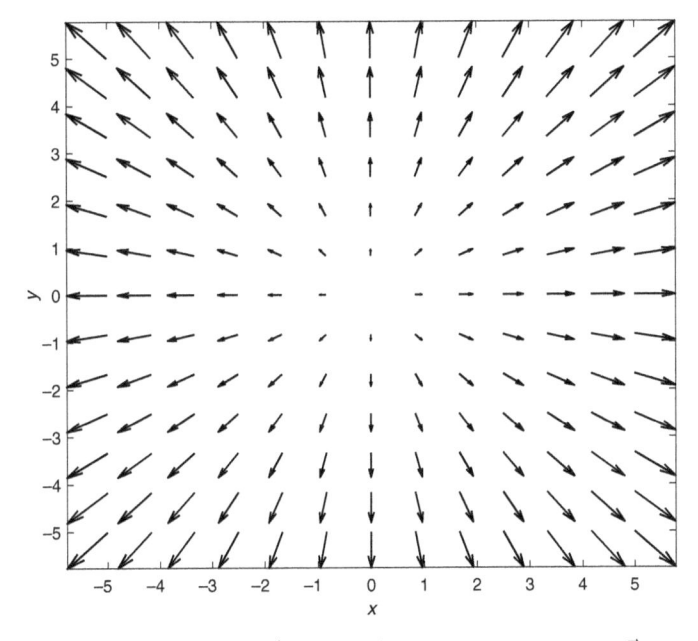

Fig. 3.20. Vector field $\vec{A} = x\hat{i} + y\hat{j}$ with a divergence div$\vec{A} = 2$.

space. Figure 3.21 shows an example of a vector field $\vec{A} = y\hat{i} - x\hat{j}$ representing a rotational flow, with a constant negative curl (clockwise rotation).

It quantifies how the vector field "swirls" around a point. The concept of the curl of a vector field finds applications in various areas of physics and engineering. Some of the key applications include electromagnetism and fluid dynamics. For example, when the curl of a vector field is zero, it implies that the field is conservative, and we can introduce a potential function U such that the force corresponding to this potential function is defined as $\vec{F} = -\vec{\nabla}U$.

Example 3.12. Find the divergence and the curl of the vector field

$$\vec{A} = x^2 y\hat{i} + xy^2\hat{j} + xyz\hat{k}.$$

From (3.51), the divergence of $\vec{A}$ is given by

$$\vec{\nabla} \cdot \vec{A} = 2xy + 2xy + xy = 5xy.$$

From (3.52), the curl of $\vec{A}$ is given by

$$\vec{\nabla} \times \vec{A} = xz\hat{i} - yz\hat{j} + (y^2 - x^2)\hat{k}.$$

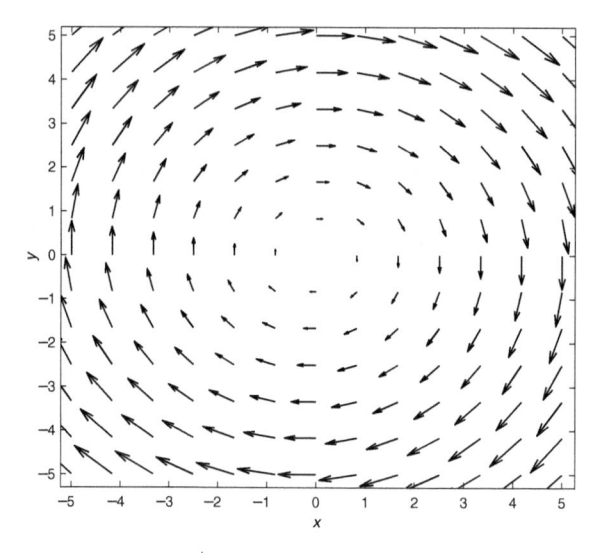

Fig. 3.21. Vector field $\vec{A} = y\hat{i} - x\hat{j}$ with a curl curl$\vec{A} = 0\hat{i} + 0\hat{j} - 2\hat{k}$.

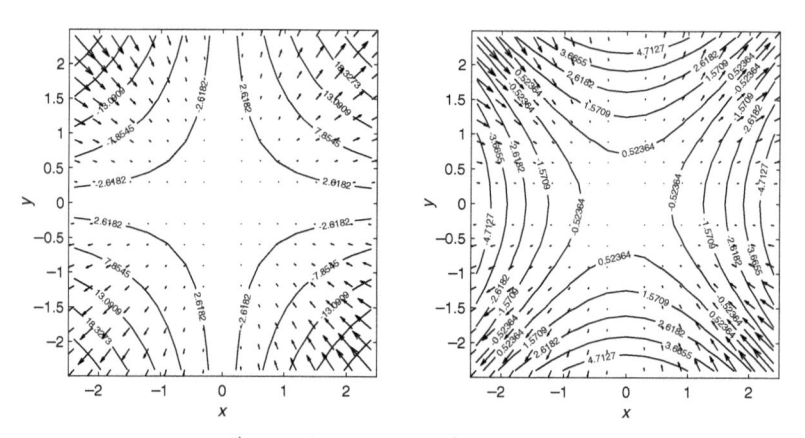

Fig. 3.22. Vector field $\vec{A} = x^2y\hat{i} + xy^2\hat{j} + xyz\hat{k}$ plotted together with the gradient $\vec{\nabla} \cdot \vec{A} = 5xy$ (left) and the curl $\vec{\nabla} \times \vec{A} = xz\hat{i} - yz\hat{j} + (y^2 - x^2)\hat{k}$ (right).

Both the divergence and the curl are functions of the space variables (x, y). Figure 3.22 shows the vector field $\vec{A}$ together with contour plots for the divergence (on the left) and z-component of the curl (on the right). ∎

We have identified three fundamental operators. The first one, the gradient, operates on scalar fields, yielding vector fields. The other two, divergence and curl, operate on vector fields. The divergence operator generates a scalar field corresponding to the original vector field, while the curl creates a new vector field from a vector field. It is possible that these operators may act sequentially, one after the other. There are nine possible combinations of grad, div, and curl. However, four combinations are do not make sense since grad can only act on scalar fields, and div and curl only act on vector fields. The four "no sense" combinations are:

(1) grad(gradφ) — grad operates on scalar fields but (gradφ) is already a vector field.
(2) grad(curl$\vec{A}$) — again, grad operates on scalar fields but (curl$\vec{A}$) is a vector field.
(3) div(div$\vec{A}$) — div operates on vector fields but (div$\vec{A}$) is a scalar field.
(4) curl(div$\vec{A}$) — a similar reason as above since (div$\vec{A}$) is a scalar field but curl operates on vector fields.

Next, we consider the remaining five possible combinations:

$$\text{grad}(\text{div}\vec{A}) = \vec{\nabla}(\vec{\nabla} \cdot \vec{A}) = \left(\frac{\partial}{\partial x}\hat{i} + \frac{\partial}{\partial y}\hat{j} + \frac{\partial}{\partial z}\hat{k}\right)\left(\frac{\partial A_x}{\partial x} + \frac{\partial A_y}{\partial y} + \frac{\partial A_z}{\partial z}\right)$$

$$= \left(\frac{\partial^2 A_x}{\partial x^2} + \frac{\partial^2 A_y}{\partial x \partial y} + \frac{\partial^2 A_z}{\partial x \partial z}\right)\hat{i} + \left(\frac{\partial^2 A_x}{\partial x \partial y} + \frac{\partial^2 A_y}{\partial y^2} + \frac{\partial^2 A_z}{\partial y \partial z}\right)\hat{j}$$

$$+ \left(\frac{\partial^2 A_x}{\partial x \partial z} + \frac{\partial^2 A_y}{\partial y \partial z} + \frac{\partial^2 A_z}{\partial z^2}\right)\hat{k}, \quad (3.53)$$

$$\text{div}(\text{grad}\varphi) = \vec{\nabla} \cdot \vec{\nabla}\varphi = \left(\frac{\partial}{\partial x}\hat{i} + \frac{\partial}{\partial y}\hat{j} + \frac{\partial}{\partial z}\hat{k}\right) \cdot \left(\frac{\partial \varphi}{\partial x}\hat{i} + \frac{\partial \varphi}{\partial y}\hat{j} + \frac{\partial \varphi}{\partial z}\hat{k}\right)$$

$$= \frac{\partial^2 \varphi}{\partial x^2} + \frac{\partial^2 \varphi}{\partial y^2} + \frac{\partial^2 \varphi}{\partial z^2} = \left(\frac{\partial^2}{\partial x^2} + \frac{\partial^2}{\partial y^2} + \frac{\partial^2}{\partial z^2}\right)\varphi = \nabla^2 \varphi.$$
$$(3.54)$$

Now, let us stop here to consider the operator

$$\nabla^2 = \frac{\partial^2}{\partial x^2} + \frac{\partial^2}{\partial y^2} + \frac{\partial^2}{\partial z^2}. \quad (3.55)$$

This operator is called the Laplacian operator, often denoted as ∇^2 or Δ. This is a scalar differential operator that is the divergence of the gradient of a scalar field. The Laplacian finds extensive use in various fields of physics and mathematics, including the study of heat conduction, electrostatics, fluid dynamics, and more:

$$\text{div}(\text{curl}\vec{A}) = \vec{\nabla} \cdot (\vec{\nabla} \times \vec{A}) = \vec{\nabla} \cdot \left(\left(\frac{\partial A_z}{\partial y} - \frac{\partial A_y}{\partial z} \right) \hat{i} \right.$$
$$\left. + \left(\frac{\partial A_x}{\partial z} - \frac{\partial A_z}{\partial x} \right) \hat{j} + \left(\frac{\partial A_y}{\partial x} - \frac{\partial A_x}{\partial y} \right) \hat{k} \right)$$
$$= \frac{\partial^2 A_z}{\partial x \partial y} - \frac{\partial^2 A_y}{\partial x \partial z} + \frac{\partial^2 A_x}{\partial y \partial z} - \frac{\partial^2 A_z}{\partial y \partial x} + \frac{\partial^2 A_y}{\partial z \partial x} - \frac{\partial^2 A_x}{\partial z \partial y} = 0,$$
(3.56)

$$\text{curl}(\text{grad}\varphi) = \vec{\nabla} \times (\vec{\nabla}\varphi) = \vec{\nabla} \times \left(\frac{\partial \varphi}{\partial x} \hat{i} + \frac{\partial \varphi}{\partial y} \hat{j} + \frac{\partial \varphi}{\partial z} \hat{k} \right)$$
$$= \left(\frac{\partial^2 \varphi}{\partial z \partial y} - \frac{\partial^2 \varphi}{\partial y \partial z} \right) \hat{i} + \left(\frac{\partial^2 \varphi}{\partial x \partial z} - \frac{\partial^2 \varphi}{\partial z \partial x} \right) \hat{j}$$
$$+ \left(\frac{\partial^2 \varphi}{\partial y \partial x} - \frac{\partial^2 \varphi}{\partial x \partial y} \right) \hat{k} = 0.$$
(3.57)

$$\text{curl}(\text{curl}\vec{A}) = \vec{\nabla} \times (\vec{\nabla} \times \vec{A}) = \vec{\nabla}(\vec{\nabla} \cdot \vec{A}) - \vec{\nabla}^2 \vec{A}. \quad (3.58)$$

Such repeated above operation forms a useful toolbox for the analysis of scalar and vector fields.

3.8. Exercises

(1) Let $\vec{a} = 3\hat{i} - 2\hat{j}$ and $\vec{b} = 1\hat{i} + 4\hat{j}$.
Find the following vectors: (a) $-\vec{a}$, (b) $3\vec{b}$, (c) $\vec{a} + \vec{b}$, (d) $\vec{a} - 2\vec{b}$, (e) $\vec{a} \cdot \vec{b}$, and (f) $\vec{a} \times \vec{b}$.
(2) If $\vec{a} + \vec{b} = 2\hat{j} + 4\hat{i}$ and $\vec{a} - \vec{b} = 6\hat{i} + 2\hat{j}$, find $\vec{a}$ and $\vec{b}$.
(3) Find the angle between $\vec{a} = 3\hat{i} + 2\hat{j}$ and $\vec{b} = 4\hat{i} - 1\hat{j}$.
(4) Let $\vec{a} = 2\hat{i} - 2\hat{j} + 1\hat{k}$. Find a unit vector in the same direction as $\vec{a}$.
(5) Show that two vectors $\vec{a}$ and $\vec{b}$ originating at the same point $(-1, 1)$ and having ends at $(1, 2)$ and $(0, -1)$ correspondingly are orthogonal.

(6) Show that vectors $\vec{a} = 1\hat{i} + 3\hat{j} - 1\hat{k}$ and $\vec{b} = 2\hat{i} - 3\hat{j} - 7\hat{k}$ are orthogonal, and find a vector $\vec{c}$ that is perpendicular to both $\vec{a}$ and $\vec{b}$.
(7) Let $\vec{c} = \vec{a} + \vec{b}$. Show that from $\vec{c} \times \vec{c} = 0$ follows $\vec{a} \times \vec{b} = -\vec{b} \times \vec{a}$.
(8) Show that $\vec{b}|a| + \vec{a}|b|$ and $\vec{a}|b| - \vec{b}|a|$ are orthogonal.
(9) Using Lagrange identity, find the value of $(\vec{a} \times \vec{b})^2 + (\vec{a} \cdot \vec{b})^2$.
(10) The position of a particle as a function of time t is given as

$$\vec{r}(t) = \sin(t)\hat{i} - \cos(t)\hat{j} + t\hat{k}.$$

Find the velocity, speed, and acceleration of the particle. Describe the motion.
(11) The position of a particle at time t is given as

$$\vec{r}(t) = \cos^2 t\,\hat{i} + \sin^2 t\,\hat{j} + 2\cos(2t)\hat{k}.$$

Describe the motion.
(12) For the scalar function

$$\varphi = x^2 + y^2 - 4xy,$$

find in what direction does the function decrease most rapidly at the point $(1, 2)$.
(13) Find the derivative of $\varphi(x, y, z) = \sin(x)e^y z^2$ at the point $(1, -2, 2)$ in the direction $3\hat{i} + 2\hat{j} + 1\hat{k}$.
(14) Suppose a height of a hill is described as

$$h(x, y) = -3x^2 - 2y^2 + 2xy - 4x + 3y + 24.$$

Find (a) where the top of the hill is located, (b) how high is the hill, and (c) if you start from a point $(0, 0)$ and go in the direction of $\hat{i} + \hat{j}$, are you walking uphill or downhill?
(15) Find the divergence and curl of the following vector fields:
 (a) $\vec{r} = xy^2\hat{i} + 2x^2y\hat{j} + xyz\hat{k}$
 (b) $\vec{r} = x\cos(y)\hat{i} + 2ye^x\hat{j} + xz\hat{k}$.
(16) Find a force corresponding to the following scalar fields, and show that the force is conservative:
 (a) $\varphi(x, y, z) = -xy - z$
 (b) $\varphi(x, y, z) = 3xy - xy^3z - z^2$.

Answers and solutions

(1) (a) $-3\hat{i} + 2\hat{j}$, (b) $3\hat{i} + 12\hat{j}$, (c) $4\hat{i} + 2\hat{j}$, (d) $1\hat{i} - 10\hat{j}$, (e) -5, (f) $0\hat{i} + 0\hat{j} + 14\hat{k}$.

(2) $\vec{a} = 4\hat{i} + 3\hat{j}$, $\vec{b} = -2\hat{i} + 1\hat{j}$.

(3) 0.833 radians or 47.7 degrees.

(4) $\vec{u} = \frac{2}{3}\hat{i} - \frac{2}{3}\hat{j} + \frac{1}{3}\hat{k}$.

(5) $\vec{a} = 2\hat{i} + 1\hat{j}$, $\vec{b} = 1\hat{i} - 2\hat{j}$, $\vec{a} \cdot \vec{b} = 0$.

(6) $\vec{c} = -24\hat{i} + 5\hat{j} - 9\hat{k}$.

(9) $a^2 b^2$.

(10) $\vec{v}(t) = \cos(t)\hat{i} + \sin(t)\hat{j} + \hat{k}$, $v = \sqrt{2}$, a helix.

(11) It's better to move this exercise as example into the text.

(12) The velocity is $\vec{v}(t) = 2\sin(2t) * (\hat{i} + \hat{j} - 4\hat{k})$, motion along a straight line in the direction $\hat{i} + \hat{j} - 4\hat{k}$ with the speed of $v = 6\sqrt{2}\sin(2t)$.

(13) grad $\varphi = (2x - 4y)\hat{i} + (2y - 4x)\hat{j}$. At $(1, 2)$, the grad $\varphi = -6\hat{i} + 0\hat{j}$. Therefore, the direction of most rapid decrease is $6\hat{i} - 0\hat{j}$ or in the direction of $\hat{i}$.

(14) (a) $(-1/2, 1/2)$, (b) $103/4$, (c) $\vec{\nabla} h = -1$ at $(-1/2, 1/2)$ (downhill).

(15) (a) $\vec{\nabla} \cdot \vec{r} = xy + 2x^2 + y^2$, $\vec{\nabla} \times \vec{r} = xz\hat{i} - yz\hat{j} + 2xy\hat{k}$
(b) $\vec{\nabla} \cdot \vec{r} = x + \cos y + 2e^x$, $\vec{\nabla} \times \vec{r} = 0\hat{i} - z\hat{j} + (2ye^x + x\sin(y))\hat{k}$.

(16) (a) $\vec{F} = y\hat{i} + x\hat{j} + 1\hat{k}$, curl$\vec{F} = 0$
(b) $\vec{F} = (y^3 * z - 3y)\hat{i} + (3xzy^2 - 3x))\hat{j} + (xy^3 + 2z)\hat{k}$, curl$\vec{F} = 0$.

Chapter 4

Matrices

4.1. Introduction

Application of matrices goes well beyond solving systems of linear equations and finding eigenvalues and eigenvectors. Matrices have a wide range of applications in statistics and data analysis, engineering, computer graphics, economics and finance, artificial intelligence, machine learning, and many more.

We begin this chapter with a simple algebraic problem: solving a system of two linear equations. This problem will lead us to important concepts, specifically matrices and determinants.

A system of two non-homogeneous linear equations can be expressed as

$$\begin{aligned} a_{11}x_1 + a_{12}x_2 &= b_1, \\ a_{21}x_1 + a_{22}x_2 &= b_2. \end{aligned} \quad (4.1)$$

The coefficients, denoted as a_{ij}, employ subscript notation with two subscripts. The first of which indicates in which equation this coefficient is located, and the second subscript indicates at which of the unknowns it is standing. The terms b_i are the corresponding non-homogeneous terms in the ith equation. The system can be readily solved by elimination with two solutions as

$$x_1 = \frac{b_1 a_{22} - b_2 a_{12}}{a_{11} a_{22} - a_{12} a_{21}}, \quad x_2 = \frac{b_2 a_{11} - b_1 a_{21}}{a_{11} a_{22} - a_{12} a_{21}}.$$

If we invest effort and time in solving a system of three linear equations

$$a_{11}x_1 + a_{12}x_2 + a_{13} = b_1,$$
$$a_{21}x_1 + a_{22}x_2 + a_{23} = b_2, \qquad (4.2)$$
$$a_{31}x_1 + a_{32}x_2 + a_{33} = b_3,$$

then we find the solutions as

$$x_1 = \frac{b_1 a_{22} a_{33} + a_{12} a_{23} b_3 + a_{13} b_2 a_{32} - b_1 a_{23} a_{32} - a_{12} b_2 a_{33} - a_{13} a_{22} b_3}{a_{11} a_{22} a_{33} + a_{12} a_{23} a_{31} + a_{13} a_{21} a_{32} - a_{11} a_{23} a_{32} - a_{12} a_{21} a_{33} - a_{13} a_{22} a_{31}},$$

$$x_2 = \frac{a_{11} b_2 a_{33} + b_1 a_{23} a_{31} + a_{13} a_{21} b_3 - a_{11} a_{23} b_3 - b_1 a_{21} a_{33} - a_{13} a_{22} a_{31}}{a_{11} a_{22} a_{33} + a_{12} a_{23} a_{31} + a_{13} a_{21} a_{32} - a_{11} a_{23} a_{32} - a_{12} a_{21} a_{33} - a_{13} a_{22} a_{31}},$$

$$x_3 = \frac{a_{11} a_{22} b_3 + a_{12} a_{31} b_2 + a_{21} a_{32} b_1 - a_{11} a_{32} b_2 - a_{12} a_{21} b_3 - a_{22} a_{31} b_1}{a_{11} a_{22} a_{33} + a_{12} a_{23} a_{31} + a_{13} a_{21} a_{32} - a_{11} a_{23} a_{32} - a_{12} a_{21} a_{33} - a_{13} a_{22} a_{31}}.$$

We observe that the denominators for all three solutions are identical. Additionally, we observe that the numerator for the first solution can be obtained from the denominator by simply substituting the coefficients a_{i1} with the non-homogeneous terms b_i. Similarly, such substitution applies to the other two solutions as well. Consequently, we only need to determine a rule for evaluating the denominator to find solutions for linear systems of equations.

We can see that the denominators for both (4.1) and (4.2) linear systems have only a_{ij} coefficients. We represent these coefficients in the form of a square table, maintaining the order in which they appear in the system itself. For instance, in the case of the system of three linear equations, we have

$$\begin{bmatrix} a_{11} & a_{12} & a_{13} \\ a_{21} & a_{22} & a_{23} \\ a_{n1} & a_{n2} & a_{n3} \end{bmatrix}. \qquad (4.3)$$

Such a table is called a *matrix*. The first index i in a_{ij} shows which row the element is in, and the second index j indicates the column number. The denominator in the solutions is a single number corresponding to the matrix, and it is called a *determinant*.

In the following section, we explore both matrices and determinants in greater depth.

4.2. Matrices

A matrix is a rectangular table of numbers containing a certain number of m rows and a certain number of n columns:

$$\begin{pmatrix} a_{11} & a_{12} & a_{13} & \cdots & a_{1n} \\ a_{21} & a_{22} & a_{23} & \cdots & a_{2n} \\ \cdots & \cdots & \cdots & \cdots & \cdots \\ a_{m1} & a_{m2} & a_{m3} & \cdots & a_{mn} \end{pmatrix}. \qquad (4.4)$$

It is common to use a capital letter as a shorter version for writing a matrix. For example, for the matrix above, we can use A as the shorter notation, more specifically as $A = [a_{ij}], i = 1, 2, \ldots, n$, $j = 1, 2, \ldots, n$. In the equivalent notation, the square brackets $[a_{ij}]$ are replaced by round brackets (a_{ij}).

If $m = n$, then the matrix is called a *square matrix*, and the number $m = n$ is called its *order* or *dimension* of a matrix. In this chapter, we primarily focus on square matrices and also on matrices having just one column, which are commonly referred to as vectors:

$$x = \begin{pmatrix} b_1 \\ b_2 \\ \cdots \\ b_n \end{pmatrix}. \qquad (4.5)$$

For square matrices, we use concepts of *main and secondary diagonals*. The main diagonal of a matrix is the diagonal going from the upper left corner of this matrix to its lower right corner as $a_{11}a_{22}\ldots a_{nn}$. The side diagonal of the same matrix is the diagonal going from the lower left corner to the upper right corner as $a_{n1}a_{n-1,2}\ldots a_{1n}$.

4.2.1. Basic operations on matrices

Before introducing basic operations on matrices, it is essential to introduce the concept of equality between two matrices. Two matrices A and B are said to be equal $A = B$ if these matrices have the same order, and all their corresponding elements are all the same, i.e., $a_{ij} = b_{i,j}$.

Matrix addition: For two matrices $A = [a_{ij}], i, j = 1, 2 \ldots, n$ and $B = [b_{ij}], i, j = 1, 2 \ldots, n$ of the same order n, we can define a sum $C = A + B$, where the elements c_{ij} are

$$c_{ij} = a_{ij} + b_{ij} \quad (i = 1, 2 \ldots, n; j = 1, 2 \ldots, n), \tag{4.6}$$

or it can be written as

$$\begin{bmatrix} a_{11} & a_{12} & \cdots & a_{1n} \\ a_{21} & a_{22} & \cdots & a_{2n} \\ \cdots & \cdots & \cdots & \cdots \\ a_{n1} & a_{n2} & \cdots & a_{nn} \end{bmatrix} + \begin{bmatrix} b_{11} & b_{12} & \cdots & b_{1n} \\ b_{21} & b_{22} & \cdots & b_{2n} \\ \cdots & \cdots & \cdots & \cdots \\ b_{n1} & b_{n2} & \cdots & b_{nn} \end{bmatrix}$$
$$= \begin{bmatrix} a_{11} + b_{11} & a_{12} + b_{12} & \cdots & a_{1n} + b_{1n} \\ a_{21} + b_{21} & a_{22} + b_{22} & \cdots & a_{2n} + b_{2n} \\ \cdots & \cdots & \cdots & \cdots \\ a_{n1} + b_{n1} & a_{n2} + b_{n2} & \cdots & a_{nn} + b_{nn} \end{bmatrix}. \tag{4.7}$$

Using definition (4.6) we can define a zero matrix O such that all elements of the zero matrix are equal to zero, namely

$$o_{ij} = 0 \quad (i = 1, 2 \ldots, n; j = 1, 2 \ldots, n).$$

In this case, we have

$$A + O = A.$$

From the definition of the sum of matrices (4.6), it follows that the operation of matrix addition has the same properties as the addition of numbers, namely
commutative property

$$A + B = B + A$$

and *associative* property

$$(A + B) + C = A + (B + C).$$

The difference between the two matrices is defined in the same way, namely for $C = A - B$, the elements of the matrix C are

$$c_{ij} = a_{ij} - b_{ij} \quad (i = 1, 2 \ldots, n; j = 1, 2 \ldots, n). \tag{4.8}$$

Note that the addition and difference of matrices A and B with different orders are not defined.

Multiplication of a matrix by a constant: Matrix A multiplied by a number μ is a matrix $C = \mu A$, where the elements of c_{ij} are calculated as

$$c_{ij} = \mu a_{ij} \quad (i = 1, 2 \ldots, n; j = 1, 2 \ldots, n). \tag{4.9}$$

It is easy to see that the difference C of two matrices C and B as in (4.8) can be obtained as

$$C = A + (-1)B.$$

Matrix multiplication: Multiplication of matrix $A = [a_{ij}]$ ($i = 1, 2, \ldots, m; j = 1, 2, \ldots, n$) on matrix $B = [b_{ij}]$ ($i = 1, 2, \ldots, n; j = 1, 2, \ldots, l$), i.e., AB, is defined as

$$c_{ij} = \sum_{k=1}^{n} a_{ik} b_{kj} \quad (i = 1, 2, \ldots, m; j = 1, 2, \ldots, p). \tag{4.10}$$

From the definition above, it follows that the matrix A cannot be multiplied by just any matrix B, unless the number of columns in matrix A is equal to the number of rows in matrix B. In the same way, we define BA product. Specifically, both the products AB and BA can be calculated only when the number of columns in matrix A matches the number of rows in matrix B, and the number of rows in matrix A matches the number of columns in matrix B. When these conditions are met, both matrices AB and BA will be square matrices, but their orders will typically differ. For two square matrices to be multiplied, they have to be of the same order n. And a matrix of order b can be multiplied by a column vector of length n.

Example 4.1. As an illustration of the application the multiplication rule, we present the formula for multiplying second-order square matrices:

$$\begin{bmatrix} a_{11} & a_{12} \\ a_{21} & a_{22} \end{bmatrix} \cdot \begin{bmatrix} b_{11} & b_{12} \\ b_{21} & b_{22} \end{bmatrix} = \begin{bmatrix} (a_{11}b_{11} + a_{12}b_{21}) & (a_{11}b_{12} + a_{12}b_{22}) \\ (a_{21}b_{11} + a_{22}b_{21}) & (a_{21}b_{12} + a_{22}b_{22}) \end{bmatrix}.$$

■

Basic properties of matrix multiplication: Matrix multiplication is *associative*, which means that when you have three matrices

A, B, and C, the order in which you perform the multiplication does not matter:

$$(AB)C = A(BC).$$

Matrix multiplication is *distributive* over addition, which means that when you multiply a matrix by the sum of two other matrices, you can distribute the multiplication across the sum

$$A(B + C) = AB + AC.$$

For both products AB and BA to not only be well defined but also have the same order, it is necessary and sufficient that both matrices A and B are square matrices of the same order. Matrix multiplication generally is not a *commutative* operation, which means that the order of multiplication matters. In general, for matrices A and B, the product AB is not necessarily equal to the product BA, i.e., generally

$$AB \neq BA.$$

Two matrices whose product has the commutative property are usually called *commuting*.

Example 4.2. Let us consider AB and BA as

$$\begin{bmatrix} 1 & 2 \\ 0 & 3 \end{bmatrix} \cdot \begin{bmatrix} 4 & 0 \\ 5 & 6 \end{bmatrix} = \begin{bmatrix} 14 & 12 \\ 15 & 18 \end{bmatrix} \text{ and } \begin{bmatrix} 4 & 0 \\ 5 & 6 \end{bmatrix} \cdot \begin{bmatrix} 1 & 2 \\ 0 & 3 \end{bmatrix} = \begin{bmatrix} 4 & 8 \\ 5 & 28 \end{bmatrix}.$$

As we can see, $AB \neq BA$. ■

However, there are important special cases in which the commutative property holds. We consider a class of square matrices where all elements located outside the main diagonal are zero. These square matrices are referred to as *diagonal matrices*. Each diagonal matrix of order n has the form

$$D = \begin{bmatrix} d_{11} & 0 & \ldots & 0 \\ 0 & d_{22} & \ldots & 0 \\ \ldots & \ldots & \ldots & \ldots \\ 0 & 0 & \ldots & d_{nn} \end{bmatrix}. \quad (4.11)$$

From the definition of matrix multiplication (4.34), we can note that if all diagonal elements are equal to each other, i.e., $d_{11} = d_{22} = \cdots = d_{nn} = d$, then any square matrix A of order n commutes with matrix D, i.e., $AD = DA$. Indeed, from (4.34) follows for $C = AD$

$$c_{ij} = a_{ij}d_{jj} = a_{ij}d,$$

and for $C' = DA$

$$c'_{ij} = d_{jj}a_{ij} = da_{ij},$$

i.e., $c_{ij} = c'_{ij}$. Two diagonal matrices A and B commute because multiplication of diagonal matrices only involves element-wise multiplication.

Among all diagonal matrices (4.11) with equal elements $d_{11} = d_{22} = \cdots = d_{nn} = d$, two matrices play a particularly important role. The first of these matrices is obtained when $d = 1$, which is known as the *identity matrix* of the nth order and is denoted by the symbol I or I_n (where n represents the size or order of the matrix). The second matrix is obtained when $d = O$, which is called the zero matrix of the nth order and is denoted by the symbol O or O_n. Thus,

$$I = \begin{bmatrix} 1 & 0 & \cdots & 0 \\ 0 & 1 & \cdots & 0 \\ \cdots & \cdots & \cdots & \cdots \\ 0 & 0 & \cdots & 1 \end{bmatrix}, \quad O = \begin{bmatrix} 0 & 0 & \cdots & 0 \\ 0 & 0 & \cdots & 0 \\ \cdots & \cdots & \cdots & \cdots \\ 0 & 0 & \cdots & 0 \end{bmatrix}. \quad (4.12)$$

For the identity matrix, we have

$$AI = IA = A \quad (4.13)$$

and for the zero matrix, we have

$$AO = OA = 0, \quad A + O = A.$$

4.2.2. *Types of square matrices*

There are a few types of square matrices essential for science and engineering applications, such matrices as symmetric, Hermitian, transpose, and inverse.

An *inverse matrix* A^{-1} is a matrix with the property

$$AA^{-1} = I. \qquad (4.14)$$

While there are many ways to derive the inverse matrix A^{-1}, it is instructive to do it from the definition, using a matrix of dimension $n = 2$. Let $X = A^{-1}$ be the inverse matrix of matrix A, then

$$\begin{bmatrix} a_{11} & a_{12} \\ a_{21} & a_{22} \end{bmatrix} \cdot \begin{bmatrix} x_{11} & x_{12} \\ x_{21} & x_{22} \end{bmatrix} = \begin{bmatrix} 1 & 0 \\ 0 & 1 \end{bmatrix}.$$

Sine the matrix multiplication AX is equal to I matrix, we obtain a system of four linear equations for finding unknown elements x_{ij}:

$$a_{11}x_{11} + a_{12}x_{21} = 1,$$
$$a_{11}x_{12} + a_{12}x_{22} = 0,$$
$$a_{21}x_{11} + a_{22}x_{21} = 0,$$
$$a_{21}x_{12} + a_{22}x_{22} = 1.$$

Solving the system above produces the following solutions:

$$x_{11} = \frac{a_{22}}{a_{11}a_{22} - a_{12}a_{21}}, \quad x_{21} = \frac{-a_{21}}{a_{11}a_{22} - a_{12}a_{21}},$$

$$x_{12} = \frac{-a_{12}}{a_{11}a_{22} - a_{12}a_{21}}, \quad x_{22} = \frac{a_{11}}{a_{11}a_{22} - a_{12}a_{21}},$$

or in matrix form

$$X = A^{-1} = \frac{1}{a_{11}a_{22} - a_{12}a_{21}} \begin{bmatrix} a_{22} & -a_{12} \\ -a_{21} & a_{11} \end{bmatrix}.$$

Now, we show that a right inverse matrix is equal to the left inverse matrix as $AA^{-1} = A^{-1}A$. Indeed,

$$A^{-1} = A^{-1}I = A^{-1}(AA^{-1}),$$

then using the associative property of matrix multiplication

$$A^{-1}(AA^{-1}) = (A^{-1}A)A^{-1}),$$

thus it must be $A^{-1}A = I$, or A and A^{-1} are commute.

The *trace* of a square matrix Tr(a) is the sum of its diagonal elements, i.e.,

$$\text{Tr}(a) = a_{11} + a_{22} + \cdots + a_{nn} = \sum_{i=1}^{n} a_{ii}. \quad (4.15)$$

The trace of a matrix is number (a scalar value), and it has several important properties and applications, in particular, in finding eigenvalues, which we discuss in Section 4.5.

The *transpose* of a matrix A^T is a new matrix formed by interchanging its rows and columns, such as

$$a_{ij}^T = a_{ji}.$$

A *symmetric matrix* is a matrix when it is equal to its transpose

$$A = A^T.$$

An *orthogonal matrix* is a matrix when

$$A^T = A^{-1} \quad \text{or} \quad A^T A = I.$$

A *complex conjugate matrix* A^* is a matrix when we apply complex conjugation to every element of matrix A, i.e., a_{ij}^*.

A Hermitian conjugate for a complex matrix is defined as

$$A^\dagger = (A^*)^T = (A^T)^*.$$

A *Hermitian matrix* is a matrix when its Hermitian conjugate is equal to the original matrix A:

$$A^\dagger = A \quad \text{if} \quad a_{ij} = a_{ja}^*.$$

4.3. Determinants

In Section 4.1, we previously observed that determinants provide solutions to systems of two and three linear equations. Moreover, determinants of square matrices have important applications in finding eigenvalues and eigenvectors (see Section 4.5), calculating the inverse of a matrix, changing variables in multiple integrals, and many more.

So, for the second-order square matrix

$$\begin{bmatrix} a_{11} & a_{12} \\ a_{21} & a_{22} \end{bmatrix}, \tag{4.16}$$

the determinant is a number

$$\Delta = \det A = \begin{vmatrix} a_{11} & a_{12} \\ a_{21} & a_{22} \end{vmatrix} = a_{11}a_{22} - a_{12}a_{21}. \tag{4.17}$$

And for the third-order square matrix

$$\begin{bmatrix} a_{11} & a_{12} & a_{13} \\ a_{21} & a_{22} & a_{23} \\ a_{31} & a_{32} & a_{33} \end{bmatrix}, \tag{4.18}$$

the determinant is

$$\Delta = \det A = a_{11}a_{22}a_{33} + a_{12}a_{23}a_{31} + a_{13}a_{21}a_{32}$$
$$- a_{11}a_{23}a_{32} - a_{12}a_{21}a_{33} - a_{13}a_{22}a_{31}. \tag{4.19}$$

We wish to evaluate a determinant for a square matrix on any order n. By analyzing determinants (4.17) and (4.17) for matrices of the second- and the third-order, we can derive procedures for evaluating determinants for square matrices of any order.

4.3.1. *Calculating determinants by permutations*

For example, for the third-order matrix, we see that the determinant consists of six terms, and each of its terms is the product of three elements of matrix (4.18), and the elements of each row and each column are presented in this product. Indeed, these products look like

$$a_{1i}a_{2j}a_{3k},$$

where $i = 1, 2, 3$, $j = 1, 2, 3$, and $k = 1, 2, 3$ are arranged in some specific order. There are six possible permutations, namely

$$1, 2, 3; \ 2, 3, 1; \ 3, 1, 2; \ 1, 3, 2; \ 2, 1, 3; \ 3, 2, 1.$$

Thus, we obtain all six terms or 3! terms in (4.19). However, some terms are with a plus sign and others with a minus sign. Then, we need to find the rule according to which the sign must be chosen. The sign is actually determined by a number of permutations needed to bring i, j, k numbers to $1, 2, 3$ sequence. Thus, we get $+$ sign for $a_{11}a_{22}a_{33}$ (zero permutations), $a_{12}a_{23}a_{31}$ (two permutations from 231 to 213 and to 123), and $a_{13}a_{21}a_{32}$ (two permutations from 312 to 132 to 123), and we get $-$ for $a_{11}a_{23}a_{32}$ (one permutation from 132 to 123), $a_{12}a_{21}a_{33}$ (one permutation from 213 to 123), and $a_{13}a_{22}a_{31}$ (three permutations from 321 to 231 to 213 to 123).

For a matrix of order n, its determinant has $n!$ terms, and each term is a product of n numbers a_{ij}, such as each product contains one number from each row and one from each column. These products will look like

$$a_{1k_1}a_{1k_2} \ldots a_{nk_n},$$

where $k_1, k_2, \ldots, k_n$ are numbers arranged in some order. Let $[k_1, k_2, \ldots, k_n]$ be the number of steps needed to bring $k_1, k_2, \ldots, k_n$ numbers in $1, 2, 3, \ldots, n$ form. Thus, a determinant can be written as

$$\det A = \sum_{k_1, k_2, \ldots, k_n} (-1)^{[k_1, k_2, \ldots, k_n]} a_{1k_1} a_{1k_2} \ldots a_{nk_n}. \quad (4.20)$$

Here, the summation extends to all possible permutations of the second indexes, i.e., to all possible permutations $k_1, k_2, \ldots, k_n$. While it is possible to evaluate determinants using the above formula, it is not very practical. Therefore, we consider another more practical way to evaluate determinants.

4.3.2. *Expansion by minors*

We can rewrite determinant (4.19) as

$$\det A = a_{11} * (a_{22}a_{33} - a_{23}a_{32}) - a_{12}(a_{21}a_{33} - a_{23}a_{31})$$
$$+ a_{13}(a_{21}a_{32} - a_{22}a_{31})$$

or as

$$\det A = a_{11}\begin{vmatrix} a_{22} & a_{23} \\ a_{32} & a_{33} \end{vmatrix} - a_{12}\begin{vmatrix} a_{21} & a_{23} \\ a_{31} & a_{33} \end{vmatrix} + a_{13}\begin{vmatrix} a_{21} & a_{22} \\ a_{31} & a_{32} \end{vmatrix}.$$

Thus, the first term is a product of a_{11} on a determinant of a second-order matrix that is obtained from the original matrix by removing the first row and first column containing a_{11} element. The second term has a product of $(-a_{12})$ and the determinant of a matrix obtained from A by removing the first row and second column. And the third term is a_{13} multiplied by the determinant of a matrix obtained from A by removing the first row and the third column. By induction, we can extend this approach to any square matrix of order n. By first, we introduce a concept of a minor. We call the *minor* of an element a_{ij} of a matrix of order n the determinant of order $n-1$ corresponding to the matrix that is obtained from matrix (4.4) by deleting the ith row and jth column (that row and the column at the intersection of which element a_{ij} is located). The minor of the element a_{ij} is denoted by the symbol M_{ij}. It is convenient to also introduce a *cofactor*, defined as

$$C_{ij} = (-1)^{i+j} M_{ij}. \tag{4.21}$$

Then, the determinant of a square matrix of order n can be written as

$$\Delta = \det A = \begin{vmatrix} a_{11} & a_{12} & \cdots & a_{1n} \\ a_{21} & a_{22} & \cdots & a_{2n} \\ \cdots & \cdots & \cdots & \cdots \\ a_{n1} & a_{n2} & \cdots & a_{nn} \end{vmatrix} = \sum_{j=1}^{n} (-1)^{1+j} a_{1j} M_{1j} = \sum_{j=1}^{n} a_{1j} C_{1j}. \tag{4.22}$$

The formula above demonstrates a rule for evaluating a determinant using the elements of the first row. Nevertheless, this rule is applicable to any row. In such cases, for any row number $i (i = 1, 2, \ldots, n)$, we can express it as follows:

$$\Delta = \det A = \begin{vmatrix} a_{11} & a_{12} & \cdots & a_{1n} \\ a_{21} & a_{22} & \cdots & a_{2n} \\ \cdots & \cdots & \cdots & \cdots \\ a_{n1} & a_{n2} & \cdots & a_{nn} \end{vmatrix} = \sum_{j=1}^{n} (-1)^{i+j} a_{ij} M_{ij} = \sum_{j=1}^{n} a_{ij} C_{ij}, \tag{4.23}$$

which is referred as an expansion of the determinant using the ith row. It is also possible to expand the nth-order determinant in any of its columns as

$$\Delta = \det A = \sum_{i=1}^{n} (-1)^{i+j} a_{ij} M_{ij} = \sum_{i=1}^{n} a_{ij} C_{ij}. \tag{4.24}$$

It is useful to note that if we utilize minors from any other row (or column), then the resulting sum is equal to zero:

$$\sum_{j=1}^{n} a_{ij} C_{kj} = 0 \quad (k \neq i), \quad \sum_{i=1}^{n} a_{ij} C_{ik} = 0 \quad (k \neq j). \quad (4.25)$$

Example 4.3. Find the determinant for the matrix

$$\begin{pmatrix} 3 & 2 & 1 \\ -2 & 1 & -1 \\ 5 & 2 & 4 \end{pmatrix}.$$

Expanding over the first row gives

$$\Delta = 3 \cdot (-1)^2 \begin{vmatrix} 1 & -1 \\ 2 & 4 \end{vmatrix} + 2 \cdot (-1)^3 \begin{vmatrix} -2 & -1 \\ 5 & 4 \end{vmatrix} + 1 \cdot (-1)^4 \begin{vmatrix} -2 & -1 \\ 5 & 2 \end{vmatrix} = 15.$$

∎

While the expansion by minors looks as a straightforward way to calculate determinants, it is not practical for calculations with large matrices. For example, using the minor expansion for a matrix of order n requires $n!$ multiplication. Thus, for a matrix of order $n = 20$, it will take about one month to calculate its determinant on a teraflop computer (10^{12} floating-point operation per second). In the present calculations in science and engineering, it is common to deal with matrices where n is counted in many thousands. Therefore, in practical calculations, many problems (like solving systems of linear equations, or finding eigenvalues and eigenvectors for large matrices) are calculated either directly (without using determinants) or using iterative techniques.

4.3.3. *Properties of determinants*

Determinants have a number properties that are useful both for calculating determinants and in various applications. These properties can be easily tested for second- and third-order matrices and proved by induction for n-order matrices. Here, we list some of most practical properties.

(1) *Interchange rows or columns*: When two rows (or two columns) are interchanged, the determinant does not change its absolute value,

but it does change its sign. For a second-order determinant, this property immediately follows from (4.17), namely

$$\begin{vmatrix} a_{11} & a_{12} \\ a_{21} & a_{22} \end{vmatrix} = - \begin{vmatrix} a_{21} & a_{22} \\ a_{11} & a_{12} \end{vmatrix}, \quad \begin{vmatrix} a_{11} & a_{12} \\ a_{21} & a_{22} \end{vmatrix} = - \begin{vmatrix} a_{12} & a_{11} \\ a_{22} & a_{21} \end{vmatrix}.$$

For $n > 2$, this property can be derived either by induction or by expanding minors in (4.23) or (4.24) until we are left only with minors of the second order. But, as we demonstrated above, the second-order determinant changes its sign if rows (or columns) are interchanged. From the interchange rows (or columns) property follows that a determinant with two identical rows (or columns) is equal to zero.

(2) *Scalar multiplication*: Multiplying all elements of a row (or a column) by a number μ is equivalent to multiplying the determinant by this number μ. Indeed, let us consider the determinant, where all elements of the first row are multiplied by a constant μ:

$$\Delta_\mu = \begin{vmatrix} \mu a_{11} & \mu a_{12} & \cdots & \mu a_{1n} \\ a_{21} & a_{22} & \cdots & a_{2n} \\ \cdots & \cdots & \cdots & \cdots \\ a_{n1} & a_{n2} & \cdots & a_{nn} \end{vmatrix}.$$

Then, according to expansion (4.22), we obtain

$$\Delta_\mu = \sum_{j=1}^{n} (-1)^{1+j} \mu a_{1j} M_{1j} = \mu \sum_{j=1}^{n} (-1)^{1+j} a_{1j} M_{1j} = \mu \Delta.$$

From this property follows that if all elements of a row (or some column) are equal to zero, then the determinant itself is equal to zero (it is the same as setting $\mu = 0$). And it also follows that if the elements of two rows (or two columns) of a matrix are proportional, then the determinant is equal to zero.[1]

(3) *Linear property*: If elements of a row (or a column) are sums of an equal number of terms, then the determinant is equal to the sum

---

[1] Having the common factor out, we obtain a determinant with two identical rows/columns.

of the determinants in which the elements of the line (column) are replaced by individual terms, i.e., for the second-order determinant, we have

$$\begin{vmatrix} a_{11} & a_{12} + b_{12} \\ a_{21} & a_{22} + b_{22} \end{vmatrix} = a_{11}(a_{22} + b_{22}) - (a_{12} + b_{12})a_{21}$$

$$= a_{11}a_{22} - a_{12}a_{21} + a_{11}b_{22} - a_{21}b_{12}.$$

As we can note, the four terms above are corresponding to the sum of two determinants, such that we can write

$$\begin{vmatrix} a_{11} & a_{12} + b_{12} \\ a_{21} & a_{22} + b_{22} \end{vmatrix} = \begin{vmatrix} a_{11} & a_{12} \\ a_{21} & a_{22} \end{vmatrix} + \begin{vmatrix} a_{11} & b_{12} \\ a_{21} & b_{22} \end{vmatrix}.$$

In fact, this property directly follows from the linearity of determinants. For the minor expansion by a row, we can write

$$\Delta_{ab} = \sum_{j=1}^{n} (-1)^{i+j}(a_{ij} + b_{ij})M_{ij}$$

$$= \sum_{j=1}^{n} (-1)^{i+j}a_{ij}M_{ij} + \sum_{j=1}^{n} (-1)^{i+j}b_{ij}M_{ij} = \Delta_a + \Delta_b,$$

and for expansion by a column, we have

$$\Delta_{ab} = \sum_{i=1}^{n} (-1)^{i+j}(a_{ij} + b_{ij})M_{ij}$$

$$= \sum_{i=1}^{n} (-1)^{i+j}a_{ij}M_{ij} + \sum_{i=1}^{n} (-1)^{i+j}b_{ij}M_{ij} = \Delta_a + \Delta_b.$$

(4) *Row or column addition*: If to the elements of a row (or a column) we add the corresponding elements of another row (another column), multiplied by an arbitrary factor μ, then the value of the determinant will not change. This property is based on three other properties, namely the linear property, scalar multiplication property, and that a determinant with two identical rows (columns) is equal to zero. For example, for a third-order determinant, if we

add the first row multiplied by a factor μ to the third row, then, we obtain

$$\begin{vmatrix} a_{11} & a_{12} & a_{13} \\ a_{21} & a_{22} & a_{23} \\ a_{31} & a_{32} & a_{33} \end{vmatrix} = \begin{vmatrix} a_{11} & a_{12} & a_{13} \\ a_{21} & a_{22} & a_{23} \\ a_{31}+\mu a_{11} & a_{32}+\mu a_{12} & a_{33}+\mu a_{13} \end{vmatrix}.$$

The property of row (or column) addition can be formulated in a more general way — if we add to the elements of a row a linear combination of several other rows, then the value of the determinant does not change. This is one of most practical properties for both calculating determinants and solving systems of linear equations.

(5) *Linear dependence of rows or columns*: Necessary and sufficient condition for a determinant to be equal to zero is linear dependence of its rows (columns). We demonstrate this property using a third-order determinant:

$$\begin{vmatrix} a_{11} & a_{12} & a_{13} \\ a_{21} & a_{22} & a_{23} \\ a_{31} & a_{32} & a_{33} \end{vmatrix}.$$

Let us assume that the rows of the determinant are linear dependent, i.e., we can write for the first row elements

$$a_{1,k} = \alpha a_{2,k} + \beta a_{3k} \quad (k=1,2,3).$$

Next, we want to evaluate the determinant

$$\begin{vmatrix} \alpha a_{21}+\beta a_{31} & \alpha a_{22}+\beta a_{33} & \alpha a_{23}+\beta a_{33} \\ a_{21} & a_{22} & a_{23} \\ a_{31} & a_{32} & a_{33} \end{vmatrix}.$$

Using the row addition property, we can write that the above determinant can be written as a sum of two determinants:

$$\begin{vmatrix} \alpha a_{21} & \alpha a_{22} & \alpha a_{23} \\ a_{21} & a_{22} & a_{23} \\ a_{31} & a_{32} & a_{33} \end{vmatrix} + \begin{vmatrix} \beta a_{31} & \beta a_{33} & \beta a_{33} \\ a_{21} & a_{22} & a_{23} \\ a_{31} & a_{32} & a_{33} \end{vmatrix}.$$

Using the scalar multiplication property, we obtain for the above determinants

$$\alpha \begin{vmatrix} a_{21} & a_{22} & a_{23} \\ a_{21} & a_{22} & a_{23} \\ a_{31} & a_{32} & a_{33} \end{vmatrix} + \beta \begin{vmatrix} a_{31} & a_{33} & a_{33} \\ a_{21} & a_{22} & a_{23} \\ a_{31} & a_{32} & a_{33} \end{vmatrix}.$$

But the both determinants are equal to zero since the have equal rows (the first one — rows one and two, and the second one — rows one and three). The proof can be generalized for a determinant of any order n.

(6) *Transpose of a matrix*: The determinant of a matrix A is equal to the determinant of its transpose, A^T, or the value of the determinant does not change when replacing rows with columns, or replacing columns with rows, i.e., $\det A = \det A^T$. This property follows directly the minor expansions (4.24) and (4.23), specifically, when we observe that the expansion of the determinant $\det A$ along its first column is identical to the expansion of the determinant $\det A^T$ along its first row.

Example 4.4. Suppose we need to calculate the determinant of the following matrix:

$$\Delta = \begin{vmatrix} 6 & 3 & 2 & 1 \\ 0 & 3 & 9 & 0 \\ 12 & 2 & 14 & 6 \\ 4 & 7 & 27 & 2 \end{vmatrix}.$$

We use the property of "row or column addition". Subtracting twice the last column from the first column gives

$$\Delta = \begin{vmatrix} 4 & 3 & 2 & 1 \\ 0 & 3 & 9 & 0 \\ 0 & 2 & 14 & 6 \\ 0 & 7 & 27 & 2 \end{vmatrix}.$$

Next, it is natural to expand the determinant using the first column. As a result, we obtain

$$\Delta = 4 \begin{vmatrix} 3 & 9 & 0 \\ 2 & 14 & 6 \\ 7 & 27 & 2 \end{vmatrix}.$$

Next, subtract three times the first column from the second column:

$$\Delta = 4 \begin{vmatrix} 3 & 0 & 0 \\ 2 & 8 & 6 \\ 7 & 6 & 2 \end{vmatrix}.$$

Then, expanding the last third-order determinant over the first row, we finally obtain

$$\Delta = 4 \cdot 3 \begin{vmatrix} 8 & 6 \\ 6 & 2 \end{vmatrix} = 12(16 - 36) = -240.$$

∎

Example 4.5. Let us calculate the determinant of a triangular matrix of order 4, in which all elements lying below the main diagonal are equal to zero:

$$\Delta = \begin{vmatrix} a_{11} & a_{12} & a_{13} & a_{14} \\ 0 & a_{22} & a_{23} & a_{24} \\ 0 & 0 & a_{33} & a_{34} \\ 0 & 0 & 0 & a_{44} \end{vmatrix}.$$

Using the minor expansion over the first column (4.24), we obtain

$$\Delta = a_{11} \begin{vmatrix} a_{22} & a_{23} & a_{24} \\ 0 & a_{33} & a_{34} \\ 0 & 0 & a_{44} \end{vmatrix}.$$

Continuing the minor expansion over the first column in the above new matrix we come to

$$\Delta = a_{11} a_{22} \begin{vmatrix} a_{33} & a_{34} \\ 0 & a_{44} \end{vmatrix},$$

and eventually we obtain

$$\Delta = a_{11} a_{22} a_{33} a_{44}.$$

We can derive the same product result for a matrix of any order n. Thus, the determinant of a triangular matrix is the product of the elements on its main diagonal:

$$\Delta = \prod_{i=1}^{n} a_{ii}.$$

∎

4.4. Linear systems of equations

Systems of linear equations hold a special place in physics, chemistry, and engineering. In fact, multiple problems in science and technology can be mathematically expressed as a linear system of equations.

In general, a system of n linear equations with n unknowns has the following form[2]:

$$\begin{aligned} a_{11}x_1 + a_{12}x_2 + a_{13}x_3 + \cdots + a_{1n}x_n &= b_1, \\ a_{21}x_1 + a_{22}x_2 + a_{23}x_3 + \cdots + a_{2n}x_n &= b_2, \\ &\cdots\cdots\cdots, \\ a_{n1}x_1 + a_{n2}x_2 + a_{n3}x_3 + \cdots + a_{nn}x_n &= b_n, \end{aligned} \quad (4.26)$$

where x_j $(j = 1, 2, \ldots, n)$ are unknowns, a_{ij} $(i, j = 1, 2, \ldots, n)$ are the coefficients of the linear system, and b_i $(i = 1, 2, \ldots, n)$ are the non-homogeneous terms. The first subscript i identifies the row of the equation and the second subscript j identifies the column of the system of equations.

The system called *homogeneous* if all of its b_i coefficients are equal to zero. If at least one of the coefficients b_i is different from zero, then such system is called *non-homogeneous*.

A solution to system (4.26) is a set of numbers $x_1, x_2, \ldots, x_n$ which, when substituted into system (4.26), all the equations of this system become identities. A system of linear equations is said to be *consistent* if it has at least one solution, and it is said to be *inconsistent* if it has no solution.

Using the rule of matrix multiplication (4.10), it is very practical to write system (4.26) in a matrix form as

$$\begin{pmatrix} a_{11} & a_{12} & \cdots & a_{1n} \\ a_{21} & a_{22} & \cdots & a_{2n} \\ \cdots & \cdots & \cdots & \cdots \\ a_{n1} & a_{n2} & \cdots & a_{nn} \end{pmatrix} \begin{pmatrix} x_1 \\ x_2 \\ \cdots \\ x_n \end{pmatrix} = \begin{pmatrix} b_1 \\ b_2 \\ \cdots \\ b_n \end{pmatrix}, \quad (4.27)$$

or even in the compact form as

$$Ax = b. \quad (4.28)$$

---

[2] In this section, we concentrate on systems of linear equations when the number of equations is equal to the number of unknowns.

Before attempting to solve a system of linear equations, it is essential to determine whether the system is consistent or not. A system of linear equations is said to be consistent if it has at least one solution. We start with a system of homogeneous equations:

$$\begin{aligned} a_{11}x_1 + a_{12}x_2 + a_{13}x_3 + \cdots + a_{1n}x_n &= 0, \\ a_{21}x_1 + a_{22}x_2 + a_{23}x_3 + \cdots + a_{2n}x_n &= 0, \\ &\cdots\cdots\cdots\cdots\cdots\cdots\cdots\cdots\cdots\cdots\cdots\cdots, \\ a_{n1}x_1 + a_{n2}x_2 + a_{n3}x_3 + \cdots + a_{nn}x_n &= 0. \end{aligned} \quad (4.29)$$

While this system is always consistent, because there is always a so-called trivial ($x_1 = x_2 = \cdots = x_n = 0$) solution, we are interested in non-trivial solutions. Note that the existence of a non-trivial solution to system (4.29) is equivalent to a linear dependence of the columns of matrix A. However, from the property of linear dependence (see property 5), it means that the determinant of matrix A is equal to zero, i.e., $\det A = 0$.

4.4.1. Cramer's rule

Next, we analyze the non-homogeneous system (4.27). First, we assume that the system has non-trivial solutions, and for clarity, we start our analysis with a system with three equations:

$$\begin{aligned} a_{11}x_1 + a_{12}x_2 + a_{13}x_3 &= b_1, \\ a_{21}x_1 + a_{22}x_2 + a_{23}x_3 &= b_2, \\ a_{31}x_1 + a_{32}x_2 + a_{33}x_3 &= b_3. \end{aligned}$$

Next, we multiply each row of the system correspondingly by cofactors of the first column, i.e., C_{11}, C_{21}, and C_{31}, such as

$$\begin{aligned} a_{11}x_1 C_{11} + a_{12}x_2 C_{11} + a_{13}x_3 C_{11} &= b_1 C_{11}, \\ a_{21}x_1 C_{21} + a_{22}x_2 C_{21} + a_{23}x_3 C_{21} &= b_2 C_{21}, \\ a_{31}x_1 C_{31} + a_{32}x_2 C_{31} + a_{33}x_3 C_{31} &= b_3 C_{31}. \end{aligned}$$

By adding all the equations together we obtain

$$x_1(a_{11}C_{11} + a_{21}C_{21} + a_{31}C_{31}) + x_2(a_{12}C_{11} + a_{22}C_{21} + a_{32}C_{31}) \\ + x_3(a_{13}C_{11} + a_{23}C_{21} + a_{33}C_{31}) = b_1 C_{11} + b_2 C_{21} + b_3 C_{31}.$$

We can observe that the coefficient associated with x_1 is the determinant of the matrix A evaluated by expansion by minors along the

first column, i.e.,
$$a_{11}C_{11} + a_{21}C_{21} + a_{31}C_{31} = \Delta.$$

However, the coefficients for x_2 and x_3 are both equal to zero, as indicated in (4.25), given that we employ minors from column one for the expansion by columns two and three. Then, we obtain
$$x_1 \Delta = b_1 C_{11} + b_2 C_{21} + b_3 C_{31}.$$

The right-hand side of the equation above looks as a determinant of a matrix where the elements a_{11}, a_{21}, and a_{31} are replaced by the non-homogeneous terms b_1, b_2, and b_3 accordingly. By denoting such determinant as Δ_1, we have the first solution x_1 as
$$x_1 = \frac{\Delta_1}{\Delta}.$$

In the same way, we can find solutions x_2 and x_3.

Generalization for a system on n equations gives *Cramer's rule*, named after the Swiss mathematician Gabriel Cramer:
$$x_i = \frac{\Delta_i}{\Delta} \quad (i = 1, 2, \ldots, n). \tag{4.30}$$

Cramer's rule can be formulated as follows: if the determinant of system (4.26) is non-zero, then this system possesses a unique solution (4.30). Each of the unknowns is expressed as a ratio of two determinants: The denominator contains the determinant of the system, while the numerator contains the determinant obtained by replacing the coefficients of the unknown with their corresponding non-homogeneous terms.

Example 4.6. Use Cramer's rule to solve the system of equations
$$1x_1 - 2x_2 + 3x_3 = 6,$$
$$-2x_1 + 4x_2 - 1x_3 = 3,$$
$$1x_1 - 3x_2 + 4x_3 = 7.$$

The determinants are
$$\Delta = \begin{vmatrix} 1 & -2 & 3 \\ -2 & 4 & -1 \\ 1 & -3 & 4 \end{vmatrix} = 5, \quad \Delta_1 = \begin{vmatrix} 6 & -2 & 3 \\ 3 & 4 & -1 \\ 7 & -3 & 4 \end{vmatrix} = 5,$$

$$\Delta_2 = \begin{vmatrix} 1 & 6 & 3 \\ -2 & 3 & -1 \\ 1 & 7 & 4 \end{vmatrix} = 10, \quad \Delta_3 = \begin{vmatrix} 6 & -2 & 6 \\ 3 & 4 & 3 \\ 7 & -3 & 7 \end{vmatrix} = 15.$$

Then, from Cramer's rule follows

$$x_1 = \frac{\Delta_1}{\Delta} = 1, \quad x_2 = \frac{\Delta_2}{\Delta} = 2, \quad x_3 = \frac{\Delta_3}{\Delta} = 3. \qquad \blacksquare$$

Cramer's rule is particularly useful only for small system of equations as evaluating determinants can become impractical for large systems. Since Cramer's rule is based on evaluations of determinants, it needs about $n!$ multiplications and divisions for solving a system of n equations. Thus, solving a system of only 30 equations would take as much 600 times the age of the universe on a teraflop computer!

4.4.2. Solutions by the inverse matrix

Earlier in Section 4.2.2, we introduced an inverse matrix A^{-1}, such that

$$A^{-1}A = A^{-1}A = I.$$

Multiplying both side of a linear system (4.29) by the inverse matrix results in

$$A^{-1}(Ax) = A^{-1}b, \quad (A^{-1}A)x = A^{-1}b.$$

And finally, we have

$$x = A^{-1}b. \qquad (4.31)$$

Thus, having the inverse matrix immediately provides solutions to non-homogeneous systems of equations. For a matrix A of order n, its inverse matrix can be calculated as

$$A^{-1} = \frac{1}{\det A} \begin{pmatrix} C_{11} & C_{21} & \ldots & C_{n1} \\ C_{12} & C_{22} & \ldots & C_{n2} \\ \ldots & \ldots & \ldots & \ldots \\ C_{13} & C_{2n} & \ldots & C_{nn} \end{pmatrix}, \qquad (4.32)$$

where C_{ij} are co-factors of the original matrix A. Please note that the matrix of co-factors is transposed, i.e., in a place corresponding to a_{ij} element, we place C_{ji} co-factor. We can see that calculating an inverse matrix would take about the same time as using Cramer's rule. However, using an inverse matrix has benefits when we need to solve multiple times a system of linear equations with the same coefficients a_{ij} but various sets of non-homogeneous terms b_i.

Example 4.7. Solve the following system of equation by applying the inverse matrix:

$$3x_1 + 2x_2 + 1x_3 = 1,$$
$$-2x_1 + 1x_2 - 1x_3 = -3,$$
$$5x_1 + 2x_2 + 4x_3 = 3.$$

First, we need to find the inverse matrix by using (4.32). The determinant of this matrix was evaluated earlier in Example 4.3 as $\Delta = 15$. Using the definition of co-factors (4.21) we obtain

$$A^{-1} = \frac{1}{15} \begin{pmatrix} 6 & -6 & -3 \\ 3 & 7 & 1 \\ -9 & 4 & 7 \end{pmatrix}.$$

Then, the solution of the system is

$$\begin{pmatrix} x_1 \\ x_2 \\ x_3 \end{pmatrix} = \frac{1}{15} \begin{pmatrix} 6 & -6 & -3 \\ 3 & 7 & 1 \\ -9 & 4 & 7 \end{pmatrix} \begin{pmatrix} 1 \\ -3 \\ 3 \end{pmatrix} = \begin{pmatrix} 1 \\ -1 \\ 0 \end{pmatrix}. \quad \blacksquare$$

4.4.3. Direct elimination methods

Elimination methods use a simple idea that is well known from courses of algebra: a system of two equations worked out formally by solving one of the equations. Let's say we solve the first equation for the unknown x_1 in terms of the other unknown x_2. Substituting the solution for x_1 into the second equation gives us a single equation for one unknown x_2, thus x_1 is eliminated from the second equation. After x_2 is found, the other x_1 unknown can be found by back substitution.

In the general case of n linear equations, the elimination process involves operations on rows of linear equations that do not alter the solution. These operations include the following:

Scaling — multiplying any equation by a constant,
Pivoting — interchanging the order of equations,
Elimination — replacing any equation with a linear combination of that equation and any other equation.

For a better understanding of basic techniques of direct elimination, it is beneficial to apply the elimination method to a system of three linear equations:

$$a_{11}x_1 + a_{12}x_2 + a_{13}x_3 = b_1,$$
$$a_{21}x_1 + a_{22}x_2 + a_{23}x_3 = b_2, \qquad (4.33)$$
$$a_{31}x_1 + a_{32}x_2 + a_{33}x_3 = b_3.$$

Step 1a: Subtracting the first equation multiplied by a_{21}/a_{11} from the second equation and multiplied by a_{31}/a_{11} from the third equation gives

$$a_{11}x_1 + \qquad a_{12}x_2 + \qquad a_{13}x_3 = b_1,$$
$$(a_{21} - a_{21})x_1 + (a_{22} - \frac{a_{21}}{a_{11}}a_{12})x_2 + (a_{23} - \frac{a_{21}}{a_{11}}a_{13})x_3 = b_2 - \frac{a_{21}}{a_{11}}b_1,$$
$$(a_{31} - a_{31})x_1 + (a_{32} - \frac{a_{31}}{a_{11}}a_{12})x_2 + (a_{33} - \frac{a_{31}}{a_{11}}a_{13})x_3 = b_3 - \frac{a_{31}}{a_{11}}b_1.$$
$$(4.34)$$

One can see that the coefficients by the unknown x_1 in the second and the third rows of the new system are zero:

$$a_{11}x_1 + a_{12}x_2 + a_{13}x_3 = b_1,$$
$$0 + a'_{22}x_2 + a'_{23}x_3 = b'_2, \qquad (4.35)$$
$$0 + a'_{32}x_2 + a'_{33}x_3 = b'_3,$$

where $a'_{ij} = a_{ij} - \frac{a_{i1}a_{1j}}{a_{11}}$ and $b'_i = b_i - \frac{a_{i1}}{a_{11}}b_1$. Thus, we eliminated the first unknown x_1 from the second and third equations.

Step 1b: Now, let's subtract the modified second equation multiplied by a'_{32}/a'_{22} from the third equation in (4.35):

$$0 + (a'_{32} - a'_{32})x_2 + \left(a'_{33} - a'_{23}\frac{a'_{32}}{a'_{22}}\right)x_3 = b'_3 - b'_2\frac{a'_{32}}{a'_{22}}. \quad (4.36)$$

After the two eliminations we have a new form for system (4.33):

$$\begin{aligned} a_{11}x_1 + a_{12}x_2 + a_{13}x_3 &= b_1, \\ 0 + a'_{22}x_2 + a'_{23}x_3 &= b'_2, \\ 0 + 0 + a''_{33}x_3 &= b''_3, \end{aligned} \quad (4.37)$$

with $a''_{ij} = a'_{ij} - \frac{a'_{i2}a'_{2j}}{a'_{22}}$ and $b''_i = b'_i - \frac{a'_{i2}}{a'_{22}}b_2$. Thus, the original system $Ax = b$ is reduced to triangular form.

Step 2: The last equation in (4.37) can be solved for x_3, and then the second for x_2, and finally the first for x_1:

$$\begin{aligned} x_3 &= b''_3/a''_{33}, \\ x_2 &= (b'_2 - a'_{23}x_3)/a'_{22}, \\ x_1 &= (b_1 - a_{12}x_2 - a_{13}x_3)/a_{11}. \end{aligned} \quad (4.38)$$

The basic elimination algorithm for solving a system of n linear equations can be formulated as follows:

Step 1: Apply the elimination procedure to every column k ($k = 1, 2, \ldots, n-1$) for rows i ($i = k+1, k+2, \ldots, n$) to create zeros in column k below the pivot element $a_{k,k}$:

$$a_{i,j} = a_{i,j} - (a_{i,k}/a_{k,k})\,a_{k,j} \quad (i,j = k+1, k+2, \ldots, n), \quad (4.39)$$

$$b_i = b_i - (a_{i,k}/a_{k,k})\,b_k \quad (i,j = k+1, k+2, \ldots, n). \quad (4.40)$$

Step 2: The solutions of the reduced triangular system can then be found using the backward substitution:

$$x_n = (b_n/a_{n,n}) \quad (4.41)$$

$$x_j = \frac{1}{a_{i,i}}\left(b_i - \sum_{j=i+1}^{n} a_{i,j}x_j\right) \quad (i = n-1, n-2, \ldots, 1). \quad (4.42)$$

Example 4.8. Using the elimination method solve the system
$$x_1 - 3x_2 + 4x_3 = 7,$$
$$x_1 - 2x_2 + 3x_3 = 6,$$
$$-2x_1 + 4x_2 - 1x_3 = 3.$$

First, we subtract the first equation from the second, and then we add the first equation multiplied by two to the third equation. Then, we obtain
$$x_1 - 3x_2 + 4x_3 = 7,$$
$$0 + x_2 - x_3 = -1,$$
$$0 - 2x_2 + 7x_3 = 17.$$

Next, we multiple the new second equation by two and add it to the third equation:
$$x_1 - 3x_2 + 4x_3 = 7,$$
$$0 + x_2 - x_3 = -1,$$
$$0 - 0 + 5x_3 = 15.$$

From the third equation, we obtain $x_3 = 3$. Then, from the second equation, we have $x_2 = -1 + x_3 = 2$, and finally, $x_1 = 7 + 3x_2 - 4x_3 = 1$. ∎

The total number of multiplications and divisions done by the basic elimination algorithm for a system of n equations is about $O(n^3)$. The back substitution takes approximately $O(n^2)$ multiplication and divisions. As we remember, using Cramer's rule needs about $n!$ multiplications and divisions. Thus, for a system of order 10, the elimination method would need about 3,000 times fewer operations, and for a system of order 20, it would take about $3 \cdot 10^{14}$ times less time!

These are several modifications to the basic elimination method, including Gauss elimination, Gauss–Jordan elimination, and LU factorization. A detailed description of these methods falls outside the scope of this book.

4.4.4. Iterative methods

While methods based on elimination significantly reduce the number of operations compared to applying Cramer's rule by a factor of approximately $n!/n^3$, for very large linear systems, even when using computers, this reduction may not be sufficient. Iterative methods make it possible to approach a solution as a set of iterations. Iterative methods, as their name signifies, allow us to approach a solution through a series of iterations. While they may not achieve the same level of accuracy as elimination methods for arbitrary matrix A, they offer control over the number of iterations, thus allowing us to balance accuracy and computation time. Additionally, for systems of linear equations with many a_{ij} elements equal to zero, or close to zero (so-called *sparse* systems), iterative methods can demonstrate remarkable speed and efficiency.

The concept behind the iterative solution of a linear system is based on assuming an initial (trial) solution that can be iteratively refined to achieve an improved solution. This process continues until convergence is reached, with an accepted level of accuracy. It is important to note that for an iterative method to succeed or converge, the linear system of equations must exhibit diagonal dominance, such as

$$|a_{i,i}| > \sum_{j \neq i} |a_{i,j}|. \tag{4.43}$$

Let us consider a system of linear equations:

$$\sum_{j=1}^{n} a_{i,j} x_j = b_i \quad (i = 1, 2, \ldots, n). \tag{4.44}$$

Every equation can be formally solved for a diagonal element:

$$x_i = \frac{1}{a_{i,i}} \left(b_i - \sum_{j=1}^{i-1} a_{i,j} x_j - \sum_{j=i+1}^{n} a_{i,j} x_j \right) \quad (i = 1, 2, \ldots, n). \tag{4.45}$$

Choosing an initial solution as zero approximation $x_1^{(0)}, x_2^{(0)} \ldots x_n^{(0)}$ we may calculate the next iteration:

$$x_i^{(1)} = \frac{1}{a_{i,i}} \left(b_i - \sum_{j=1}^{i-1} a_{i,j} x_j^{(0)} - \sum_{j=i+1}^{n} a_{i,j} x_j^{(0)} \right) \quad (i = 1, 2, \ldots, n). \tag{4.46}$$

Equation (4.46) can be rewritten in the iterative form:

$$x_i^{(k+1)} = x_i^{(k)} + \frac{1}{a_{i,i}} \left(b_i - \sum_{j=1}^{n} a_{i,j} x_j^{(k)} \right) \quad (i = 1, 2, \ldots, n), \quad (4.47)$$

where k is an iterative number. Equation (4.47) defines the Jacobi iterative method, which is also called the method of simultaneous iterations. The method converges well for diagonally dominant matrices A. The number of iterations is either predetermined by a maximum number of allowed iterations or by one of conditions for absolute errors:

$$\max_{1 \leq i \leq n} \left| x_i^{(k+1)} - x_i^{(k)} \right| \leq \varepsilon, \text{ or } \sum_{i=1}^{n} \left| x_i^{(k+1)} - x_i^{(k)} \right| \leq \varepsilon,$$

$$\text{or } \left(\sum_{i=1}^{n} \left(x_i^{(k+1)} - x_i^{(k)} \right)^2 \right)^{1/2} \leq \varepsilon, \quad (4.48)$$

where ε is a tolerance.

In the Jacobi method, all values of $x_i^{(k+1)}$ are calculated using $x_i^{(k)}$ values. In the Gauss–Seidel method, the most recently computed values of $x_i^{(k+1)}$ are used in calculations for $j > i$ solutions:

$$x_i^{(k+1)} = \frac{1}{a_{i,i}} \left(b_i - \sum_{j=1}^{i-1} a_{i,j} x_j^{(k+1)} - \sum_{j=i+1}^{n} a_{i,j} x_j^{(k)} \right) \quad (i = 1, 2, \ldots, n) \quad (4.49)$$

or

$$x_i^{(k+1)} = x_i^{(k)} + \frac{1}{a_{i,i}} \left(b_i - \sum_{j=1}^{i-1} a_{i,j} x_j^{(k+1)} - \sum_{j=i}^{n} a_{i,j} x_j^{(k)} \right) \quad (i = 1, 2, \ldots, n). \quad (4.50)$$

The Gauss–Seidel iterations generally converge faster than Jacobi iterations.

Quite often, the iterative solution to a linear system approaches the true solution from the same direction. Then, it is possible to accelerate the iterative process by introducing the over-relaxing factor ω:

$$x_i^{(k+1)} = x_i^{(k)} + \omega \frac{1}{a_{i,i}} \left(b_i - \sum_{j=1}^{i-1} a_{i,j} x_j^{(k+1)} - \sum_{j=i}^{n} a_{i,j} x_j^{(k)} \right)$$
$$\times (i = 1, 2, \ldots, n). \tag{4.51}$$

For $\omega = 1$, system (4.51) is the Gauss–Seidel method, for $1.0 < \omega < 2.0$, the system is over-relaxed, and for $\omega < 1.0$, the system is under-relaxed. The optimum value of ω depends on the size of the system and the nature of the equations. The iterative process (4.51) is called the successive over-relaxation (SOR) method.

4.5. Eigenvalues and eigenvectors

The eigenvalue problem has multiple applications in practically every field of science. It is also widely used in engineering and data analysis. Eigenvalues are a fundamental concept in linear algebra and they are associated with square matrices.

Given a square matrix A, an eigenvalue λ and its corresponding eigenvector x are defined as follows:

$$Ax = \lambda x. \tag{4.52}$$

Equation (4.52) can be written as a system of linear equations:

$$\begin{aligned} a_{11}x_1 + a_{12}x_2 + a_{13}x_3 + \cdots + a_{1n}x_n &= \lambda x_1, \\ a_{21}x_1 + a_{22}x_2 + a_{23}x_3 + \cdots + a_{2n}x_n &= \lambda x_2, \\ &\cdots\cdots\cdots, \\ a_{n1}x_1 + a_{n2}x_2 + a_{n3}x_3 + \cdots + a_{nn}x_n &= \lambda x_n. \end{aligned} \tag{4.53}$$

System (4.53) looks like a regular linear system of equations. However, there is substantial difference between the eigenvalue problem and solving a linear system of equations. For the eigenvalue problem, the scalars λ's are unknown, and solutions for system (4.53) exist only for specific values of λ. These values are called eigenvalues.

And a set of solutions $x_1, x_2, \ldots, x_n$ corresponding to an eigenvalue λ is called an eigenvector.

Regrouping terms in system (4.53) gives a system of homogeneous linear equations:

$$\begin{pmatrix} a_{11} - \lambda & a_{12} & \cdots & a_{1n} \\ a_{21} & a_{22} - \lambda & \cdots & a_{2n} \\ \cdots & \cdots & \cdots & \cdots \\ a_{n1} & a_{n2} & \cdots & a_{nn} - \lambda \end{pmatrix} \begin{pmatrix} x_1 \\ x_2 \\ \cdots \\ x_n \end{pmatrix} = \begin{pmatrix} 0 \\ 0 \\ \cdots \\ 0 \end{pmatrix}. \quad (4.54)$$

Using a unit matrix I, which is

$$I = \begin{pmatrix} 1 & 0 & \cdots & 0 \\ 0 & 1 & \cdots & 0 \\ \cdots & \cdots & \cdots & \cdots \\ 0 & 0 & \cdots & 1 \end{pmatrix}, \quad (4.55)$$

the system of linear equations (4.54) may also be written as

$$(A - \lambda I)x = 0. \quad (4.56)$$

Non-trivial solutions for system (4.54) exist if and only if the determinant of the matrix $(A - \lambda I)$ is zero, that is,

$$\det |A - \lambda I| = 0. \quad (4.57)$$

When expanded, determinant (4.57) reveals itself as a characteristic polynomial of degree n in λ. It has n roots, or n eigenvalues λ_i $(i = 1, 2, \ldots, n)$, some of which may occur as repeated roots. Since a polynomial can have not only real but complex roots also, the eigenvalues can be real or complex. The general solution for finding the roots of a polynomial of any degree analytically exists for degrees up to 4.[3] For polynomials of higher degrees (5 or greater), it is generally necessary to use numerical methods.

It is important to note that for every eigenvalue λ there is a corresponding eigenvector $x_1, x_2, \ldots x_n$. To differentiate eigenvectors associated with distinct eigenvalues λ, we shall employ the notation $x_i^{(\lambda)}$.

---

[3]The Abel–Ruffini theorem demonstrates the non-existence of a general solution for polynomials of degree 5 or higher using only arithmetic operations and radicals.

Eigenvalues and eigenvectors have several important properties when associated with matrices. Here are some of most practical properties:

Eigenvalues and the trace of a matrix: The sum of the eigenvalues of a matrix is equal to the sum of its diagonal entries, which is also equal to the trace of the matrix:

$$\sum_{i=1}^{n} \lambda_i = \text{tr}(A) = \sum_{i=1}^{n} a_{ii}. \tag{4.58}$$

Eigenvalues and determinants: The product of the eigenvalues is equal to the determinant of the matrix:

$$\prod_{i=1}^{n} \lambda_i = \det(A). \tag{4.59}$$

Eigenvalues of transpose matrix: The eigenvalues of a matrix A are the same as the eigenvalues of its transpose matrix A^T,

Eigenvalues of symmetric matrices: For symmetric matrices (with real numbers), all eigenvalues are real numbers, and the eigenvectors corresponding to distinct eigenvalues are orthogonal.

Linear independence of eigenvectors: Eigenvectors corresponding to distinct eigenvalues of a matrix are linearly independent. This means that if a matrix A has n distinct eigenvalues, then its eigenvectors form a linearly independent set of n vectors. If a matrix has repeated eigenvalues, then the eigenvectors corresponding to the repeated eigenvalues may not be linearly independent.

Example 4.9. Show that a real symmetric second-order matrix has only real eigenvalues. Let's write such a matrix as

$$A = \begin{pmatrix} a & g \\ g & b \end{pmatrix}.$$

The characteristic equation for finding eigenvalues is

$$\Delta = \begin{vmatrix} a - \lambda & g \\ g & b - \lambda \end{vmatrix} = (a - \lambda)(b - \lambda) - g^2 = \lambda^2 + (a + b)\lambda + ab - g^2.$$

The solutions of the equation are

$$\lambda_{1,2} = \frac{1}{2}\left[(a+b) \pm \sqrt{(a+b)^2 - 4ab + 4g^2}\right]$$

$$= \frac{1}{2}\left[(a+b) \pm \sqrt{(a-b)^2 + 4g^2}\right].$$

Since $\sqrt{(a-b)^2 + 4g^2}$ is always positive, then both eigenvalues of the above symmetric matrix are real. ∎

Example 4.10. Find eigenvalues and eigenvectors of matrix A, given that

$$A = \begin{pmatrix} 4 & 2 \\ 2 & 1 \end{pmatrix}.$$

This is a symmetric matrix, and we expect only real eigenvalues. First, we find eigenvalues using condition (4.57):

$$\Delta = \begin{vmatrix} 4-\lambda & 2 \\ 2 & 1-\lambda \end{vmatrix} = (4-\lambda)(1-\lambda) - 4 = \lambda^2 - 5\lambda = 0.$$

This equation has two solutions, namely $\lambda_1 = 0$ and $\lambda_2 = 5$. We can note that the sum of the eigenvalues is equal to the trace of the matrix $\lambda_1 + \lambda_2 = \text{tr}A$ (the sum of elements on the main diagonal), and the product of the eigenvalues is equal to the matrix's determinant which is $\lambda_1 \lambda_2 = \det(A) = 0$.

Next, we need to solve a system of linear equations:

$$(4 - \lambda_i)x_1 + 2x_2 = 0,$$
$$2x_1 + (1 - \lambda_i)x_2 = 0,$$

for every value of λ. For $\lambda_1 = 0$, we obtain

$$4x_1 + 2x_2 = 0,$$
$$2x_1 + x_2 = 0.$$

We can see that the two equations are homogeneous, the determinant of their coefficients is equal to zero, and the two equations are linearly dependent (we can obtain the first equation by multiplying the second by a factor of two). Discarding the first equation we have *one* equation with *two* unknowns x_1 and x_2:

$$2x_1 + x_2 = 0.$$

It means that we can assign an arbitrary value to one of the unknowns. Let us set $x_1 = 1$, then $x_2 = -2$. Repeating calculations for $\lambda_2 = 5$ we obtain $x_1 = 2$ and $x_2 = 1$. Thus, the matrix has two eigenvectors:

$$\begin{pmatrix} x_1^{(1)} \\ x_2^{(1)} \end{pmatrix} = \begin{pmatrix} 1 \\ -2 \end{pmatrix} \quad \text{and} \quad \begin{pmatrix} x_1^{(2)} \\ x_2^{(2)} \end{pmatrix} = \begin{pmatrix} 2 \\ 1 \end{pmatrix}.$$

Both eigenvectors are defined within an arbitrary multiplicative factor, i.e., any eigenvector multiplied by a constant is still the same eigenvector. How do we choose the arbitrary factor? There are two popular methods. In the first method, the smallest component is set to one. That is exactly what we did above. In physics, it is common to normalize eigenvectors on its length, like $L = \sqrt{x_1^2 + x_2^2}$. ∎

Hermitian matrices: A Hermitian matrix is a special type of square matrix with wide applications in physics and other fields. A matrix is Hermitian if it is equal to its conjugate transpose

$$\begin{pmatrix} a_{11} & a_{12} & \cdots & a_{1n} \\ a_{21} & a_{22} & \cdots & a_{2n} \\ \cdots & \cdots & \cdots & \cdots \\ a_{n1} & a_{n2} & \cdots & a_{nn} \end{pmatrix} = \begin{pmatrix} a_{11} & a_{21}^* & \cdots & a_{n1}^* \\ a_{12}^* & a_{22} & \cdots & a_{n2}^* \\ \cdots & \cdots & \cdots & \cdots \\ a_{1n}^* & a_{2n}^* & \cdots & a_{nn} \end{pmatrix}, \quad (4.60)$$

with the elements a_{ii} on its leading diagonal all real. For example, the set of Pauli matrices

$$\sigma_1 = \begin{pmatrix} 0 & 1 \\ 1 & 0 \end{pmatrix}, \quad \sigma_2 = \begin{pmatrix} 0 & -i \\ i & 0 \end{pmatrix}, \quad \sigma_3 = \begin{pmatrix} 1 & 0 \\ 0 & -1 \end{pmatrix}$$

is used to represent the spin angular momentum components in quantum mechanics. The Pauli matrices are also used as fundamental gates in quantum computing.

Hermitian matrices possess important properties, including the fact that all their eigenvalues are real numbers, and the eigenvectors associated with different eigenvalues are mutually orthogonal. The scalar product of two eigenvectors corresponding to two eigenvalues λ_1 and λ_2 is defined as

$$\sum_{i=1}^{n} x_i^{(\lambda_1)*} x_i^{(\lambda_2)}, \quad (4.61)$$

where $x_i^{(\lambda_1)*}$ are the complex conjugate components of the first vector.

In quantum physics, Hermitian matrices represent Hermitian operators, which directly relate to observable physical quantities. The real eigenvalues of these operators correspond to the measurable values of physical observables.

Example 4.11. Prove that a Hermitian matrix of order two has only real eigenvalues.

Let us consider a Hermitian matrix

$$\begin{pmatrix} d_1 & g+ih \\ g-ih & d_2 \end{pmatrix},$$

where d_1, d_2, g, and h are all real numbers. For finding the eigenvalues, we need to solve for λ:

$$\begin{vmatrix} d_1 - \lambda & g+ih \\ g-ih & d_2 - \lambda \end{vmatrix} = (d_1 - \lambda)(d_2 - \lambda) - (g+ih)(g-ih)$$
$$= \lambda^2 - \lambda(d_1 + d_2) + d_1 d_2 - (g^2 + h^2) = 0.$$

This is a quadratic equation with respect to λ. For a quadratic equation to have only real roots, we must have[4]

$$(d_1 + d_2)^2 - 4(d_1 d_2 - (g^2 + h^2)) \geq 0.$$

Rewriting the above equation we obtain

$$(d_1 - d_2)^2 + 4(g^2 + h^2) \geq 0,$$

which is non-negative for any real d_1, d_2, g, h numbers. ∎

Example 4.12. Find eigenvalues and eigenvectors of the following matrix:

$$\begin{pmatrix} 3 & 1-i \\ 1+i & 2 \end{pmatrix}.$$

---

[4] Roots of a quadratics equation $ax^2 + bx + c = 0$ are

$$x_{1,2} = \frac{-b \pm \sqrt{b^2 - 4ac}}{2a}.$$

The corresponding determinant for finding the eigenvalues is

$$\begin{vmatrix} 3 - \lambda & 1 - i \\ 1 + i & 2\lambda \end{vmatrix} = (3 - \lambda)(2 - \lambda) - (1 - i)(1 + i) = \lambda^2 - 5\lambda + 4 = 0,$$

with the solutions $\lambda_1 = 4$ and $\lambda_2 = 1$. The system of linear equations for finding the eigenvectors is

$$(3 - \lambda_i)x_1 + (1 - i)x_2 = 0,$$
$$(1 + i)x_1 + (2 - \lambda_i)x_2 = 0,$$

with solutions

$$x^{(\lambda_1)} = \begin{pmatrix} 1 - i \\ 1 \end{pmatrix} \quad \text{and} \quad x^{(\lambda_2)} = \begin{pmatrix} 1 \\ -1 - i \end{pmatrix}.$$

As we see, while the eigenvalues of the matrix are real, its eigenvectors can be complex. A quick check of the orthogonality shows

$$x^{(\lambda_1)} \cdot x^{(\lambda_2)} = x_1^{(\lambda_1)*}x_1^{(\lambda_2)} + x_2^{(\lambda_1)*}x_2^{(\lambda_2)} = (1 + i)1 + 1(-1 - i) = 0.$$

Thus, the two eigenvectors are orthogonal. ■

4.6. Exercises

(1) Find $A + B$, $A - B$, AB, BA, A^2, and B^2 for the following matrices:

(a) $A = \begin{pmatrix} 2 & -4 \\ 4 & 3 \end{pmatrix}$, $B = \begin{pmatrix} -2 & 1 \\ 2 & -1 \end{pmatrix}$

(b) $A = \begin{pmatrix} 1 & 2 \\ 3 & -5 \end{pmatrix}$, $B = \begin{pmatrix} 10 & 4 \\ -5 & -2 \end{pmatrix}$.

(2) The Pauli spin matrices in quantum mechanics are

$$A = \begin{pmatrix} 0 & 1 \\ 1 & 0 \end{pmatrix}, \quad B = \begin{pmatrix} 0 & -i \\ i & 0 \end{pmatrix}, \quad C = \begin{pmatrix} 1 & 0 \\ 0 & -1 \end{pmatrix}$$

(a) Show that $A^2 = B^2 = C^2 = I$
(b) Show that any of these matrices anticommute, e.g., $AB = -BA$, $AC = -CA$, $BC = -CB$
(c) Show that $AB - BA = 2iC$.

(3) Find the inverse matrix of

(a) $A = \begin{pmatrix} 4 & 3 \\ 1 & 1 \end{pmatrix}$, (b) $B = \begin{pmatrix} 2 & 3 \\ 4 & 5 \end{pmatrix}$.

(4) Evaluate the determinants of the following matrices

(a) $\begin{pmatrix} 1 & 2 & 3 \\ 4 & 5 & 6 \\ 7 & 8 & 9 \end{pmatrix}$, (b) $\begin{pmatrix} 3 & 1 & -2 \\ 0 & 4 & 5 \\ -1 & 2 & 3 \end{pmatrix}$.

(5) Solve the following systems of linear equations:

(a) $2x + y = 5$
$3x - 2y = 8$

(b) $2x + y + z = 2$
$-x + y - z = 3$
$x + 2y + 3z = -10$

(c) $3x + 2y + 4z = 4$
$2x - 3y + z = 2$
$x + y + 2z = 3$.

(6) Find the eigenvalues and eigenvectors of the following symmetric matrices

(a) $\begin{pmatrix} 2 & 1 \\ 1 & 2 \end{pmatrix}$ (b) $\begin{pmatrix} 0 & -1 \\ 2 & 3 \end{pmatrix}$

(c) $\begin{pmatrix} 4 & 0 & 2 \\ 0 & 2 & 0 \\ 2 & 0 & 4 \end{pmatrix}$ (d) $\begin{pmatrix} 4 & 2 & 2 \\ 2 & 6 & 0 \\ 2 & 0 & 6 \end{pmatrix}$.

(7) Find eigenvalues and eigenvectors of the following Hermitian matrices. Show that the eigenvectors are orthogonal:

(a) $\begin{pmatrix} 1 & i \\ -i & 1 \end{pmatrix}$ (b) $\begin{pmatrix} 2 & 2+i \\ 2-i & 6 \end{pmatrix}$.

Answers and solutions

(1) (a)
$$A+B = \begin{pmatrix} 0 & -3 \\ 6 & 2 \end{pmatrix}, \quad A-B = \begin{pmatrix} 4 & -5 \\ 2 & 4 \end{pmatrix}$$

$$AB = \begin{pmatrix} -12 & 6 \\ -2 & 1 \end{pmatrix}, \quad BA = \begin{pmatrix} 0 & 11 \\ 0 & -11 \end{pmatrix}$$

$$A^2 = \begin{pmatrix} -12 & -20 \\ 20 & -7 \end{pmatrix}, \quad B^2 = \begin{pmatrix} 6 & -3 \\ -6 & 3 \end{pmatrix}.$$

(2) Use basic operations with matrices.

(3) (a) $A^{-1} = \begin{pmatrix} 1 & -3 \\ -1 & 4 \end{pmatrix}$, **(b)** $B^{-1} = \dfrac{1}{2}\begin{pmatrix} -5 & 3 \\ 4 & -2 \end{pmatrix}.$

(4) (a) 0, **(b)** -7.

(5) (a) $x = 18/7,\ y = -1/7$
   (b) $x = 3,\ y = 1,\ z = -5$
   (c) $x = -2,\ y = -1,\ z = 3$.

(6) (a) $\lambda_1 = 1,\ v_1 = (-1, 1),\ \lambda_2 = 3,\ v_2 = (1, 1)$
   (b) $\lambda_1 = 1,\ v_1 = (-1, 1),\ \lambda_2 = 2,\ v_2 = (-1, 2)$
   (c) $\lambda_1 = 6,\ v_1 = (1, 0, 1),\ \lambda_2 = 2,\ v_2 = (-1, 0, 1),\ \lambda_3 = 2,\ v_3 = (0, 1, 0)$
   (d) $\lambda_1 = 2,\ v_1 = (-2, 1, 1),\ \lambda_2 = 6,\ v_2 = (0, -1, 1),\ \lambda_3 = 8,\ v_3 = (1, 1, 1).$

(7) (a) $\lambda_1 = 0,\ v_1 = (-i, 1),\ \lambda_2 = 2,\ v_2 = (i, 1)$
   (b) $\lambda_1 = 1,\ v_1 = (-2 - i, 1),\ \lambda_2 = 7,\ v_2 = (2 + i, 5).$

Chapter 5

Partial Differentiation

5.1. Complicated real world

In the real world, not everything can be described by functions of a single variable, $f(x)$! Such a simple example as describing a temperature in a room needs three space variables x, y, and z. And when we deal with physical processes developing in time, we add time t as one more independent variable. Then, a temperature distribution is a functions of four variables $T(x, y, z, t)$.

There are numerous instances where functions involving multiple variables are essential for accurately representing and understanding real-world phenomena. For example, in physics, the equations of motion in classical mechanics involve multiple variables, such as time, position, velocity, and acceleration. Few-body problem in quantum mechanics is still at the top of most challenging problems. Wave motion in two and three dimensions, water waves, and acoustic waves all need functions of multiple variables. Structural engineering involves analyzing the behavior of a structure under the influence of multiple forces, such as loads, material properties, and geometric dimensions. In macroeconomics, variables such as Gross Domestic Product (GDP) are influenced by multiple factors, including government spending, consumption, investment, and net exports. Modeling economic relationships requires functions with multiple variables to analyze their complex interactions.

Most common operations with functions of a single variable $f(x)$ include differentiation, integration, and solving differential equations.[1] This is the same situation we have with functions of many variables. In this chapter, we concentrate on differentiation (called partial differentiation) with applications to studying stationary points. We also identify how to change variables used in differential operators.

Almost all principal ideas, definitions, and properties for functions of multiple variables can be studied and understood by analyzing functions of two variables. We may say that a function of two variables $f(x,y)$ has all the complexity we need at this level. The idea of a function of two variables is quite simple — for every pair of two numbers x and y, there is one specific value $z = f(x,y)$.

5.2. The partial derivatives

Let a function $f(x,y)$ be a function of two independent variables x and y. If we assign a constant value to y and let x to be changed, then $f(x,y)$ will be a function of just one variable x. We can write a derivative of the function $f(x,y)$ over x as the limit

$$\lim_{\Delta x \to 0} \frac{f(x+\Delta x, y) - f(x,y)}{\Delta x} = \frac{\partial f}{\partial x}. \qquad (5.1)$$

This derivative is called the *partial derivative* of the function $f(x,y)$ with respect to x at the point x. Since such a derivative is evaluated by assuming that y remains constant, then partial derivatives can be evaluated using the same methods and properties that we use for evaluating derivatives of functions of single variables. When you evaluate a derivative for one variable, you treat all other variables as constant coefficients. Then, the evaluation of the partial derivative itself is essentially nothing new compared to the evaluation of the

---

[1]Differential equations for a function of a single variable are called *ordinary differential equations*. Differential equations involving partial derivatives of a function of two or more independent variables are called *partial differential equations*.

Partial Differentiation

ordinary derivative. There are various notations for partial derivatives,[2] namely

$$\frac{\partial f}{\partial x} \equiv \frac{\partial f(x,y)}{\partial x} \equiv f_x(x,y) \equiv f'_x(x,y).$$

In the same way, we define a partial derivative of $f(x,y)$ with respect to y assuming that x does not change:

$$\lim_{\Delta y \to 0} \frac{f(x, y + \Delta y) - f(x,y)}{\Delta y} = \frac{\partial f}{\partial y}. \tag{5.2}$$

Example 5.1. Find partial derivatives of $f(x,y) = x^2 y^3 + 2xe^y$:

$$\frac{\partial f}{\partial x} = 2xy^3 + 2e^y, \quad \frac{\partial f}{\partial y} = 3x^2 y^2 + 2xe^y. \qquad \blacksquare$$

Note that an ordinary derivative is a ratio of two differentials. Assume that $dy/dx = h(x)$, then we can write $dy = h(x)dx$. This is not allowed with partial derivatives because there is no partial differential ∂y.

A partial derivative depends on a choice of "fixed" variables. For example, potential energy stored in a capacitor can be written in two equivalent forms:

$$U = \frac{1}{2}CV^2 = \frac{Q^2}{2C},$$

since $Q = CV$. Then,

$$\left.\frac{\partial U}{\partial C}\right|_V = \frac{1}{2}V^2 = \frac{U}{C}, \quad \left.\frac{\partial U}{\partial C}\right|_Q = -\frac{Q^2}{2C^2} = -\frac{U}{C} \quad \text{or} \quad \left.\frac{\partial U}{\partial C}\right|_V = -\left.\frac{\partial U}{\partial C}\right|_Q.$$

---

[2]It was C.G. Jacobi who suggested to use the symbol ∂ instead of d for notations of partial derivatives.

5.3. The total differential and total derivative

For a function of a single variable $f(x)$, a small change dx of the variable x results in a small change in the function $f(x)$:

$$f(x+dx) - f(x) = f'(x)dx. \qquad (5.3)$$

Here, we disregard terms of order $(dx)^2$ since dx is small. For a function of two variables $f(x,y)$, the function change is a result of variations in *both* variables x and y as

$$f(x+dx, y+dy) - f(x,y).$$

We are going to show that such a change has two components: one component is proportional to dx and the second one is proportional to dy. We disregard terms of the second order $(dx)^2$, $(dy)^2$, and $(dxdy)$ since dx and dy are small. We can rewrite the above equation as

$$f(x+dx, y+dy) - f(x,y) = f(x+dx, y+dy) - f(x+dx, y)$$
$$+ f(x+dx, y) - f(x,y).$$

Disregarding terms $\sim (dx)^2$ the difference of the last two terms can be written as

$$f(x+dx, y) - f(x,y) = \left.\frac{\partial f(x,y)}{\partial x}\right|_y dx,$$

where $\left.\frac{\partial f(x,y)}{\partial x}\right|_y$ denotes a partial derivative of the function $f(x,y)$ evaluated under assumption that y is constant. Now, for the first two terms (disregarding terms $\sim (dy)^2$), we have

$$f(x+dx, y+dy) - f(x+dx, y) = \left.\frac{\partial f(x+dx, y)}{\partial y}\right|_{x+dx} dy.$$

Thus,

$$f(x+dx, y+dy) - f(x,y) = \left.\frac{\partial f(x+dx, y)}{\partial y}\right|_{x+dx} dy + \left.\frac{\partial f(x,y)}{\partial x}\right|_y dx.$$

Since dx is small, we can write

$$\left.\frac{\partial f(x+dx, y)}{\partial y}\right|_{x+dx} dy \sim \left.\frac{\partial f(x,y)}{\partial y}\right|_x dy + \alpha dx,$$

where α is a constant. Then,

$$\left[\left.\frac{\partial f(x,y)}{\partial y}\right|_x + \alpha dx\right] dy + \left.\frac{\partial f(x,y)}{\partial x}\right|_y dx$$

$$= \left.\frac{\partial f(x,y)}{\partial y}\right|_x dy + \left.\frac{\partial f(x,y)}{\partial x}\right|_y dx + \alpha dx dy.$$

Finally, since $\alpha dx dy$ is as small as $(dx)^2, (dy)^2$, then we can define *the total differential* as

$$df(x,y) = f(x+dx, y+dy) - f(x,y) = \left.\frac{\partial f(x,y)}{\partial x}\right|_y dx + \left.\frac{\partial f(x,y)}{\partial y}\right|_x dy,$$

or in a more concise form,

$$df(x,y) = \frac{\partial f(x,y)}{\partial x} dx + \frac{\partial f(x,y)}{\partial y} dy. \tag{5.4}$$

We can say that for small values of dx and dy, the total differential $df(x,y)$ gives an approximate value of the total increment $f(x+dx, y+dy) - f(x,y)$, corresponding to two increments of the independent variables dx and dy. Figure 5.1 help in grasping the idea of a total differential for a function of two variables.

Formula (5.4) shows a very important property of functions of several *independent* variables, which can be called the property of

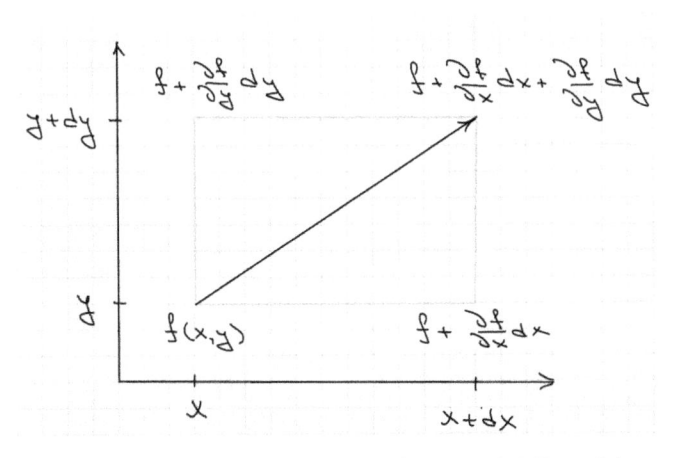

Fig. 5.1. The geometrical view of the total differential.

superposition of small changes, namely the joint effect of several small changes of dx and dy can be replaced with sufficient accuracy by the sum of the effects of each small change separately.

So far we considered x and y as two independent variables. Let the two variables $x(t)$ and $y(t)$ be functions of a parameter t. For example, in physics, a trajectory of an object moving in x, y space is actually a function $f(x, y) = f(x(t), y(t))$ of a single variable t (time). The total differential for variables depending on a parameter can be written as

$$df(x,y) = \frac{\partial f(x,y)}{\partial x} dx + \frac{\partial f(x,y)}{\partial y} dy = \frac{\partial f(x,y)}{\partial x} \frac{dx}{dt} dt + \frac{\partial f(x,y)}{\partial y} \frac{dy}{dt} dt.$$

Then, we can write a derivative along a line (a trajectory in Figure 5.2) as

$$\frac{df}{dt} = \frac{\partial f}{\partial x} \frac{dx}{dt} + \frac{\partial f}{\partial y} \frac{dy}{dt}, \tag{5.5}$$

which is the *chain rule for partial differentiation*.

Example 5.2. For $f(x,y) = xe^{-y}$, where $x(t) = 1 + at$, $y(t) = bt^2$,

$$\frac{df}{dt} = \frac{\partial f}{\partial x} \frac{dx}{dt} + \frac{\partial f}{\partial y} \frac{dy}{dt} = e^{-y} a - (xe^{-y}) 2bt = e^{-y}(a - 2btx)$$
$$= e^{-bt^2}(a - 2bt(1 + at)). \qquad \blacksquare$$

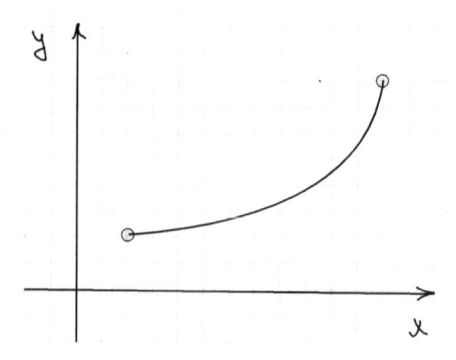

Fig. 5.2. A trajectory of a particle in 2D space.

If the variable y is given as a function of x, i.e., $y(x)$, then

$$\frac{df}{dx} = \frac{\partial f}{\partial x} + \frac{\partial f}{\partial y}\frac{dy}{dx}, \qquad (5.6)$$

which is called *the total derivative* of $f(x, y)$ with respect to x.

Example 5.3. Here is an example from physics. Assume that function $T(x, y, t)$ describes the temperature of a point with gas flowing along a trajectory so that $x(t), y(t)$. Then, $\frac{dT}{dt}$ is the rate of change of the temperature along the trajectory:

$$\frac{dT}{dt} = \frac{\partial T}{\partial t} + \frac{\partial T}{\partial x}\frac{dx}{dt} + \frac{\partial T}{\partial y}\frac{dy}{dt}.$$

Unlike in the projectile motion example, here the point has a local characteristic explicitly dependent on time and the temperature change has x and y components. If, however, T does not depend explicitly on time, then $\frac{\partial T}{\partial t} = 0$ and we have a stationary temperature field. ∎

5.4. Second partial derivatives

Second partial derivatives are introduced in practically the same way we define second-order ordinary derivatives. Thus, we write

$$\frac{\partial}{\partial x}\left(\frac{\partial f}{\partial x}\right) = f_{xx}, \quad \frac{\partial}{\partial x}\left(\frac{\partial f}{\partial y}\right) = f_{xy},$$
$$\frac{\partial}{\partial y}\left(\frac{\partial f}{\partial x}\right) = f_{yx}, \quad \frac{\partial}{\partial y}\left(\frac{\partial f}{\partial y}\right) = f_{yy}, \qquad (5.7)$$

where f_{xy} and f_{yx} are *mixed* partial derivatives.

Example 5.4. Evaluate second partial derivatives for $x(x, y) = x^2y^3 + xe^y$. The first derivative was already evaluated in the first example of this chapter:

$$\frac{\partial f}{\partial x} = 2xy^3 + e^y \quad \frac{\partial f}{\partial y} = 3x^2y^2 + xe^y.$$

For second derivatives, we get

$$f_{xx} = \frac{\partial}{\partial x}\left(2xy^3 + e^y\right) = 2y^3, \quad f_{yy} = \frac{\partial}{\partial y}\left(3x^2y^2 + xe^y\right) = 6x^2y + xe^y,$$

and for mixed derivatives, we have

$$f_{xy} = \frac{\partial}{\partial x}\left(3x^2y^2 + xe^y\right) = 6xy^2 + e^y,$$

$$f_{yx} = \frac{\partial}{\partial y}\left(2xy^3 + e^y\right) = 6xy^2 + e^y. \quad \blacksquare$$

We can see from the example above that the second mixed partial derivatives are equal. We show that this is not a coincidence. We evaluate mixed partial derivatives near a point x_0, y_0 as seen in Figure 5.3, and where h and k are small. We start with the f_{yx} derivative as

$$\left.\frac{\partial}{\partial y}\left(\frac{\partial f}{\partial x}\right)\right|_{\substack{x=x_0\\y=y_0}} = \frac{1}{2k}\left(\left.\frac{\partial f}{\partial x}\right|_{\substack{x=x_0\\y=y_0+k}} - \left.\frac{\partial f}{\partial x}\right|_{\substack{x=x_0\\y=y_0-k}}\right),$$

where

$$\left.\frac{\partial f}{\partial x}\right|_{\substack{x=x_0\\y=y_0+k}} = \frac{f(x_0+h, y_0+k) - f(x_0-h, y_0+k)}{2h}$$

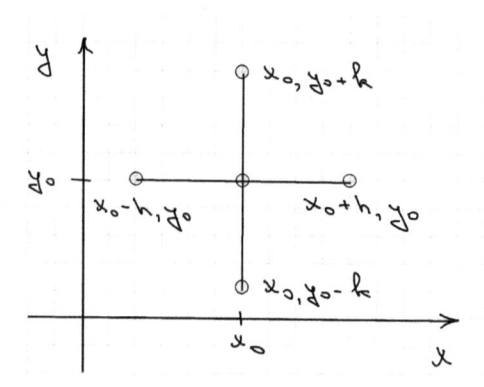

Fig. 5.3. Five points in x, y-space.

and
$$\left.\frac{\partial f}{\partial x}\right|_{\substack{x=x_0 \\ y=y_0-k}} = \frac{f(x_0+h, y_0-k) - f(x_0-h, y_0-k)}{2h}.$$

Then,
$$\left.\frac{\partial}{\partial y}\left(\frac{\partial f}{\partial x}\right)\right|_{\substack{x=x_0 \\ y=y_0}} = \frac{f(x_0+h, y_0+k) - f(x_0-h, y_0+k)}{4hk}$$
$$- \frac{f(x_0+h, y_0-k) - f(x_0-h, y_0-k)}{4hk}.$$

Now, we work with the other mixed partial derivative, namely f_{xy}
$$\left.\frac{\partial}{\partial x}\left(\frac{\partial f}{\partial y}\right)\right|_{\substack{x=x_0 \\ y=y_0}} = \frac{1}{2h}\left(\left.\frac{\partial f}{\partial y}\right|_{\substack{x=x_0+h \\ y=y_0}} - \left.\frac{\partial f}{\partial y}\right|_{\substack{x=x_0-h \\ y=y_0}}\right),$$

where
$$\left.\frac{\partial f}{\partial y}\right|_{\substack{x=x_0+h \\ y=y_0}} = \frac{f(x_0+h, y_0+k) - f(x_0+h, y_0-k)}{2k}$$

and
$$\left.\frac{\partial f}{\partial y}\right|_{\substack{x=x_0-h \\ y=y_0}} = \frac{f(x_0-h, y_0+k) - f(x_0-h, y_0-k)}{2k}.$$

Then,
$$\left.\frac{\partial}{\partial x}\left(\frac{\partial f}{\partial y}\right)\right|_{\substack{x=x_0 \\ y=y_0}} = \frac{f(x_0+h, y_0+k) - f(x_0+h, y_0-k)}{4hk}$$
$$- \frac{f(x_0-h, y_0+k) - f(x_0-h, y_0-k)}{4hk}$$

or
$$\left.\frac{\partial}{\partial y}\left(\frac{\partial f}{\partial x}\right)\right|_{\substack{x=x_0 \\ y=y_0}} = \left.\frac{\partial}{\partial x}\left(\frac{\partial f}{\partial y}\right)\right|_{\substack{x=x_0 \\ y=y_0}}.$$

Thus, we derived that the two mixed partial derivatives are equal at some point (x_0, y_0). Since this point can be chosen arbitrary, then mixed partial derivatives are equal at any point (x, y)

where these derivatives are defined and continuous. This is called Clairaut's theorem:

$$\frac{\partial}{\partial x}\left(\frac{\partial f}{\partial y}\right) = \frac{\partial}{\partial y}\left(\frac{\partial f}{\partial x}\right). \tag{5.8}$$

We see that the result of a double mixed partial differentiation does not depend on the order of differentiation. This property can be easily generalized to derivatives of any order, again, subject to the appropriate continuity conditions.

The total differential $df(x, y)$ of a function of two variables (5.4) is in turn a function of the same variables, so we can define the total differential of such a function. Thus, we get a second-order differential $d^2 f(x, y)$ of the original function $f(x, y)$, which will also be a function of the same variables, and its total differential will lead us to a third-order differential $d^3 f(x, y)$ of the original function, and so on.

The second total differential can be written as

$$\begin{aligned} d^2 f(x, y) &= d\left(\frac{\partial f(x, y)}{\partial x} dx + \frac{\partial f(x, y)}{\partial y} dy\right) \\ &= d\left(\frac{\partial f(x, y)}{\partial x} dx\right) + d\left(\frac{\partial f(x, y)}{\partial y} dy\right). \end{aligned}$$

Since the differentials of dx and dy of the independent variables must be considered as constants, we have

$$d^2 f(x, y) = dx\, d\left(\frac{\partial f(x, y)}{\partial x}\right) + dy\, d\left(\frac{\partial f(x, y)}{\partial y}\right).$$

Applying (5.4) to each term we have

$$\begin{aligned} d^2 f(x, y) &= dx \left(\frac{\partial^2 f(x, y)}{\partial^2 x} dx + \frac{\partial^2 f(x, y)}{\partial x \partial y} dy\right) \\ &+ dy \left(\frac{\partial^2 f(x, y)}{\partial y \partial x} dx + \frac{\partial^2 f(x, y)}{\partial^2 y} dy\right), \end{aligned}$$

and finally,

$$d^2 f(x, y) = \frac{\partial^2 f(x, y)}{\partial^2 x} dx^2 + 2\frac{\partial^2 f(x, y)}{\partial x \partial y} dx dy + \frac{\partial^2 f(x, y)}{\partial^2 y} dy^2. \tag{5.9}$$

In the same way, we can derive the third total differential as

$$d^3 f(x,y) = \frac{\partial^3 f(x,y)}{\partial^3 x} dx^3 + 3\frac{\partial^3 f(x,y)}{\partial x^2 \partial y} dx^2 dy + 3\frac{\partial^3 f(x,y)}{\partial x \partial y^2} dx dy^2$$
$$+ \frac{\partial^3 f(x,y)}{\partial^3 y} dy^3. \qquad (5.10)$$

Formulas (5.9) and (5.10) can be generalized to a total differential of any order as

$$d^n f(x,y) = \left(\frac{\partial}{\partial x} dx + \frac{\partial}{\partial y} dy\right)^{(n)} f(x,y). \qquad (5.11)$$

This formula should be interpreted as the sum in parentheses must be raised to the power of n, then n for $\partial/\partial x$ and $\partial/\partial y$ must be considered as the order of derivatives with respect to x and y of the function $f(x,y)$.

5.5. Taylor series for functions of two variables

Taylor series are extensively utilized in analyzing both single-variable and two-variable functions and serve as a fundamental basis for a multitude of numerical techniques. For example, the Taylor series expansion for functions of two variables has practical applications in approximating complicated functions, error analysis, optimization, computer-aided design, and geographical mapping.

Earlier in Chapter 1 we discussed the Taylor series for a function of one variable:

$$f(x) = f(a) + \frac{f'(a)}{1!}(x-a) + \frac{f''(a)}{2!}(x-a)^2$$
$$+ \cdots \frac{f^{(n)}(a)}{n!}(x-a)^n + R_{n+1}(x), \qquad (5.12)$$

where $R_{n+1}(x)$ is a remainder:

$$R_{n+1}(x) = \frac{1}{n!}\int_a^x f^{(n+1)}(t)(x-t)^n dt.$$

Now, we consider a function of two variables $f(x,y)$ near a point (a,b). Then, the "zeroth" approximation to the function is

$$f(x,y) = f(a,b).$$

It is clear that this approximation is not very practical. The first-order approximation can be written as

$$f(x,y) = f(a,b) + A(x-a) + B(y-b),$$

where A and B are some constants. We can easily find these constants by partial differentiation of $f(x,y)$:

$$f_x(x,y) = A, \quad f_y(x,y) = B.$$

Since the coefficients A and B are constants (with no dependence on x or/and y), we set $A = f_x(a,b), B = f_y(a,b)$, and finally, the first approximation to the Taylor series is

$$f(x,y) = f(a,b) + f_x(a,b)(x-a) + f_y(a,b)(y-b). \qquad (5.13)$$

In the same way, we can find the second-order approximation to the Taylor series. It is instructive to assume that it can be written as

$$f(x,y) = f(a,b) + [A(x-a) + B(y-b)]$$
$$+ \left[C(x-a)^2 + D(x-a)(y-b) + E(y-b)^2\right],$$

where A, B, C, D, E are some constant coefficients. Evaluating partial derivatives we get

$$f_x(x,y) = A + 2C(x-a) + D(y-b)$$
$$f_y(x,y) = B + D(x-a) + 2E(y-b)$$
$$f_{xx}(x,y) = 2C, \quad f_{xy}(x,y) = f_{yx}(x,y) = D, \quad f_{yy}(x,y) = 2E.$$

Choosing $x=a, y=b$, we immediately get

$$A = f_x(a,b), \quad B = f_y(a,b), \quad C = (1/2)f_{xx}(a,b),$$
$$D = f_{xy}(a,b), \quad E = (1/2)f_{yy}(a,b).$$

Then,

$$f(x,y) = f(a,b) + [f_x(a,b)(x-a) + f_y(a,b)(y-b)]$$
$$+ \frac{1}{2}\left[f_{xx}(a,b)(x-a)^2 + 2f_{xy}(a,b)(x-a)(y-b)\right.$$
$$\left. + f_{yy}(a,b)(y-b)^2\right]. \qquad (5.14)$$

Note that if a function does not depend on y, then all partial derivatives f_y, f_{xy}, f_{yy} are equal to zero, and we come to the Taylor series for a function of a single variable $f(x)$. Thus, equation (5.14) is a natural extension of the standard Taylor series to functions of two variables.

The next order of the series can be derived in the same way. We can rewrite the Taylor series for a function of two variables in a compact form. First, we replace the notation in (5.14), namely we replace x and y on $x + dx$ and $y + dy$, and then we replace a and b on x and y. In particular, $x - a$ and $x - b$ will become dx and dy accordingly. Then, (5.14) can be rewritten as

$$f(x+dx, y+dy) = f(x,y) + \left(\frac{\partial}{\partial x}dx + \frac{\partial}{\partial y}dy\right)f(x,y)$$
$$+ \left(\frac{\partial^2}{\partial^2 x}dx^2 + 2\frac{\partial^2}{\partial x \partial y}dxdy + \frac{\partial^2}{\partial^2 y}dy^2\right)f(x,y).$$

Using formula (5.11) for the total differential, we write

$$f(x+dx, y+dy) = f(x,y) + \frac{1}{1!}df(x,y) + \frac{1}{2!}d^2f(x,y) + \cdots$$
$$+ \frac{1}{n!}d^n f(x,y) + R_{n+1}, \qquad (5.15)$$

where R_{n+1} is the remainder, which can be written as

$$R_{n+1} = \frac{1}{(n+1)!}d^{n+1}f(x+\theta dx, y+\theta dy) \quad (0 < \theta < 1). \qquad (5.16)$$

We should note that terms higher than second order in the Taylor series for functions of two (or more variables) are rarely used in practical applications.

5.6. Stationary points of a function of two variables

From introductory calculus, we know that a function of a single variable $f(x)$ has a stationary point at $x = a$ if its derivative $f'(a) = 0$ at that point. More specifically, at that point, $f(x)$ has a maximum if its second derivative $f''(a) < 0$, or it has a minimum if $f''(a) > 0$. And if $f''(a) = 0$, then the next derivative should be explored.

Let a function $f(x, y)$ be continuous at a point (a, b) and in some neighborhood of it. Similar to the case of one independent variable, i.e., $f(x)$, we say that a function $f(x, y)$ of two independent variables reaches its *relative maximum* at the point (a, b) if the value of $f(a, b)$ is not less than all neighboring values of the function, namely

$$f(a + \Delta x, b + \Delta y) - f(a, b) \leq 0,$$

for all Δx and Δy sufficiently small in absolute value. In the same way, we say that a function $f(x, y)$ has a *relative minimum* at $x = a$, $y = b$ if

$$f(a + \Delta x, b + \Delta y) - f(a, b) \geq 0.$$

Let $f(x, y)$ have a maximum or minimum at $x = a$ and $y = b$, i.e., as in Figure 5.4, where $a = 0$ and $b = 0$. Next, we consider the function $f(x, b)$ of one independent variable x, where the value of the

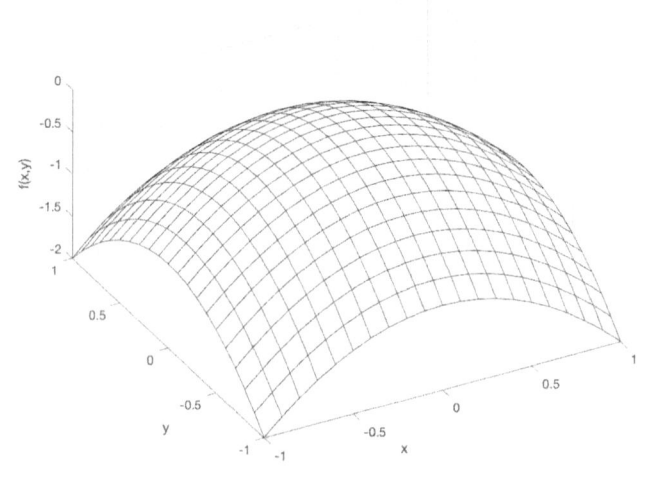

Fig. 5.4. The function $f(x, y) = -x^2 - y^2$ with a maximum at $a = 0$, $b = 0$.

Partial Differentiation

y variable is fixed at $y = b$. Then, the function $f(x, b)$ must reach a maximum or minimum at $x = a$, and therefore its derivative with respect to x at $x = a$ must vanish:

$$\left[\frac{d}{dx}f(x,b)\right]\bigg|_{x=a} = 0.$$

By the same reasoning, we can verify that the derivative of the function $f(a, y)$ with respect to y must vanish at $y = b$:

$$\left[\frac{d}{dy}f(a,y)\right]\bigg|_{y=b} = 0.$$

Thus, we have the following *necessary conditions* for the existence of a maximum or minimum: Function $f(x, y)$ of two independent variables can attain a maximum or minimum only for those values of x and y where the first partial derivatives vanish:

$$f_x(x, y) = 0, \quad f_y(x, y) = 0. \tag{5.17}$$

Since the first-order differential (5.4)

$$df(x, y) = f_x(x, y)dx + f_y(x, y)dy$$

is the sum of the partial derivatives with respect to the independent variables, we can therefore conclude that when a function $f(x, y)$ has a maximum or a minimum, the first-order differential $df(x, y)$ must vanish.[3]

Similarly, changing only x, or only y, and by analyzing the second partial derivatives, we can conclude that a *necessary condition for a maximum* is

$$f_{xx} \leq 0 \quad \text{and} \quad f_{yy} \leq 0, \tag{5.18}$$

and a *necessary condition for a minimum* is

$$f_{xx} \geq 0 \quad \text{and} \quad f_{yy} \geq 0. \tag{5.19}$$

Unlike for a function of a single variable, a *sufficient condition* for a maximum or a minimum of a function of two variables $f(x, y)$ is much more complicated. Let us consider two examples. For

---

[3]This form of the necessary condition is convenient because the first differential, also known as the total differential, remains unaffected by the choice of variables. It represents the linear approximation of the function's change with respect to all its variables.

function $f(x, y) = -x^2 - y^2$ in Figure 5.4, condition (5.17) gives $2x = 0, 2y = 0$, or the "suspected" point for maximum or minimum is a point at the origin where $x = 0, y = 0$. Indeed, here we have a maximum, since at this point the function is equal to zero $f(0, 0) = 0$, and it is negative $f < 0$ at all other points. Regarding the second partial derivatives in this example, they are constant, and $f_{xx} = -2$, $f_{yy} = -2$.

For the function $f(x, y) = -x^2 + y^2$ in Figure 5.5, we still have $f_x = 0$ at $x = 0$ and $f_y = 0$ at $y = 0$. However, there is no maximum or minimum at $x = 0, y = 0$, instead the function has a maximum along $x = 0$ as $f(0, y) = -y^2$ and a minimum along $y = 0$ as $f(x, 0) = x^2$. Such a case is called "a saddle point". Note that at a saddle point, second partial derivatives f_{xx} and f_{yy} have opposite signs.

Hence, besides maxima and minima, a function of two variables $f(x, y)$ may have *saddle* points. Therefore, we need to carry out additional analysis for identifying the nature of points, where $f_x(x, y) = 0$ and $f_y(x, y) = 0$.

By setting the first-order partial derivatives to zero (5.17), we obtain a system of two equations. Solutions of the system provide values of x and y where the function $f(x, y)$ can reach a maximum

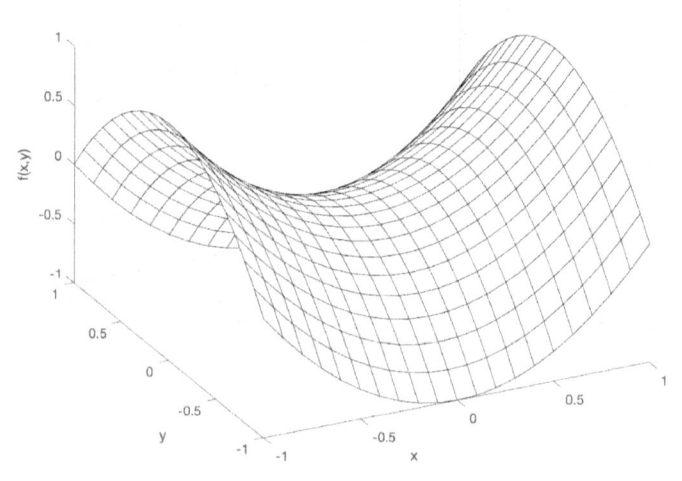

Fig. 5.5. The function $f(x, y) = x^2 - y^2$ with a saddle at $a = 0, b = 0$.

or a minimum. For a complete solution of the problem, it is also necessary to analyze second partial derivatives in order to decide whether the function reaches a maximum or minimum for these values of the independent variables, and if it does, then what exactly is a maximum or a minimum or a saddle point. Figure 5.6 illustrates a function having all three types of the stationary points.

We start our analysis with the necessary condition (5.17) satisfied for a point (a, b), i.e., $f_x(a, b) = 0$ and $f_y(a, b) = 0$. Then, for the Taylor series (5.14), we can write

$$f(x, y) \approx f(a, b) + \frac{1}{2}\left[f_{xx}(a, b)(x - a)^2 + 2f_{xy}(a, b)(x - a)(y - b) + f_{yy}(a, b)(y - b)^2\right].$$

Of course, in this equation, we neglect third-order terms. However, in studying stationary points, these terms are much less substantial, comparing to second-order terms. We can rewrite the above equation as

$$f(x, y) - f(a, b) = \frac{1}{2}(y - b)^2 \left[\frac{(x - a)^2}{(y - b)^2} f_{xx}(a, b) + 2\frac{(x - a)}{(y - b)} f_{xy}(a, b) + f_{yy}(a, b)\right].$$

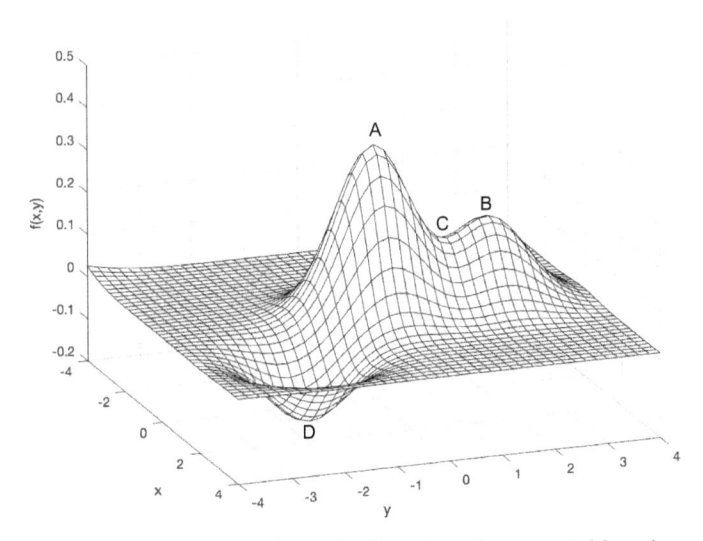

Fig. 5.6. Various stationary points of a function of two variables. A maximum at A, a second local maximum at B, a saddle point at C, and a minimum at D.

Now, introducing $t = (x-a)/(y-b)$, we have

$$f(x,y) - f(a,b) = \frac{1}{2}(y-b)^2 \left[f_{xx}(a,b)t^2 + 2f_{xy}(a,b)t + f_{yy}(a,b) \right], \tag{5.20}$$

where the right-hand side defines the sign of the difference $f(x,y) - f(a,b)$ in the vicinity of a,b. If the difference is positive, then we have a minimum, and if the difference is negative, then we have a maximum. If the difference changes the sign in the vicinity of a,b, then we have a saddle point (also called a *minimax* point).

The terms in the square brackets in (5.20) look like an equation of a parabola $g(t) = at^2 + bt + c$ relative to a variable t. From basic algebra follows that if $b^2 - 4ac > 0$, then the parabola has both positive and negative values as a function of t, and it crosses the t-axis at two points (that are the roots of equation $at^2 + bt + c = 0$). So, this is a saddle point case. If $b^2 - 4ac < 0$, then the parabola does not change its sign being either positive or negative at all values of t. If we set $t = 0$, then the coefficient a defines the sign of the parabola.

Applying this analysis to (5.20), and keeping in mind the necessary conditions for second derivatives (equations (5.18) and (5.19)), we can analyze the nature of points when $f_x(a,b) = 0$ and $f_y(a,b) = 0$ as follows:

- If $f_{xy}^2(a,b) - f_{xx}(a,b) f_{yy}(a,b) < 0$[4] and
  if $f_{xx}(a,b) > 0$ and $f_{yy}(a,b) > 0$, then $f(x,y)$ has a minimum at (a,b),
  if $f_{xx}(a,b) < 0$ and $f_{yy}(a,b) < 0$, then $f(x,y)$ has a maximum at (a,b).
- If $f_{xy}^2(a,b) - f_{xx}(a,b) f_{yy}(a,b) > 0$, then $f(x,y)$ has a saddle point at (a,b) independently if $f_{xx}(a,b)$ and $f_{yy}(a,b)$ have the same or opposite signs.
- If $f_{xy}^2(a,b) - f_{xx}(a,b) f_{yy}(a,b) = 0$, then further investigation is required (including the analysis of third-order derivatives).

---

[4]We can write the conditions for $f_{xy}^2(a,b) - f_{xx}(a,b) f_{yy}(a,b)$ by using the Hessian matrix H with it's determinant $D(x,y)$:

$$H = \begin{pmatrix} f_{xx} & f_{xy} \\ f_{yx} & f_{yy} \end{pmatrix} \quad D(x,y) = f_{xx} f_{yy} - f_{xy}^2.$$

In our analysis, we essentially use $-D(x,y)$.

The above analysis can be extended to functions of n variables. Finding stationary points for a function of multiple variables plays an important role in solving many practical problems in science and engineering.

Example 5.5. Find stationary points of $f(x, y) = x^3 - y^3 - 3xy + 4$. Condition (5.17) on first partial derivatives gives

$$f_x = x^2 - y = 0,$$
$$f_y = -y^2 - x = 0.$$

The system of equations has two solutions: $x = 0$, $y = 0$ and $x = -1$, $y = 1$. The second partial derivatives are $f_{xx} = 6x$, $f_{yy} = -6y$, $f_{xy} = -3$. Since for the first point $x = 0$, $y = 0$ we have $f_{xy}^2(0,0) - f_{xx}(0,0)f_{yy}(0,0) > 0$, this is a saddle point. For the second point $x = -1$, $y = 1$, $f_{xx}(-1,1) = -6 < 0$, $f_{yy}(-1,1) = -6 < 0$ and $f_{xy}^2(-1,1) - f_{xx}(-1,1)f_{yy}(-1,1) < 0$, so this is a maximum as shown in Figure 5.7. ∎

Example 5.6. There are n points in a plane (x, y) with given masses m_k and positions (a_k, b_k). Find a point (x_0, y_0) so that the rotational inertia of the system is the smallest relative to this point.

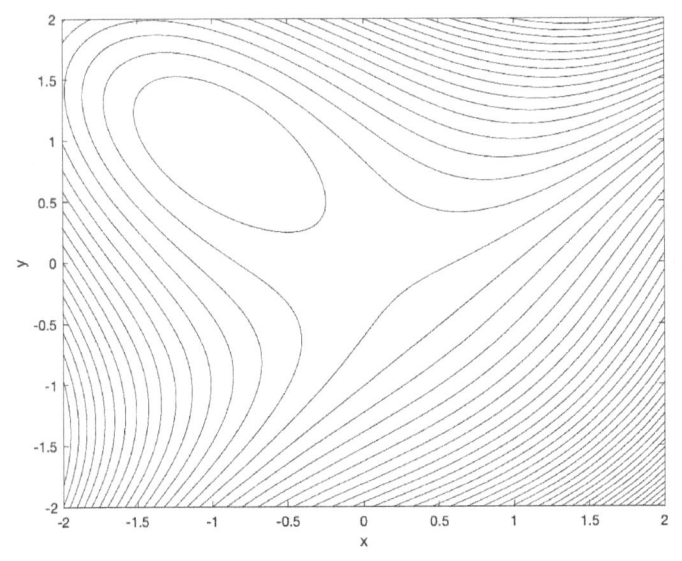

Fig. 5.7. Contour plot of $f(x, y) = x^3 - y^3 - 3xy + 4$ with a saddle point at $x = 0$, $y = 0$ and a local maximum at $x = -1, y = 1$.

The rotational inertia in a plane is defined as

$$I(x,y) = \sum_{k=1}^{n} m_k \left[(x - a_k)^2 + (y - b_k)^2 \right].$$

The condition for stationary points (5.17) gives

$$\frac{\partial I}{\partial x} = 2 \sum_{k=1}^{n} m_k (x_0 - a_k) = 0, \quad \frac{\partial I}{\partial y} = 2 \sum_{k=1}^{n} m_k (y_0 - b_k) = 0.$$

Solving these equations for x_0 and y_0 we have

$$x_0 = \frac{m_1 a_1 + m_2 a_2 + \cdots m_n a_n}{m_1 + m_2 + \cdots + m_n}, \quad y_0 = \frac{m_1 b_1 + m_2 b_2 + \cdots m_n b_n}{m_1 + m_2 + \cdots + m_n}.$$

Since

$$\frac{\partial^2 I}{\partial x^2} = 2 \sum_{k=1}^{n} m_k > 0, \quad \frac{\partial^2 I}{\partial x^2} = 2 \sum_{k=1}^{n} m_k > 0, \quad \frac{\partial^2 I}{\partial x \partial y} = 0,$$

and $I_{xy}^2 - I_{xx} I_{yy} < 0$, then $I(x, y)$ has a minimum at (x_0, y_0). Note that the point (x_0, y_0) is the positions for the center of mass. ∎

5.7. Stationary points with constrains

So far we considered problems of finding stationary values of a function $f(x, y)$ of two variables when both x and y could be independently varied. However, in physics (like much in life), most optimization problems have constraints.[5] For example, we need to find a maximum value for a function $f(x, y)$ on a line that is a subject of a constraint $g(x, y) = c$ so that x and y are not independent.

If a constraint has a relatively simple connection between x and y, for example, we can explicitly solve $g(x, y) = c$ as $y(x)$, then substituting $y(x)$ into $f(x, y)$ results in a problem for a function of a single variable.

---

[5]Finding stationary points for multi-variable functions with constraints which is a subset of optimization problems.

Example 5.7. Find stationary points $f(x,y) = x^2 + y^2$ with the constraint $x + y - 1 = 0$. Thus, we are looking for stationary points not within the whole $x - y$ plane but on the line defined by the condition $x + y - 1 = 0$. Here, we can simplify the problem by reducing the search for a conditional extremum of a two-variable function to a standard problem of finding an extremum of a single-variable function. From the constraint, it follows $y = 1 - x$. Substituting it into the original function gives $f(x,y) \equiv f(x) = x^2 + (1-x)^2 = 2x^2 - 2x + 1$. The unconditional extremum can be easily found since $f'(x) = 4x - 2 = 0$ and $f''(x) = 4$. Then, the function $2x^2 - 2x + 1$ has a minimum $f(x) = 1/2$ at $x = 1/2$. And the original function $f(x,y) = x^2 + y^2$ has a conditional minimum (or a minimum with a constraint) at $x = 1/2, y = 1/2$. Note that the unconditional minimum for $f(x,y) = x^2 + y^2$ is located at $x = 0, y = 0$ and $f(0,0) = 0$. ■

In the above method, the symmetry with respect to variables is violated: some of them are treated as independent (e.g., x in the example above), some as dependent (e.g., y in the example above), some differentials are excluded, and others are preserved. Very often it is practically difficult, if not impossible, to apply such an approach.

The *method of Lagrange undetermined multipliers* is considered as the most efficient and elegant method for solving problems with constraints. The method treats all variables equally by maintaining its symmetry. It works for any number of variables with any number of constraints. However, we need to remember that for problems with more than three variables, it is better to use a computer instead of working analytically.

Let us find a maximum or a minimum of a function $f(x,y)$ under the condition that x and y are connected by a condition:

$$g(x,y) = c. \tag{5.21}$$

This means that values of the function $f(x,y)$ are only analyzed for points connected by condition (5.21), i.e., x and y are connected as $y(x)$ or $x(y)$. If we assume that $y(x)$ is an implicit function of x defined in (5.21), then we are looking for a maximum or a minimum of a function of a single variable $f(x,y(x))$ with the necessary condition for an extremum as $f'(x) = 0$. Earlier, we derived a total derivative for $f(x, y)$ with respect to x when y is a function of x (see equation (5.6)), then we can write

$$\frac{df}{dx} = \frac{\partial f}{\partial x} + \frac{\partial f}{\partial y}\frac{dy}{dx} = 0, \qquad (5.22)$$

where dy/dx is a derivative of an implicit function $y(x)$ defined by (5.21). Differentiating (5.21) gives

$$\frac{dg}{dx} = \frac{\partial g}{\partial x} + \frac{\partial g}{\partial y}\frac{dy}{dx} = 0, \quad \text{or} \quad \frac{dy}{dx} = -\frac{\partial g/\partial x}{\partial g/\partial y}.$$

Substituting it into (5.22) gives that at the point of interest we have

$$\frac{\partial f/\partial x}{\partial g/\partial x} = \frac{\partial f/\partial y}{\partial g/\partial y} = -\lambda, \qquad (5.23)$$

where λ is an unknown parameter, called the Lagrange multiplier. The equation above can be written as a system of two equations:

$$\begin{aligned}\frac{\partial f}{\partial x} + \lambda\frac{\partial g}{\partial x} &= 0, \\ \frac{\partial f}{\partial y} + \lambda\frac{\partial g}{\partial y} &= 0.\end{aligned} \qquad (5.24)$$

Equations (5.24) together with constrain (5.21) form a system of three equations with three unknowns x, y, and λ.

The nature of the Lagrange multiplier can be explored by considering x, y and $f(x, y)$ as functions of c from (5.21). Then, we can write

$$\frac{df}{dc} = f_x\frac{dx}{dc} + f_y\frac{dy}{dc}$$

and

$$\frac{dg}{dc} = g_x\frac{dx}{dc} + g_y\frac{dy}{dc} = 1.$$

Using the above two equations together with (5.24) gives

$$\frac{df}{dc} = -\lambda,$$

i.e., the Lagrange multiplier is equal to the rate of change of $f(x, y)$ when changing the parameter c in condition (5.21).

The method does not provide information if we have found a maximum or a minimum (the analysis for the second-order partial derivatives from Section 5.6 cannot be applied here). The most straightforward way to identify the nature of stationary points is to calculate values of $f(x,y)$ at these points. Having some knowledge about your problem is very helpful.

Example 5.8. Find the maximum of $f(x,y) = 1 + xy$ on the circle $g(x,y) = x^2 + y^2 = 1$. Using (5.26), we generate three equations:

$$\left(\frac{\partial f}{\partial x} + \lambda \frac{\partial g}{\partial x}\right) = y + 2\lambda x = 0,$$

$$\left(\frac{\partial f}{\partial y} + \lambda \frac{\partial g}{\partial y}\right) = x + 2\lambda y = 0,$$

$$x^2 + y^2 = 1.$$

This is a nonlinear system, but it can be solved analytically. There are four solutions of the system, i.e., four stationary points:

$$x_1 = +\frac{1}{\sqrt{2}}, \quad y_1 = -\frac{1}{\sqrt{2}}; \quad x_2 = -\frac{1}{\sqrt{2}}, \quad y_2 = +\frac{1}{\sqrt{2}};$$

$$x_3 = +\frac{1}{\sqrt{2}}, \quad y_3 = +\frac{1}{\sqrt{2}}; \quad x_4 = -\frac{1}{\sqrt{2}}, \quad y_4 = -\frac{1}{\sqrt{2}}.$$

Which one is the maximum? Evaluating $f(x,y) = 1 + xy$ at these four points tells that there are two maxima located at (x_3, y_3) and (x_4, y_4). ∎

In case of three (or more variables), the generalization of the Lagrange method (5.26) is very straightforward:

$$\frac{\partial f}{\partial x} + \lambda \frac{\partial g}{\partial x} = 0, \quad \frac{\partial f}{\partial y} + \lambda \frac{\partial g}{\partial y} = 0, \quad \frac{\partial f}{\partial z} + \lambda \frac{\partial g}{\partial z} = 0, \quad g(x,y,z) = c.$$
(5.25)

The method of Lagrange multipliers works for two (or more) constraints. Assume that we are looking for a stationary point (maximum or minimum) of a function $f(x,y)$ with two constraints

$g(x,y) = c_1$ and $h(x,y) = c_2$. Then, we have a system of four equations:

$$\frac{\partial f}{\partial x} + \lambda \frac{\partial g}{\partial x} + \mu \frac{\partial h}{\partial x} = 0,$$
$$\frac{\partial f}{\partial y} + \lambda \frac{\partial g}{\partial y} + \mu \frac{\partial h}{\partial y} = 0, \qquad (5.26)$$
$$g(x,y) = c_1$$
$$h(x,y) = c_2,$$

where λ and μ are two unidentified Lagrange multipliers.

Example 5.9. Find the stationary point of $f(x,y,z) = x^2 - y^2 + z^2 - 4x$ when $x + y - z = 0$ and $x + 2y - 2 = 0$.

By using (5.25), but for three variables x, y, and z, we obtain the system of equations:

$$2x - 4 + \lambda_1 + \lambda_2 = 0,$$
$$-2y + \lambda_1 + 2\lambda_2 = 0,$$
$$2z - \lambda_1 = 0,$$
$$x + y - z = 0,$$
$$x + 2y - 2 = 0.$$

Solutions of this system of linear equations are

$$x = 1, \quad y = \frac{1}{2}, \quad z = \frac{3}{2}, \quad \lambda_1 = 3, \quad \lambda_2 = -1. \qquad \blacksquare$$

5.8. Change of variables

Changing variables is a powerful method for solving many problems. More often we change variables by changing coordinate systems. Most popular coordinate systems include Cartesian and polar for two-dimensional problems, and Cartesian, spherical, cylindrical, and parabolic for three-dimensional problem. The essence of the problem can be formulated in the following way. Suppose that we are given an expression having independent variables x, y and a function $f(x,y)$:

$$F\left(x, y, \frac{\partial f}{\partial x}, \frac{\partial f}{\partial y}, \ldots\right).$$

Partial Differentiation

Instead of independent variables x, y and the function $f(x,y)$ we introduce new independent variables u and v such that $u = u(x,y)$ and $v = v(x,y)$. Then, we need to transform the original expression to the new variables.

According to the chain rule for partial differentiation, we have

$$\frac{\partial f}{\partial u} = \frac{\partial f}{\partial x}\frac{\partial x}{\partial u} + \frac{\partial f}{\partial y}\frac{\partial y}{\partial u}, \qquad (5.27)$$

$$\frac{\partial f}{\partial v} = \frac{\partial f}{\partial x}\frac{\partial x}{\partial v} + \frac{\partial f}{\partial y}\frac{\partial y}{\partial v}, \qquad (5.28)$$

or writing symmetrically

$$\frac{\partial f}{\partial x} = \frac{\partial f}{\partial u}\frac{\partial u}{\partial x} + \frac{\partial f}{\partial v}\frac{\partial v}{\partial x}, \qquad (5.29)$$

$$\frac{\partial f}{\partial y} = \frac{\partial f}{\partial u}\frac{\partial u}{\partial y} + \frac{\partial f}{\partial v}\frac{\partial v}{\partial y}. \qquad (5.30)$$

The above equations provide a way to change variables in differential operators.

Now, we are going to show that the total differential

$$df = \frac{\partial f}{\partial u}du + \frac{\partial f}{\partial v}dv$$

does not depend on the choice of variables:

$$df = \left(\frac{\partial f}{\partial x}\frac{\partial x}{\partial u} + \frac{\partial f}{\partial y}\frac{\partial y}{\partial u}\right)du + \left(\frac{\partial f}{\partial x}\frac{\partial x}{\partial v} + \frac{\partial f}{\partial y}\frac{\partial y}{\partial v}\right)dv$$

$$= \frac{\partial f}{\partial x}\left(\frac{\partial x}{\partial u}du + \frac{\partial x}{\partial v}dv\right) + \frac{\partial f}{\partial y}\left(\frac{\partial y}{\partial u}du + \frac{\partial y}{\partial v}dv\right).$$

Since

$$dx = \left(\frac{\partial x}{\partial u}du + \frac{\partial x}{\partial v}dv\right), \quad dy = \left(\frac{\partial y}{\partial u}du + \frac{\partial y}{\partial v}dv\right),$$

then

$$df = \frac{\partial f}{\partial x}dx + \frac{\partial f}{\partial y}dy = \frac{\partial f}{\partial u}du + \frac{\partial f}{\partial v}dv.$$

This property can be easily used to find total differentials for a sum, a product, and a division of a function of a few variables. For example,

$$d(uv) = \frac{\partial(uv)}{\partial u} du + \frac{\partial(uv)}{\partial v} dv = vdu + udv.$$

Thus, we have

$$d(u+v) = du + dv, \quad d(uv) = vdu + udv, \quad d\left(\frac{u}{v}\right) = \frac{vdu - udv}{v^2}, \tag{5.31}$$

where u and v are functions of independent variables.

Very many physics phenomena can be described as solutions of second-order partial differential equations (that include second partial derivatives). Assume that $f(x, y)$ describes a physics problem in (x, y) variables, and $g(u, v)$ describes the same problem but in (u, v) variables. Although f and g may appear distinct, they both describe the same underlying physics. Then, the second derivative is

$$\frac{\partial^2 f}{\partial x^2} = \frac{\partial}{\partial x}\left(\frac{\partial}{\partial x} f\right) = \frac{\partial}{\partial x}\left(\frac{\partial}{\partial x}\right) f.$$

Since

$$\frac{\partial}{\partial x} = \frac{\partial}{\partial u}\frac{\partial u}{\partial x} + \frac{\partial}{\partial v}\frac{\partial v}{\partial x},$$

then

$$\frac{\partial^2}{\partial x^2} = \left(\frac{\partial}{\partial u}\frac{\partial u}{\partial x} + \frac{\partial}{\partial v}\frac{\partial v}{\partial x}\right)\left(\frac{\partial}{\partial u}\frac{\partial u}{\partial x} + \frac{\partial}{\partial v}\frac{\partial v}{\partial x}\right). \tag{5.32}$$

In the same way, we can write

$$\frac{\partial^2}{\partial y^2} = \left(\frac{\partial}{\partial u}\frac{\partial u}{\partial y} + \frac{\partial}{\partial v}\frac{\partial v}{\partial y}\right)\left(\frac{\partial}{\partial u}\frac{\partial u}{\partial y} + \frac{\partial}{\partial v}\frac{\partial v}{\partial y}\right). \tag{5.33}$$

Example 5.10. Changing variables for the Laplace operator is one of the best examples to illustrate the technique. In the Cartesian coordinates, the operator is defined as

$$\Delta f = \frac{\partial^2 f}{\partial x^2} + \frac{\partial^2 f}{\partial y^2}.$$

Now, we want to write the Laplace operator in polar coordinates, where

$$x = \rho \cos \varphi, \quad \rho = (x^2 + y^2)^{1/2},$$
$$y = \rho \sin \varphi, \quad \varphi = \tan^{-1}\left(\frac{y}{x}\right).$$

According to (5.32),

$$\frac{\partial^2}{\partial x^2} = \left(\frac{\partial}{\partial \rho}\frac{\partial \rho}{\partial x} + \frac{\partial}{\partial \varphi}\frac{\partial \varphi}{\partial x}\right)\left(\frac{\partial}{\partial \rho}\frac{\partial \rho}{\partial x} + \frac{\partial}{\partial \varphi}\frac{\partial \varphi}{\partial x}\right).$$

From the definitions for the polar coordinates follows

$$\frac{\partial \rho}{\partial x} = \frac{x}{(x^2 + y^2)^{1/2}} = \cos \varphi, \quad \frac{\partial \varphi}{\partial x} = -\frac{y}{x^2 + y^2} = -\frac{\sin \varphi}{\rho},$$

$$\frac{\partial \rho}{\partial y} = \frac{y}{(x^2 + y^2)^{1/2}} = \sin \varphi, \quad \frac{\partial \varphi}{\partial y} = \frac{1/x}{1 + (y/x)^2} = \frac{\cos \varphi}{\rho}.$$

Then, we have

$$\frac{\partial^2}{\partial x^2} = \left(\cos \varphi \frac{\partial}{\partial \rho} - \frac{\sin \varphi}{\rho}\frac{\partial}{\partial \varphi}\right)\left(\cos \varphi \frac{\partial}{\partial \rho} - \frac{\sin \varphi}{\rho}\frac{\partial}{\partial \varphi}\right)$$

$$= \cos^2 \varphi \frac{\partial^2}{\partial \rho^2} - \cos \varphi \frac{\partial}{\partial \rho}\left(\frac{\sin \varphi}{\rho}\frac{\partial}{\partial \varphi}\right) - \frac{\sin \varphi}{\rho}\frac{\partial}{\partial \varphi}\left(\cos \varphi \frac{\partial}{\partial \rho}\right)$$
$$+ \frac{\sin \varphi}{\rho}\frac{\partial}{\partial \varphi}\left(\frac{\sin \varphi}{\rho}\frac{\partial}{\partial \varphi}\right)$$

$$= \cos^2 \varphi \frac{\partial^2}{\partial \rho^2} + \cos \varphi \sin \varphi \frac{1}{\rho^2}\frac{\partial}{\partial \varphi} - \cos \varphi \sin \varphi \frac{1}{\rho}\frac{\partial^2}{\partial \rho \partial \varphi}$$
$$+ \sin^2 \varphi \frac{1}{\rho}\frac{\partial}{\partial \rho} - \cos \varphi \sin \varphi \frac{1}{\rho}\frac{\partial^2}{\partial \rho \partial \varphi} + \cos \varphi \sin \varphi \frac{1}{\rho^2}\frac{\partial}{\partial \varphi}$$
$$+ \sin^2 \varphi \frac{1}{\rho^2}\frac{\partial^2}{\partial \varphi^2}$$

$$= \cos^2 \varphi \frac{\partial^2}{\partial \rho^2} + 2\cos \varphi \sin \varphi \frac{1}{\rho^2}\frac{\partial}{\partial \varphi} - 2\cos \varphi \sin \varphi \frac{1}{\rho}\frac{\partial^2}{\partial \rho \partial \varphi}$$
$$+ \sin^2 \varphi \frac{1}{\rho}\frac{\partial}{\partial \rho} + \sin^2 \varphi \frac{1}{\rho^2}\frac{\partial^2}{\partial \varphi^2}.$$

Similarly, we derive

$$\frac{\partial^2}{\partial y^2} = \sin^2\varphi \frac{\partial^2}{\partial \rho^2} - 2\cos\varphi\sin\varphi \frac{1}{\rho^2}\frac{\partial}{\partial \varphi} + 2\cos\varphi\sin\varphi \frac{1}{\rho}\frac{\partial^2}{\partial \rho \partial \varphi}$$
$$+ \cos^2\varphi \frac{1}{\rho}\frac{\partial}{\partial \rho} + \cos^2\varphi \frac{1}{\rho^2}\frac{\partial^2}{\partial \varphi^2}.$$

And finally, we arrive at

$$\frac{\partial^2 x}{\partial x^2} + \frac{\partial^2 f}{\partial y^2} = \frac{\partial^2 g}{\partial \rho^2} + \frac{1}{\rho}\frac{\partial}{\partial \rho} + \frac{1}{\rho^2}\frac{\partial^2}{\partial \varphi^2}. \qquad (5.34)$$

This example provides a good starting point for deriving the Laplace operator in spherical and cylindrical coordinates. ∎

5.9. Implicit functions

Partial differentiation can be useful to find derivatives of functions of single variables given in implicit form. As a reminder, note that a function is called implicit when we have no direct analytic expression for it in terms of independent variables but only an equation relating its values to those of the independent variables. For instance, if a variable y is related to a variable x by the equation

$$y^3 + 3axy + c^3 = 0, \qquad (5.35)$$

then y is an implicit function of the independent variable x.

And here is another example of an implicit function:

$$(x+y)^3 - b^2(x-y) = 0. \qquad (5.36)$$

Quite often finding $y(x)$ in an explicit form can be difficult if not impossible. Furthermore, an explicit form may be very complicated to deal with analytically, thus making it very challenging to evaluate even first derivatives. For example, for the implicit form (5.35), the function $y(x)$ can be written as

$$y(x) = \left(-\frac{c^3}{2} + \left(\frac{c^6}{4} + a^3 x^3\right)^{1/2}\right)^{1/3} + \left(-\frac{c^3}{2} - \left(\frac{c^6}{4} + a^3 x^3\right)^{1/2}\right)^{1/3}.$$

However, implicit functions can be considered as functions of two variables, namely $F(x,y) = 0$. In this section, we show how to evaluate first and second derivatives for implicit functions with the help of partial differentiation. We are going to consider three forms of functions given in implicit form. For the first two (rather special) forms, we can work without using partial derivatives. But in more general case, we cannot proceed without them.

Case 1 — The simplest form: In this case, it is much easier to find x as a function of y. For our first Example (5.35), we can easily write

$$x = -\frac{c^3 + y^3}{3ay}.$$

But how can we find $y(x)$ for a specific value of x? From the equation for $x(y)$, we can calculate a table of values of y_n for a set of x_n. Then, using interpolation (even linear if the values x_n are close enough, or function $y(x)$ is smooth enough), we can find a new value of y for $y(x)$ when $x_{i-1} \leq x \leq x_i$.

But what if we need the first derivative dy/dx? If we assume that $x = \varphi(y)$, then we can write

$$dx = \frac{d\varphi}{dy} dy, \quad \frac{dy}{dx} = \frac{1}{(d\varphi/dy)} = \frac{1}{\varphi'(y)}.$$

However, there is a drawback. The derivative dy/dx is a function of y, not x. Nevertheless, we can solve the problem by calculating a table for derivatives like we did for the function.

What about a second derivative? Can we impulsively write in a similar manner that

$$\frac{d^2y}{dx^2} = \frac{1}{\varphi''(y)}?$$

The answer is no! A quick dimensional analysis for derivatives shows that the dimension for second derivatives must be $\frac{d^2y}{dx^2} \sim \frac{y}{x^2}$ but our guess gives $\frac{1}{\varphi''(y)} \sim \frac{y^2}{x}$. Here is the right way to find second derivatives:

$$\frac{d^2y}{dx^2} = \frac{d}{dx}\left(\frac{dy}{dx}\right) = \frac{dy}{dx}\frac{d}{dy}\left(\frac{dy}{dx}\right) = \frac{1}{\varphi'(y)}\frac{d}{dy}\left(\frac{1}{\varphi'(y)}\right) = -\frac{\varphi''(y)}{[\varphi'(y)]^3}$$

or

$$\frac{d^2y}{dx^2} = -\left(\frac{d^2x}{dy^2}\right) \Big/ \left(\frac{dx}{dy}\right)^2, \qquad (5.37)$$

with the correct dimension
$$\frac{d^2y}{dx^2} \sim \frac{y}{x^2}.$$

Attention: A quick note on dimensional analysis for derivatives:
$$\frac{dy}{dx} \sim \frac{y}{x}, \quad \frac{d^2y}{dx^2} = \frac{d}{dx}\left(\frac{dy}{dx}\right) \sim \frac{d}{dx}\left(\frac{y}{x}\right) \sim \left(\frac{y}{x^2}\right).$$

Note here a familiar example from physics for velocity and acceleration: The dimension of velocity is L/T or m/s, and the dimension of acceleration is L/T^2 or m/s^2.

Case 2 — A parametric dependence: Assume that we can write both x and y in a parametric form $x = x(t)$ and $y = y(t)$. Thus, for equation (5.36), if we introduce $x + y = t$, then for $(x+y)^3 - b^2(x-y) = 0$, we obtain $x - y = t^3/b^2$. Then solving the two equations we have
$$x = \frac{t}{2} + \frac{t^3}{2b^2}, \quad y = \frac{t}{2} - \frac{t^3}{2b^2},$$
with t as a parameter.

For the first derivative, we can write
$$dx = \frac{dx}{dt}dt = x'(t)dt, \quad dy = \frac{dy}{dt}dt = y'(t)dt,$$
and then
$$\frac{dy}{dx} = \frac{y'(t)}{x'(t)}. \tag{5.38}$$

For the second derivative, we have
$$\frac{df}{dx} = \frac{df}{dt}\frac{dt}{dx} = \frac{df}{dt}\frac{1}{(dx/dt)},$$
then
$$\frac{d^2y}{dx^2} = \frac{d}{dx}\left(\frac{dy}{dx}\right) = \frac{d}{dx}\left(\frac{y'(t)}{x'(t)}\right) = \frac{d}{dt}\left(\frac{y'(t)}{x'(t)}\right)\frac{dt}{dx} = \frac{1}{x'(t)}\frac{d}{dt}\left(\frac{y'(t)}{x'(t)}\right),$$
and finally,
$$\frac{d^2y}{dx^2} = \frac{y''(t)}{[x'(t)]^2} - \frac{y'(t)x''(t)}{[x'(t)]^3}. \tag{5.39}$$

Case 3 — More generality: There are situations when an implicit equation for $y(x)$ cannot be restructured as $x(y)$ or as a parametric equation for $x(t)$ and $y(t)$.

Example 5.11. Here, $y^5 + xy + x^5 - 7 = 0$ is one of those cases. However, this equation is still susceptible to a simple treatment, namely if we differentiate the equation, we get $5y^4 y' + (1 \cdot y + x \cdot y') + 5x^4 = 0$. Solving this equation for y' gives

$$y' = dy/dx = -(y + 5x^4)/(5y^4 + x).$$ ∎

Nevertheless, let us try to use partial differential to attack the problem. Assume $z = f(x, y)$, then

$$dz = \frac{\partial z}{\partial x} dx + \frac{\partial z}{\partial y} dy$$

or

$$dx = \left(\frac{\partial z}{\partial x}\right)^{-1} dz - \left(\frac{\partial z}{\partial y}\right) \left(\frac{\partial z}{\partial x}\right)^{-1} dy.$$

However,

$$dx = \frac{\partial x}{\partial z} dz + \frac{\partial x}{\partial y} dy,$$

or comparing the past two equations we get

$$\left.\frac{\partial x}{\partial z}\right|_y = \frac{1}{\partial z/\partial x|_y}, \quad \frac{\partial x}{\partial y} = -\left(\frac{\partial z}{\partial y}\right)\Big|_x \Big/ \left(\frac{\partial z}{\partial x}\right)\Big|_y. \quad (5.40)$$

Example 5.12. For $z = x^2 + px + y^2 + gy + kxy$, find $\frac{\partial x}{\partial y}$. If we wanted first to solve the equation for x, then we would deal with a quadratic equations, but instead we use (5.40), then from

$$\frac{\partial z}{\partial x} = 2x + p + ky, \quad \frac{\partial z}{\partial y} = 2y + q + kx,$$

it follows

$$\frac{\partial x}{\partial y} = -\frac{2y + q + kx}{2x + p + ky}.$$ ∎

Now, we can use the above technique to evaluate ordinary derivatives for an implicit function given as $z = f(x, y)$ where we consider that $f(x,y) = 0$. Then, the problem to find dy/dx leads to the evaluation of $\partial y/\partial x|_{z=\text{const}}$. For the example that we considered at the beginning of Case 3, we need to set $z = y^5 + xy + x^5$. Then, according to equation (5.40),

$$\frac{dy}{dx} = \frac{\partial y}{\partial x}\bigg|_z = -\frac{\partial z}{\partial x}\bigg|_y \bigg/ \frac{\partial z}{\partial y}\bigg|_x = -\frac{y + 5x^4}{5y^4 + x},$$

and we have arrived at the same answer as before.

5.10. Exercises

(1) Find $f_x, f_y, f_{xx}, f_{xy}, f_{yy}$ for the following functions:
   (a) $xe^y + ye^x$
   (b) $x \sin y$
   (c) $\sin(xy)$
   (d) $e^{-(x^2+y^2)}$
   (e) $(x^2 + y^2)^{\frac{1}{2}}$.

(2) Find the second partial derivatives for $f(x,y) = e^{-ax}\cos(by)$ (where a and b are some constants), and also verify that $f_{xy} = f_{yx}$.

(3) Show that
   (a) if $f(x,y) = xy + xe^{y/x}$, then
   $$x\frac{\partial f}{\partial x} + y\frac{\partial f}{\partial y} = xy + f(x,y)$$
   (b) if $f(x,y,z) = x + (x-y)/(y-z)$, then
   $$\frac{\partial f}{\partial x} + \frac{\partial f}{\partial y} + \frac{\partial f}{\partial z} = 1.$$

(4) Find total differentials df/dt:
   (a) $x^2 - y^2$, for $x = t + 1/t, y = t + \sqrt{t}$
   (b) e^{x-y} for $x = \sin t, y = t^2$
   (c) $x^3 + 3xy^2$ for $x = t^2, y = e^{-t}$
   (d) $\ln(x^2 - y^2)$ for $x = t^2, y = \cos 2t$.

(5) Find df/dx if $f(x,y) = \ln(x + e^y), y = x^2$.

(6) Find the two variable Maclaurin series:
   (a) $f(x,y) = e^{x+y}$
   (b) $f(x,y) = e^x \sin y$ by finding the series for e^{x+iy} and taking the imaginary part (use knowledge from complex numbers).

(7) Find the stationary points and identify their nature:
   (a) $f(x,y) = x^2 - xy + y^2 - 2x + 4y - 1$
   (b) $f(x,y) = x^2 + 3xy - 2y + 2$
   (c) $f(x,y) = 4 + x + y - x^2 - xy - \frac{1}{2}y^2$
   (d) $f(x,y) = x^3 + xy^2 - 12x - y^2$.

(8) Find the stationary point on $x + y = 1$ for which $f(x,y) = xy$, and analyze its nature (maximum or minimum).

(9) Find stationary points using the Lagrange multiplier method:
   (a) $f(x,y) = x + 2y$ if $x^2 + y^2 = 5$
   (b) $f(x,y) = x^2 + y^2$ with $x/2 + y/3 = 1$
   (c) $f(x,y) = x + y - 2$ if $x^2 + y^2 = 4$
   (d) $f(x,y,z) = 4x^2 + y^2 + z^2$ with $x + 2y + 3z = 53$.

(10) If an open rectangular box has prescribed volume V, find the dimensions if its surface area is to be the least possible.

(11) The temperature of a point (x,y,z) is given by $T(x,y,z) = 1 + xy + yz$. Find the temperature of the hottest point on a unit sphere $x^2 + y^2 + z^2 = 1$.

(12) If $x = e^u \cos\theta$ and $y = e^u \sin\theta$, show that

$$\frac{\partial^2 g}{\partial u^2} + \frac{\partial^2 g}{\partial \theta^2} = (x^2 + y^2)\left(\frac{\partial^2 f}{\partial x^2} + \frac{\partial^2 f}{\partial y^2}\right),$$

where $g(u,\theta)$ and $f(x,y)$ describe the same physics just in different variables.

(13) Rewrite the Laplace operator

$$\nabla^2 f = \frac{\partial^2 f}{\partial x^2} + \frac{\partial^2 f}{\partial y^2} + \frac{\partial^2 f}{\partial z^2}$$

in spherical polar coordinates

$$\nabla^2 f = \frac{\partial^2 f}{\partial r^2} + \frac{2}{r}\frac{\partial f}{\partial r} + \frac{1}{r^2}\frac{\partial^2 f}{\partial \theta^2} + \frac{1}{r^2}\frac{\cos(\theta)}{\sin(\theta)}\frac{\partial f}{\partial \theta} + \frac{1}{r^2 \sin^2(\theta)}\frac{\partial^2 f}{\partial \varphi^2},$$

or in the standard compact form

$$\nabla^2 f = \frac{1}{r^2}\frac{\partial}{\partial r}\left[r^2\frac{\partial f}{\partial r}\right] + \frac{1}{r^2\sin(\theta)}\frac{\partial}{\partial \theta}\left[\sin(\theta)\frac{\partial f}{\partial \theta}\right] + \frac{1}{r^2\sin^2(\theta)}\frac{\partial^2 f}{\partial \varphi^2},$$

where

$$x = r\sin(\theta)\cos(\varphi), \quad y = r\sin(\theta)\sin(\varphi), \quad z = r\cos(\theta)$$

or

$$r = \sqrt{x^2 + y^2 + z^2}, \quad \theta = \arccos\left(\frac{z}{r}\right), \quad \varphi = \arctan\left(\frac{y}{x}\right).$$

(14) Find dy/dx and d^2y/dx^2 for following functions given in implicit forms:
   (a) $x = t/2, y = t^2 + t$
   (b) $x = 2\sin^3 t, y = 2\cos^3 t$
   (c) $x = \cos t + t\sin t, y = \sin t - t\cos t$
   (d) $x = \sin t, y = \cos 2t$.

(15) Find dy/dx when
   (a) $x^2 + y^2 - 4x - 10y = -4$
   (b) $x^4 y + xy^4 - x^2 y^2 - 1 = 0$.

Comment: Equations and symbols for creating figures with Adobe Illustrator:

$$f(x,y) \quad f + \frac{\partial f}{\partial x}dx \quad f + \frac{\partial f}{\partial y}dy \quad f + \frac{\partial f}{\partial x}dx + \frac{\partial f}{\partial y}dy$$

$$x_0, y_0 \quad x_0 - h, y_0 \quad x_0 + h, y_0 \quad x_0, y_0 - k \quad x_0, y_0 + k.$$

Answers and solutions

(1) Partial derivatives
   (a) $f_x = e^y + ye^x$, $f_y = xe^y + e^x$, $f_{xx} = ye^x$, $f_{yy} = xe^y$, $f_{xy} = e^x + e^y$.
   (b) $f_x = \sin(x)$, $f_y = x\cos(y)$, $f_{xx} = 0$, $f_{yy} = -x\sin(y)$, $f_{xy} = 0$.

(c) $f_x = y\cos(xy)$, $f_y = x\cos(xy)$, $f_{xx} = -y^2\sin(xy)$, $f_{yy} = -x^2\sin(xy)$, $f_{xy} = \cos(xy) - xy\sin(xy)$.
(d) $f_x = -2xe^{-(x^2+y^2)}$, $f_y = -2ye^{-(x^2+y^2)}$, $f_{xx} = (4x^2 - 2)e^{-(x^2+y^2)}$, $f_{yy} = (4y^2 - 2)e^{-(x^2+y^2)}$, $f_{xy} = 4xye^{-(x^2+y^2)}$.
(e) $f_x = x(x^2+y^2)^{-1/2}$, $f_y = y(x^2+y^2)^{-1/2}$, $f_{xx} = y^2(x^2+y^2)^{-3/2}$, $f_{yy} = x^2(x^2+y^2)^{-3/2}$, $f_{xy} = -xy(x^2+y^2)^{-3/2}$.

(2) $f_{xx} = a^2 e^{-ax}\cos(by)$, $f_{yy} = -b^2 e^{-ax}\cos(by)$, $f_{xy} = abe^{-ax}\sin(by)$, $f_{yx} = abe^{-ax}\sin(by)$.

(4) Total differentials

(a) $df = \frac{2t^3 - 8t\sqrt{t} + 2}{t^2} dt$.

(b) $df = e^{\sin(t) - t^2}(\cos(t) - 2t)dt$.

(c) $df = 6t^5 dt + 6te^{-2t} dt - 6t^2 e^{-2t} dt$.

(d) $df = \frac{4t^3 + 4\cos(2t)\sin(2t)}{t^4 - \cos^2(2t)} dt$.

(5) $df = \frac{1 + 2xe^{x^2}}{x + e^{x^2}} dx$.

(6) Maclaurin series

(a) $f(x,y) = 1 + x + y + \frac{1}{2!}(x^2 + 2xy + y^2) + \frac{1}{3!}(x^3 + 3x^2 y + 3xy^2 + y^3) + \cdots$.

(b) $f(x,y) = y + \frac{1}{3!}y^3 + \frac{1}{5!}y^5 + \cdots$.

(7) Stationary points

(a) $(0, -2)$ minimum.
(b) $(2/3, -4/9)$ saddle point.
(c) $(0, 1)$ maximum.
(d) $(-2, 0)$ maximum, $(1, -3)$ saddle point, $(1, 3)$ saddle point, $(2, 0)$ minimum.

(8) $(1/2, 1/2)$, maximum.

(9) (a) $(-1, -2)$ and $(1, 2)$.

(b) $(18/13, 12/13)$.

(c) $(-\sqrt{2}, -\sqrt{2})$ and $(\sqrt{2}, \sqrt{2})$.

(d) $(1, 8, 12)$.

(10) $x = \sqrt[3]{2V}$, $y = \sqrt[3]{2V}$, and $z = \frac{1}{2}\sqrt[3]{2V}$.

(11) There are four points stationary points:
$$\left(\frac{1}{2}, \pm\frac{1}{\sqrt{2}}, \frac{1}{2}\right), \quad \left(-\frac{1}{2}, \pm\frac{1}{\sqrt{2}}, -\frac{1}{2}\right)$$
with maxima at
$$\left(\frac{1}{2}, \frac{1}{\sqrt{2}}, \frac{1}{2}\right), \quad \left(-\frac{1}{2}, -\frac{1}{\sqrt{2}}, -\frac{1}{2}\right) \quad T = 1 + \frac{1}{\sqrt{2}}.$$

(14) (a) $\frac{dy}{dx} = 4t + 2 = 8x + 2$, $\frac{d^2y}{dx^2} = 8$.

(b) $\frac{dy}{dx} = -\cot(t)$, $\frac{d^2y}{dx^2} = \frac{1}{6\sin^4 t \cos t}$.

(c) $\frac{dy}{dx} = \tan(t)$, $\frac{d^2y}{dx^2} = \frac{1}{\cos^3 t}$.

(d) $\frac{dy}{dx} = -4\sin t = -4x$.

(15) (a) $\frac{dy}{dx} = \frac{2-x}{y-5}$.

(b) $\frac{dy}{dx} = \frac{2xy^2 - y^4 - 4x^3 y}{x^4 + 4xy^3 - 2x^2 y}$.

Chapter 6

Line Integrals and Multiple Integrals

6.1. Introduction

Line integrals and multiple integrals have a wide range of practical applications in science, engineering, and many other fields. We mention just a few applications.

For example, by using line integrals, we can calculate the work done by a force and the electric field produced by a charged wire and determine the lift and drag forces on an airplane wing. Line integrals can also be used to solve boundary value problems. Path integrals, that are a generalization of a concept of line integrals, provide a powerful tool for studying the behavior of quantum systems.[1]

Most known examples of applications of multiple integrals include calculating areas and volumes and finding centers of mass and moments of inertia. Multiple integrals are used in every area of physics. It is a powerful tool in engineering, economics, and computer science.

6.2. Line integrals: First kind

6.2.1. *Definition*

Suppose we have a curve L in the (x, y) plane, as in Figure 6.1. This curve does not intersect itself or have overlap sections. Initially,

---

[1]R. P. Feynman and A. R. Hibbs, *Quantum Mechanics and Path Integrals.* McGraw-Hill, New York, 1965.

we assume it is not a closed curve and is bounded by the points (x_a, y_a) and (x_b, y_b). Let's assume that the direction on the curve L is an arbitrarily set (one of the two possible directions) so that the position of point M on the curve can be determined by measuring the length of the arc from the initial point a. Then, the curve can be parametrically expressed by equations of the following form:

$$x = x(t), \quad y = y(t), \quad (t_a \leq t \leq t_b). \tag{6.1}$$

Next, we suppose that a continuous function $f(x, y)$ is defined along the curve L. We partition the segment $t_a \leq t \leq t_b$ into n sub-segments $[t_{k-1}, t_k]$ ($k = 0, 1, 2, \ldots, n$). For every point k, we have a point with coordinates (x_k, y_k). Next, for every sub-segment, we choose an arbitrary point ξ_k, η_k, such that $\xi_k = x(\tau_k)$ and $\eta_k = y(\tau_k)$ with $t_{k-1} \leq \tau_k \leq t_k$. Let Δl_k be a length of kth sub-segment. We can define a sum

$$\sigma_1 = \sum_{k=1}^{n} f(\xi_k, \eta_k) \Delta l_k. \tag{6.2}$$

If there is a limit of the sum σ_1 when the largest $\Delta l_k \to 0$, then such a limit is called *a line integral of the first kind*:

$$I_1 = \int_L f(x, y) dl = \lim_{\Delta l_k \to 0} \sum_{k=1}^{n} f(\xi_k, \eta_k) \Delta l_k. \tag{6.3}$$

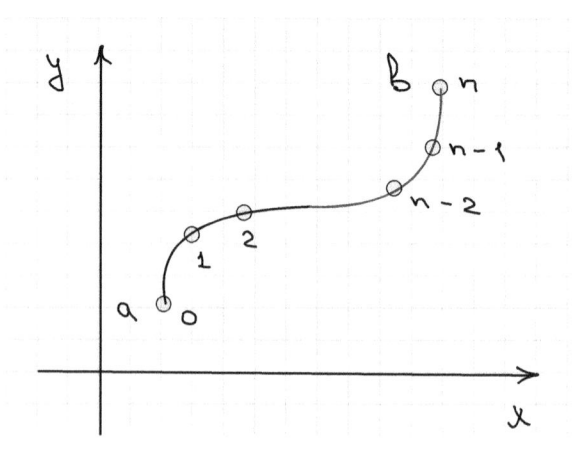

Fig. 6.1.   A curve L in xy-plane.

From the definition of the line integral (6.3), it follows that the integral is independent of the direction of integration, i.e., from a to b or from b to a.

Next, we explore the physical meaning of the line integral of the first kind. Consider a wire defined by the curve L, which has a linear density along the curve described by a function $f(x,y)$. To calculate the mass of the curve, we can divide the wire into small sections, small enough that we can assume the linear density is approximately constant within each section. The mass of each section is then approximately equal to the product of the density at that section by the length of this section. In this case, the mass of the entire wire will be approximately equal to the integral sum (6.2). The linear integral of the first kind (6.3) gives the exact mass of the wire, when the linear density along it is determined by $f(x,y)$.

6.2.2. Reduction to an ordinary definite integral

The line integral (6.3) can be expressed as a regular definite integral. For $\Delta l_k \to 0$ we have, as illustrated in Figure 6.2,

$$\Delta l_k^2 = (x_k - x_{k-1})^2 + (y_k - y_{k-1})^2.$$

Dividing both side by $\Delta t_k = t_k - t_{k-1}$, we arrive at

$$\frac{\Delta l_k}{\Delta t_k} = \left(\left[\frac{x_k - x_{k-1}}{t_k - t_{k-1}} \right]^2 + \left[\frac{y_k - y_{k-1}}{t_k - t_{k-1}} \right]^2 \right)^{1/2}.$$

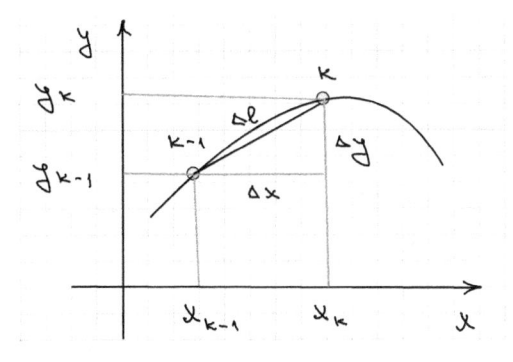

Fig. 6.2. A geometrical view of Δl.

Since for $\Delta t_k \to 0$ we can write

$$dl = \sqrt{(x'(t))^2 + (y'(t))^2}\, dt, \qquad (6.4)$$

then the first kind of line integral can be written as the usual definite integral in parametric form:

$$\int_L f(x,y)\, dl = \int_{t_a}^{t_b} f(x(t), y(t)) \sqrt{[x'(t)]^2 + [y'(t)]^2}\, dt. \qquad (6.5)$$

In the case where the path of integration is explicitly given as $y = y(x)$, the above equation (6.5) takes the following form:

$$\int_L f(x,y)\, dl = \int_{x_a}^{x_b} f(x,y) \sqrt{\left(1 + [y'(x)]^2\right)}\, dx. \qquad (6.6)$$

Line integrals of the first kind exhibit similar properties to those of regular integrals, including the following:

(1) Linearity

$$\int_L [\alpha f(x,y) + \beta g(x,y)]\, dl = \alpha \int_L f(x,y)\, dl + \beta \int_L g(x,y)\, dl. \qquad (6.7)$$

(2) Additivity

$$\int_{ab} f(x,y)\, dl = \int_{ac} f(x,y)\, dl + \int_{cb} f(x,y)\, dl. \qquad (6.8)$$

(3) The mean value theorem

$$\int_L f(x,y)\, dl = f(x^*, y^*) l, \qquad (6.9)$$

where (x^*, y^*) is some point located on L and l is the length of L.

However, the integral of the first kind does not change sign when the direction of integration is reversed, namely

$$\int_{ab} f(x,y)\, dl = \int_{ba} f(x,y)\, dl. \qquad (6.10)$$

In a similar manner to the one discussed, we can introduce the concept of the line integrals of the first kind extended to a curve in

xyz-space. Since there are no new fundamental points to consider, there is no need to delve into further details at this point, and we just write the parametric form as

$$\int_L f(x,y,z)dl = \int_{t_a}^{t_b} f(x(t),y(t),z(t))\sqrt{[x'(t)]^2 + [y'(t)]^2 + [z'(t)]^2}dt. \quad (6.11)$$

Now, we consider the case when the path L forms a closed contour, meaning the beginning point a and the end point b of the integration path coincide. Along the path L, we choose a point c that is distinct from a. With these considerations, we can write

$$\int_{a,b} f(x,y)dl = \int_{a,c} f(x,y)dl + \int_{c,b} f(x,y)dl.$$

Since the result does not depend on the choice of points a and c (do you see why?), equations (6.5) and (6.6) work for closed contours as well.

Example 6.1. Evaluate integral

$$\int_L ye^{-x}dl$$

along the path described as $x = \ln(t^2+1)$, and $y = 2\arctan t - t + 3$ between $t = 0$ and $t = 1$. Since

$$\frac{dx}{dt} = \frac{2t}{t^2+1}, \quad \frac{dy}{dt} = \frac{2}{t^2+1} - 1,$$

then $\sqrt{x'^2 + y'^2} = 1$ and we need to evaluate a regular definite integral

$$\int_0^1 \frac{2\arctan t - t + 3}{t^2+1} dt = \frac{\pi^2}{16} - \frac{1}{2}\ln 2 + \frac{3\pi}{4}. \quad \blacksquare$$

Example 6.2. A box is released from a point $(0,h)$ and slides without friction along a path L to reach a point $(2h,0)$, as shown in Figure 6.3.

We want to determine the time it takes for the box to reach the final point. While this problem can be solved using second Newton's

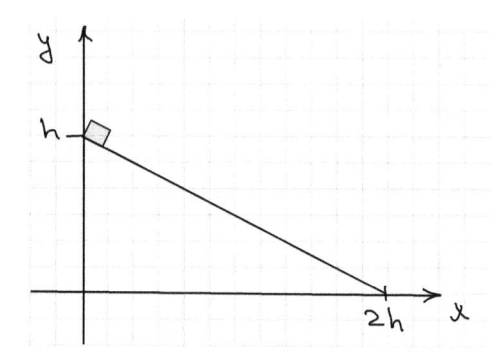

Fig. 6.3. The box sliding from $(0, h)$ to $(2h, 0)$.

law, we approach it using conservation of energy. By doing so, we can find the answer for any shape of L, not necessarily restricted to a straight line. According to conservation of energy, we have the following relationship:

$$mgh = mgy + \frac{1}{2}mv^2, \quad v = \sqrt{2g(h-y)}.$$

For a short interval, we have $dl = vdt$, then the total time is determined as the line integral:

$$T = \int_L \frac{dl}{v}.$$

Applying (6.6) with $y(x) = h - x/2$, the time to reach the ground is

$$T = \int_0^{2h} \frac{1}{\sqrt{2gx/2}} \sqrt{1 + \frac{1}{4}}\, dx = \sqrt{10 \frac{h}{g}}.$$

As you can see with line integrals of the first kind, we can we can readily obtain the traveling time by using conservation of energy. ∎

6.3. Line integrals: Second kind

6.3.1. *Definition*

Let a continuous path L be given connecting two points a and b in (x, y) plane, and let a function $f(x, y)$ be given along L. For simplicity, we assume that the path is open, meaning $a \neq b$. We compose two integral sums looking similar to the integral sum for the first kind of line integrals (6.2):

$$\sigma_x = \sum_{k=1}^{n} f(\xi_k, \eta_k)(x_k - x_{k-1}) \tag{6.12}$$

and

$$\sigma_y = \sum_{k=1}^{n} f(\xi_k, \eta_k)(y_k - y_{k-1}). \tag{6.13}$$

The distinction from the first kind of integral sum lies in the fact that we multiply the function values, denoted as $f(\xi_k, \eta_k)$, not by the length of the sub-segments Δl_k but by the projections of Δl_k onto the x and y axes. In the limit $\Delta l_k \to 0$, these integral sums are called line integrals of the second kind of $f(x,y)dx$ type

$$I_x = \int_L f(x,y)dx = \lim_{\Delta x_k \to 0} \sum_{k=1}^{n} f(\xi_k, \eta_k)\Delta x_k \tag{6.14}$$

and $f(x,y)dy$ type

$$I_y = \int_L f(x,y)dx = \lim_{\Delta y_k \to 0} \sum_{k=1}^{n} f(\xi_k, \eta_k)\Delta y_k. \tag{6.15}$$

If two functions $P(x,y)$ and $Q(x,y)$ are defined along the path L, then the integral can be expressed as follows:

$$I_2 = \int_L P(x,y)dx + \int_L Q(x,y)dy \equiv \int_L P(x,y)dx + Q(x,y)dy. \tag{6.16}$$

Integral (6.16) is *called second kind line integral of general type* or simply *second kind line integral*. From the integral sums (6.14) and (6.14) follows that changing the direction of integration from ab to ba results in changing the sign of the integral, i.e.,

$$\int_{ab} f(x,y)dx = -\int_{ba} f(x,y)dx.$$

Then, (6.16) becomes

$$\int_{ab} P(x,y)dx + Q(x,y)dy = -\int_{ba} P(x,y)dx + Q(x,y)dy. \tag{6.17}$$

Next, we analyze some physics that can be described by such integral. Consider a force $F(x,y)$ with its components $F_x(x,y) = P(x,y)$ and $F_y(x,y) = Q(x,y)$ that moves an object along a path L from point (x_a, y_a) to point (x_b, y_b). For calculating work done by this displacement, we can divide the path L into small sections. Assuming that the force changes very little within each section, we can approximate the work as two integral sums $W = W_x + W_y$, where

$$W_x = \sum_{k=1}^{n} P(\xi_k, \eta_k)(x_k - x_{k-1}), \quad W_y = \sum_{k=1}^{n} Q(\xi_k, \eta_k)(y_k - y_{k-1}).$$

Then, by taking the limit as the length of the largest section tends to zero, we can determine the exact value of this work:

$$W = \int_L P(x,y)dx + \int_L Q(x,y)dy.$$

Thus, the line integral of the second kind can evaluate work done to move an object from point (x_a, y_a) to point (x_b, y_b) along the path L under the action of a force $F(x,y)$ that has components $F_x(x,y) = P(x,y)$ and $F_y(x,y) = Q(x,y)$.

6.3.2. *Evaluating a second kind of line integral*

Now, let's explore how to evaluate the line integral of the second kind (6.16). We assume that the path L can be described by parametric equations as $x = x(t)$ and $y = y(t)$ with $t_a \leq t \leq t_b$. Similar to the line integrals of the first kind, we partition the interval $t_a \leq t \leq t_b$ into n sub-intervals $[t_{k-1}, t_k]$ ($k = 0, 1, 2, \ldots, n$). With this assumption, we can write

$$x_k - x_{k-1} = x(t_k) - x(t_{k-1}) = \int_{t_{k-1}}^{t_k} x'(t)dt.$$

Substituting it into the integral sums gives

$$\sigma_x = \sum_{k=1}^{n} P(x(\tau_k), y(\tau_k)) \int_{t_{k-1}}^{t_k} x'(t)dt \quad t_{k-1} \leq \tau_k \leq t_k.$$

As $\Delta t = t_k - t_{k-1} \to 0$,

$$\sum_{k=1}^{n} P(x(\tau_k), y(\tau_k)) \int_{t_{k-1}}^{t_k} x'(t)dt = \sum_{k=1}^{n} \int_{t_{k-1}}^{t_k} P(x(t), y(t))x'(t)dt,$$

and finally,

$$\int_L P(x,y)dx = \int_{t_a}^{t_b} P(x(t),y(t))x'(t)dt. \quad (6.18)$$

By using the same analysis for the $\int_L Q(x,y)dy$ integral, we have

$$\int_L Q(x,y)dy = \int_{t_a}^{t_b} Q(x(t),y(t))y'(t)dt. \quad (6.19)$$

Thus, the second kind of line integral can be evaluated using regular definite integrals as

$$\int_L P(x,y)dx + Q(x,y)dy$$
$$= \int_{t_a}^{t_b} \left[P(x(t),y(t))x'(t) + Q(x(t),y(t))y'(t)\right] dt. \quad (6.20)$$

If path L is given explicitly as function $y(x)$, then evaluation of the second kind of line integral (6.20) seems to be particularly simple:

$$\int_L P(x,y)dx + Q(x,y)dy = \int_{x_a}^{x_b} \left[P(x,y) + Q(x,y)y'(x)\right] dx. \quad (6.21)$$

Line integrals of the second kind, like the first kind, have the following properties: linearity (6.7), additivity (6.8), and mean value theorem (6.8). But, unlike the first kind of line integral, the second kind changes sign when the limits of integration are reversed:

$$\int_{ab} P(x,y)dx + Q(x,y)dy = -\int_{ba} P(x,y)dx + Q(x,y)dy. \quad (6.22)$$

Example 6.3. Evaluate line integral of the second order

$$I = \int_L (x^2 - 2xy)dx + (y^2 - 2xy)dy,$$

where the path L is defined explicitly as $y = x^2$ with $-1 \leq x \leq 1$.

This integral can be evaluated either by using the general (parametric form) (6.20) with $x = t$ and $y = t^2$ or by using (6.21). We proceed with the latter with $y = x^2$ and $y' = 2x$. Then, the original line integral is transformed into

$$I = \int_{-1}^{1} \left[(x^2 - 2x^3) + (x^4 - 2x^3)2x\right] dx = -\frac{14}{15}. \quad \blacksquare$$

Example 6.4. In this example, we consider two line integrals:

$$I_1 = \int_L xy\,dx + (y-x)\,dy$$

and

$$I_2 = \int_L 2xy\,dx + x^2\,dy,$$

where each is integrated from the origin $(0,0)$ to $(1,1)$ point along three different paths, namely (a) $y = x$, (b) $y = x^2$, and (c) $y = x^3$, as shown in Figure 6.4. Since the path is given explicitly as $y(x)$, we may use equation (6.21) for evaluating both integrals. For example, the first integral along path (a) gives

$$I_{1a} = \int_{x=0}^{x=1} [xx + (x-x)]\,dx = \int_0^1 x^2\,dx = \frac{1}{3}.$$

In the same manner, evaluating the first integral along the other two paths gives $I_{1b} = \frac{1}{12}$, $I_{1c} = -\frac{1}{20}$. For the second integral, we have $I_{2a} = 1$, $I_{2b} = 1$, and $I_{2c} = 1$, i.e., the same value for all three paths. ∎

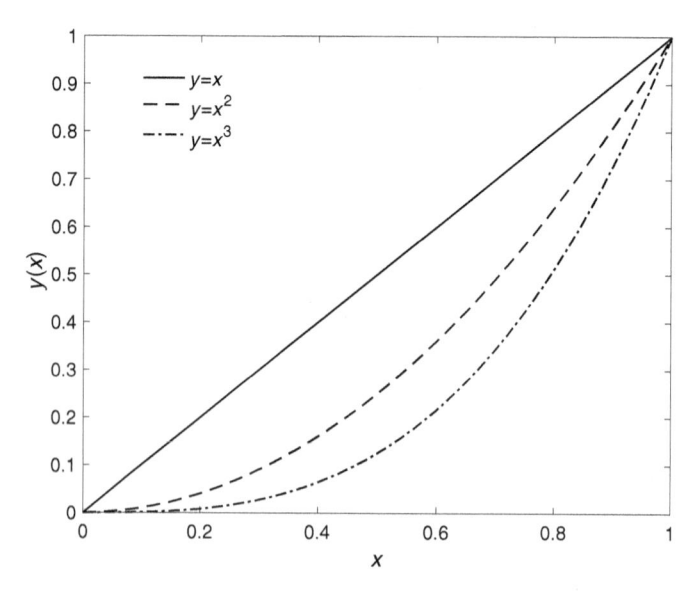

Fig. 6.4. Three paths of integration from $(0,0)$ to $(1,1)$.

6.3.3. Path independence of the line integral

It is only natural to question whether it is merely a coincidence that the second integral yields the same values regardless of the chosen paths of integration. No, it is not a coincidence! For the line integral (6.16) to be independent of the path of integration L, it is necessary and sufficient for $P(x,y)dx + Q(x,y)dy$ to be a complete differential of a function $F(x,y)$ such that

$$dF(x,y) = \frac{\partial F}{\partial x}dx + \frac{\partial F}{\partial y}dy = P(x,y)dx + Q(x,y)dy,$$

or it means that

$$P(x,y) = \frac{\partial F}{\partial x} \quad \text{and} \quad Q(x,y) = \frac{\partial F}{\partial y}. \tag{6.23}$$

When $P(x,y)dx + Q(x,y)dy$ forms a complete differential, the line integral, $\int_L P(x,y)dx + Q(x,y)dy$, can be evaluated using the function $F(x,y)$. Let us consider a path of integration L connecting two points (x_a, y_a) and (x_b, y_b), and a parametric representation for the path L, such as $x = x(t)$, $y = y(t)$ with $x_a = x(t_a)$, $y_a = y(t_a)$ and $x_b = x(t_b)$, $y_b = y(t_b)$. Using (6.20), we can write

$$\int_{t_a}^{t_b} \left[P(x(t),y(t))x'(t) + Q(x(t),y(t))y'(t) \right] dt$$

$$= \int_{t_a}^{t_b} \left[\frac{\partial F}{\partial x}x'(t) + \frac{\partial F}{\partial y}y'(t) \right] dt$$

$$= \int_{t_a}^{t_b} \frac{d}{dt} F(x(t),y(t)) dt = F(x(t),y(t))|_{t_a}^{t_b} = F(x_b, y_b) - F(x_a, x_b).$$

And finally, if $P(x,y)dx + Q(x,y)dy$ is a complete differential, then the second kind line integral can be evaluated as

$$\int_{x_a, y_a}^{x_b, y_b} P(x,y)dx + Q(x,y)dy = F(x_b, y_b) - F(x_a, x_b) = F(x,y)|_{x_a, y_a}^{x_b, y_b}, \tag{6.24}$$

where $F(x,y)$ is an anti-derivative function defined in (6.23). This formula demonstrates that the value of the line integral of the second kind in this case is independent of the specific path of integration.

However, it is important to note once again that equation (6.24) can only be used when the integrand is an exact differential.

It is practical to establish a criterion to test whether $P(x,y)dx + Q(x,y)dy$ is an exact differential or not. The answer to this question will also help clarify the conditions for the path independence of a line integral. Let us assume that $P(x,y)dx + Q(x,y)dy$ is a differential of some function $F(x,y)$, such that

$$P(x,y) = \frac{\partial F(x,y)}{\partial x} \quad \text{and} \quad Q(x,y) = \frac{\partial F(x,y)}{\partial y},$$

then

$$\frac{\partial P(x,y)}{\partial y} = \frac{\partial^2 F(x,y)}{\partial x \partial y}, \quad \frac{\partial Q(x,y)}{\partial x} = \frac{\partial^2 F(x,y)}{\partial y \partial x}.$$

Since mixed partial derivatives are independent of the order of differentiation (5.8)

$$\frac{\partial^2 F(x,y)}{\partial x \partial y} = \frac{\partial^2 F(x,y)}{\partial y \partial x},$$

then we have

$$\frac{\partial P(x,y)}{\partial y} = \frac{\partial Q(x,y)}{\partial x}. \tag{6.25}$$

Thus, this simple relation happens to be a necessary and sufficient condition for $P(x,y), dx + Q(x,y), dy$ to be an exact differential.

6.3.4. *Integration over a closed path*

So far we have considered second kind of path integrals along open paths. Now, we consider integrals taken along a closed contour L. When we integrate the second kind of line integral over a closed contour, the sign depends on the direction of integration (6.17). Of the two possible directions for integrating over a closed loop L, we call *positive* the direction of integration at which the region lying inside the contour remains on the left side with respect to the point traveling along the contour, i.e., *counterclockwise direction*. Previously, in Section 6.3.3, we examined the important case of path-independent line integrals. Now, let's explore the condition under which an integral

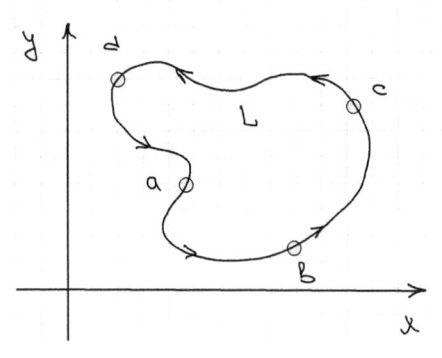

Fig. 6.5. Positive direction of integration over a closed contour — counterclockwise direction.

over a closed contour always evaluates to zero. Interestingly, this condition is the same as that for path-independent line integrals. Specifically, if an integral is path-independent, then the contour integral will always be equal to zero, and vice versa. Indeed, assume that a line integral is path independent, then integration from a to c in Figure 6.5 along the path abc is equal to integration along the path adc, i.e., $\int_{abc} = \int_{adc}$. Since $\int_{adc} = -\int_{cda}$, then

$$\int_{abcda} = \int_{abc} + \int_{cda} = \int_{abc} - \int_{adc} = 0.$$

Thus, a line integral

$$\int_L P(x,y)dx + Q(x,y)dy$$

over a closed contour L turns to zero if the integrand is a complete differential:

$$\frac{\partial P(x,y)}{\partial y} = \frac{\partial Q(x,y)}{\partial x}.$$

The same result can be derived by using Green's theorem which is considered later in this chapter (see Section 6.5.3).

6.3.5. *Calculation of surface areas using line integrals' integrals*

Linear integrals (of the second kind) can be used to calculate the areas of plane figures. Let us consider an area $ABCD$ bounded by

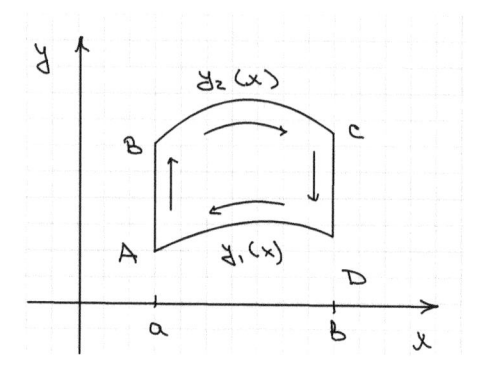

Fig. 6.6. Clockwise direction.

straight lines AB and CD parallel to the y-axis and by two lines BC and AD described by functions $y_1(x)$ and $y_2(x)$ correspondingly, as shown in Figure 6.6. The total area of the figure can be viewed as a difference of two areas, namely $aBCb$ and $aADb$:

$$S = \int_a^b y_2(x)dx - \int_a^b y_1(x)dx.$$

Since $y_2(x)$ and $y_1(x)$ are explicit functions of x, we can use (6.21) as

$$\int_{BC} y(x)dx = \int_a^b y_2(x)dx, \quad \int_{AD} y(x)dx = \int_a^b y_1(x)dx.$$

Therefore, we can write

$$S = \int_{BC} y(x)dx + \int_{DA} y(x)dx.$$

We changed the sign in front of the second integral, since we changed the direction of integration. If we add to the right side the integrals

$$\int_{AB} y(x)dx + \int_{DC} y(x)dx$$

that are equal to zero, then we arrive at

$$S = \int_{ABCD} y(x)dx, \tag{6.26}$$

where the direction of integration runs counterclockwise as $ABCD$.

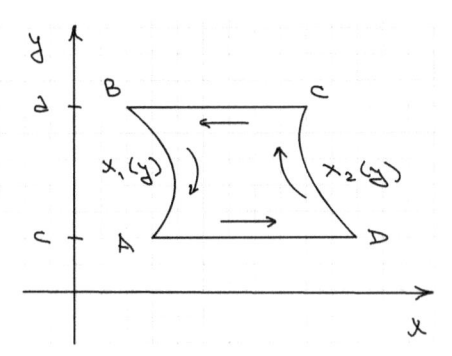

Fig. 6.7. Counterclockwise direction of integration.

Applying the same approach to an area in Figure 6.7 we can write for the area

$$S = \int_{ADCB} x(y)dy. \tag{6.27}$$

For an area of arbitrary shape, a symmetric formula applies

$$S = \frac{1}{2}\int_L xdy - ydx, \tag{6.28}$$

where the integration is carried out in the counterclockwise direction along the boundary of the area.

Example 6.5. Evaluate the area of an ellipse with semi-axes a and b.

A parametric formula for an ellipse is $x = a\cos t$ and $y = b\sin t$ with $0 \le t \le 2\pi$. Then, to evaluate the area, we can use formula (6.28) along with (6.20) for evaluating integrals of the second kind. Thus, we have

$$S = \frac{1}{2}\int_L xdy - ydx = \frac{1}{2}\int_0^{2\pi} a\cos(t)b\cos(t)dt - b\sin(t)(-a\sin(t))dt$$
$$= \frac{ab}{2}\int_0^{2\pi} dt = \pi ab.$$
∎

6.3.6. Connection between linear integrals of both kinds

Let's begin with the line integral of the second kind, which is integrated along a path L. Specifically, we have
$$\int_L P(x,y)dx + Q(x,y)dy.$$
According to (6.20), such an integral can be evaluated as a regular integral:
$$\int_{l_a}^{l_b} \left[P(x(l),y(l))x'(l) + Q(x(l),y(l))y'(l)\right] dl,$$
where l is a parameter. For our case, we consider the parameter l as the path. If we denote the angle made with the x-axis by α (Figure 6.8), which corresponds to a tangent pointed in the direction of increasing l, then, as is known,
$$\cos\alpha = \frac{dx}{dl} = x'(l), \quad \sin\alpha = \frac{dy}{dl} = y'(l).$$
Thus, the second kind of line integral can be expressed as
$$\int_{l_a}^{l_b} \left[P(x(l),y(l))\cos\alpha(l) + Q(x(l),y(l))\sin\alpha(l)\right] dl$$
$$\equiv \int_{l_a}^{l_b} \left[P(l)\cos\alpha(l) + Q(l)\sin\alpha(l)\right] dl.$$
Thus, the second kind of line integral turned out to be reduced to the first kind of line integral!

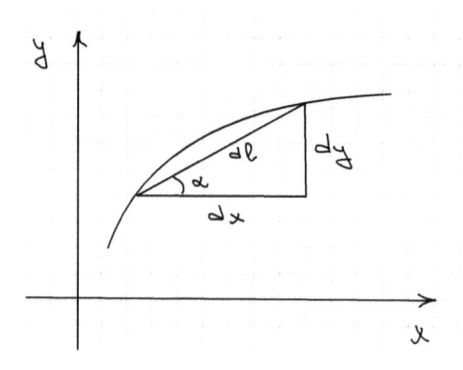

Fig. 6.8. Integration along path L.

6.4. Application of line integrals

6.4.1. Mass, center of mass, and rotational inertia of a wire

For an object located in the x, y-plane, mass, center of mass, and center of inertia can be written as line integrals of the first kind, namely for the mass as

$$M = \int_L \rho(x, y) dl, \qquad (6.29)$$

for center of mass as

$$x_{cm} = \frac{1}{M} \int_L x \rho(x, y) dl,$$

$$y_{cm} = \frac{1}{M} \int_L y \rho(x, y) dl, \qquad (6.30)$$

and for rotational inertia as

$$I_x = \int_L y^2 \rho(x, y) dl \quad \text{around } x\text{-axis},$$

$$I_y = \int_L x^2 \rho(x, y) dl \quad \text{around } y\text{-axis},$$

$$I_z = \int_L ((x - x_0)^2 + (y - y_0)^2) \rho(x, y) dl \quad \text{around } (x_0, y_0) \text{ point}.$$

$$(6.31)$$

Example 6.6. Let's consider a semicircle wire defined by equation $x^2 + y^2 = R^2$, where $y \geq 0$, as shown in Figure 6.9. This wire has a linear density described by the function $\rho(x, y)$. Our objective is to find the mass M, the center of mass (x_{cm} and y_{cm}), and the rotational inertias I_x, I_y, and I_z of the wire. The semicircle $x^2 + y^2 = R^2$ can be described using an angle θ as a parameter, ranging from 0 to π as $x = R\cos\theta$, $y = R\sin\theta$. With the derivatives $x' = -R\sin\theta$ and $y' = R\cos\theta$ we have

$$dl = \sqrt{(R^2 \cos^2\theta + R^2 \sin^2\theta)} d\theta = R d\theta,$$

then the mass, the center of mass, and the rotational inertias can be calculated using equations (6.29), (6.30), and (6.31).

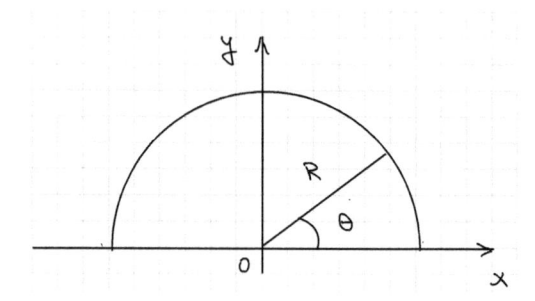

Fig. 6.9. The semicircle for $y \geq 0$.

In a case of a constant linear density, i.e., $\rho(x,y) \equiv \rho_0$, we have the mass as

$$M = \int_L \rho_0 dl = \int_0^\pi \rho_0 R d\theta = \rho_0 \pi R,$$

the center of mass as

$$x_{\text{cm}} = \frac{1}{\pi R \rho_0} \int_0^\pi \rho_0 R^2 \cos\theta d\theta = 0,$$

$$y_{\text{cm}} = \frac{1}{\pi R \rho_0} \int_0^\pi \rho_0 R^2 \sin\theta d\theta = \frac{2}{\pi} R,$$

and for the rotational inertia,

$$I_x = \int_0^\pi \rho_0 R^2 \sin^2\theta R d\theta = \frac{1}{2} MR^2.$$ ∎

6.4.2. Work done by a force

In physics, the work done by a force can be calculated as the line integral of the force along a path. For a two-dimensional case, let's consider a force vector field $\vec{F}(x,y) = \hat{i} F_x(x,y) + \hat{j} F_y(x,y)$ and a path L parametrized by $\vec{r} = \hat{i} x(t) + \hat{j} y(t)$, where $a \leq t \leq b$. The work W by the force along the path L is given by the line integral:

$$W = \int_L \vec{F} \cdot d\vec{r}, \tag{6.32}$$

where the dot product can be written as $\vec{F} \cdot d\vec{r} = F_x dx + F_y dy$. Since the path is given in the parametric form, then the line integral can be computed as

$$W = \int_a^b \vec{F} \cdot \frac{d\vec{r}}{dt} dt = \int_a^b \left(F_x \frac{dx}{dt} + F_y \frac{dy}{dt} \right) dt. \quad (6.33)$$

This integral is evaluated over the parameter range $[a, b]$ that corresponds to the path.

Example 6.7. Suppose we have a force defined as $\vec{F}(x, y) = \hat{i}2x + \hat{j}y$ and a path L defined by the parametric equations $x(t) = t^2$ and $y(t) = t$, where $0 \leq t \leq 2$. The work done by the force $\vec{F}(x, y)$ along the path L is given by (6.33):

$$W = \int_a^b \left(F_x \frac{dx}{dt} + F_y \frac{dy}{dt} \right) dt = \int_a^b \left(2x \frac{dx}{dt} + y \frac{dy}{dt} \right) dt.$$

By substituting the parametric equations and their derivatives into the integral, we obtain

$$W = \int_0^2 (2t^2 \cdot 2t + t \cdot 1) \, dt = \int_0^2 (4t^3 + t) dt = \left[\frac{t^4}{4} + \frac{t^2}{2} \right]_0^2$$
$$= (4 + 2) - (0 + 0) = 6.$$

So, in this example, the work done by the force $\vec{F}$ along the path L is 6. ∎

Both applications mentioned above are defined for a two-dimensional case. Extending them to a three-dimensional case is relatively straightforward. (Can you see how to do it?)

6.5. Multiple integrals

In introductory calculus classes, the domain of integration for the definite integral has always been limited to a straight line, such as running from x_{min} to x_{max}. However, this approach only applied to a one-dimensional world. To address this limitation, we need to generalize the concept of integration to include multi-dimensional integrals.

Let's focus on evaluating a mass of an object with volume V and variable density $f(x, y, z)$. We divide the entire volume (domain) of the object into N small elements Δv of arbitrary shape. Each element is chosen to be sufficiently small such that the density $f(x_i^*, y_i^*, z_i^*)$ can be approximated as constant within each elementary volume Δv_i. Then, the total mass can be approximated by

$$M \approx \sum_{i=1}^{N} f(x_i^*, y_i^*, z_i^*)\Delta v_i,$$

where (x_i^*, y_i^*, z_i^*) is a point inside Δv_i, and summation extends over all the elements that make up the volume V. The smaller we make the elements Δv_i and the larger the number of elements N, the more accurate the approximation becomes. In the limit $N \to \infty$, we can express the exact mass as

$$M = \lim_{N \to \infty} \sum_{i=1}^{N} f(x_i^*, y_i^*, z_i^*)\Delta v_i. \qquad (6.34)$$

By applying the same approach, we can also evaluate the electric charge of an object with a charge density $f(x, y, z)$, as well as many other physical quantities.

The sum described above in (6.34) bears a resemblance to Riemann sums, which are used as a foundation for defining definite integrals in calculus. By partitioning the domain and approximating the function over each subinterval, we can compute an approximation of the integral. The analogy is notable in the context of evaluating physical quantities, as the process involves dividing the volume or domain into smaller elements and summing their contributions, similar to how Riemann sums approximate the area under a curve.

6.5.1. Double integrals

In the following, our primary focus is on as simple multiple integrals as possible, namely double integrals. By following the examples mentioned earlier, we can evaluate masses of planar objects and other similar quantities.

Moving away from geometric concepts employed thus far, we can define the limit of a sum independent of the geometric form of the function $f(x, y)$.

Consider a bounded plane domain denoted by S and let $f(x, y)$ be a function defined within the domain. Now, we divide S into N

sub-domains with areas $\sigma_i (i = 1\ldots N)$, and let $f(x_i^*, y_i^*)$ be arbitrary points within each respective domains. We form then the sum of the products:

$$\sum_{i=1}^{N} f(x_i^*, y_i^*)\sigma_i.$$

As we increase the number of sub-domains N indefinitely and decrease the area σ_i of each sub-domain, the limit of the sum is known as *the double integral* of the function $f(x,y)$ over the domain S. This is denoted as

$$\iint_S f(x,y)d\sigma = \lim_{N\to\infty} \sum_{i=1}^{N} f(x_i^*, y_i^*)\sigma_i. \qquad (6.35)$$

The existence of the limit expressed in equation (6.35) is self-evident because, as mentioned earlier, it corresponds to the area described above. Although this argument is not rigorously proven, a rigorous analytic proof of the limit's existence is feasible for reasonably general conditions imposed on the function $f(x,y)$.

The fundamental properties of double (and multiple) integrals bear similarities to the corresponding properties of regular (one-dimensional) integrals. These properties can be straightforwardly derived from the definition (6.35) itself:

(1) The integral of the sum of two functions is equal to the sum of the integrals

$$\iint_S (f(x,y) + g(x,y))d\sigma = \iint_S f(x,y)d\sigma + \iint_S g(x,y)d\sigma.$$

(2) Constant factor can be taken outside the integral

$$\iint_S Cf(x,y)d\sigma = C\iint_S f(x,y)d\sigma \quad (C = \text{const}).$$

(3) For any partition of the area S on the parts S_1 and S_2, there will be

$$\iint_S f(x,y)d\sigma = \iint_{S_1} f(x,y)d\sigma + \iint_{S_2} f(x,y)d\sigma.$$

(4) The integral of $f(x,y) = 1$ over area is equal to the area of integration

$$S = \iint_S d\sigma.$$

(5) If the function being integrated and the variables involved have dimensions (units), the dimension of the integral is equal to the product of the dimensions of the function and the dimensions of the integrated variables. Thus, for a double integral over a surface,

$$\left[\iint_S f(x,y)d\sigma\right] = [f(x,y)] \cdot [S],$$

where the square brackets denote the dimensions. For example, consider a surface density $[\rho(x,y)]$ with units of kg $\cdot$ m^{-2}. The double integral of this density over the surface S, denoted as

$$\iint_S \rho(x,y)d\sigma,$$

would have units kg $\cdot$ m$^{-2} \cdot$ m^2 = kg. In other words, the double integral yields the mass of a planar object with a given surface density distribution.

(6) The average value of a function $\bar{f}$ on a domain is

$$\iint_\sigma \bar{f}d\sigma = \bar{f}\iint_\sigma d\sigma = \bar{f}S = \iint_\sigma f(x,y)d\sigma \quad \text{then}$$

$$\bar{f} = \frac{1}{S}\iint_\sigma f(x,y)d\sigma.$$

6.5.2. *Evaluation of double integrals*

By viewing a double integral as a sum (referring to equation (6.35)), we can establish a method for simplifying a double integral into an iterated integral. This approach involves evaluating one integral with respect to one variable and then performing another integral with respect to the remaining variable. This process allows us to compute double integrals in a more manageable and systematic manner.

A rectangular area: Consider a function $f(x,y)$ defined throughout a rectangular region, such as the one given by $[a \leq x \leq b]$

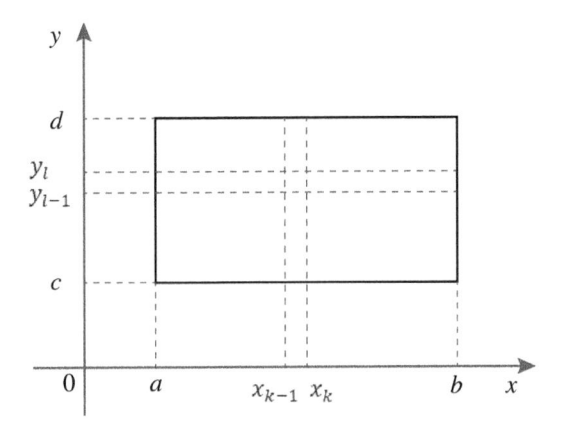

Fig. 6.10. Double integral over a rectangular area.

and $[c \leq y \leq d]$. Now, we divide the $[a, b]$ segment into n intervals with points $a = x_0 < x_1 < x_2 < \cdots < x_{n-1} < x_n = b$, and similarly we divide the $[c, d]$ segment into m intervals with points $c = y_0 < y_1 < y_2 < \cdots < y_{m-1} < y_m = d$, as shown in Figure 6.10. Let (x_k^*, y_l^*) be a point inside a box with sides $\Delta x_k = x_k - x_{k-1}, \Delta y_l = y_l - y_{l-1}$. We can now write $\Delta \sigma_{k,l} = \Delta x_k \Delta y_l$, and eventually we can define a double Riemann sum as

$$\sum_{k=1}^{n} \sum_{l=1}^{m} f(x_k^*, y_l^*) \Delta x_k \Delta y_l.$$

Then, the Riemann definition for a double integral over a rectangular area is

$$I = \lim_{n \to \infty} \lim_{m \to \infty} \sum_{k=1}^{n} \sum_{l=1}^{m} f(x_k^*, y_l^*) \Delta x_k \Delta y_l. \tag{6.36}$$

Rearranging the terms in (6.36) the sum can be also written as

$$I = \lim_{n \to \infty} \sum_{k=1}^{n} \Delta x_k \left[\lim_{m \to \infty} \sum_{l=1}^{m} f(x_k^*, y_l^*) \Delta y_l \right] \tag{6.37}$$

or equivalently as

$$I = \lim_{m \to \infty} \sum_{l=1}^{m} \Delta y_l \left[\lim_{n \to \infty} \sum_{k=1}^{n} f(x_k^*, y_l^*) \Delta x_k \right]. \tag{6.38}$$

Let's concentrate our attention on (6.37). The term in the square brackets is a regular integral if we consider x_k^* as a constant

$$I_1(x_k^*) = \int_c^d f(x_k^*, y) dy = \lim_{m \to \infty} \sum_{l=1}^{m} f(x_k^*, y_l^*) \Delta y_l,$$

then (6.37) can be rewritten as

$$I = \lim_{n \to \infty} \sum_{k=1}^{n} \Delta x_k I_1(x_k^*) = \int_a^b I_1(x) dx = \int_a^b dx \int_c^d f(x, y) dy.$$

We observe that the first integration with respect to y, while keeping x constant, corresponds to summing over rectangles composed of strips parallel to the OY axis. These rectangles all have the same width dx, which is placed outside the integral sign for the first integration. The second integration, with respect to x, involves adding up these individual sums obtained from the first integration.

Thus, by employing definition (6.36), a double integral can be expressed as an iterated integral, given by

$$I = \int_a^b dx \int_c^d f(x, y) dy = \int_c^d dy \int_a^b f(x, y) dx. \tag{6.39}$$

The quantity $d\sigma = dxdy$ is called an elementary area in rectangular coordinates.

Attention: If the first integration in equation (6.39) is performed with respect to y, then x is treated as a constant within that integral. Conversely, if we integrate first with respect to x, we treat y within the dx integral as a constant. This approach allows us to effectively evaluate the double integral by iteratively integrating with respect to one variable while treating the other variable as a constant.

In special cases, when a function $f(x, y)$ can be written as a product of two function of independent variables $f(x, y) = g(x)h(y)$, the integral (6.39) is a product of two integrals:

$$I = \int_a^b dx \int_c^d f(x, y) dy = \int_a^b g(x) dx \int_c^d h(y) dy.$$

An area with an arbitrary shape: A result similar to (6.39) can be derived for a more general case when the domain S has an arbitrary shape, but lines parallel to the axes intersect the boundary of

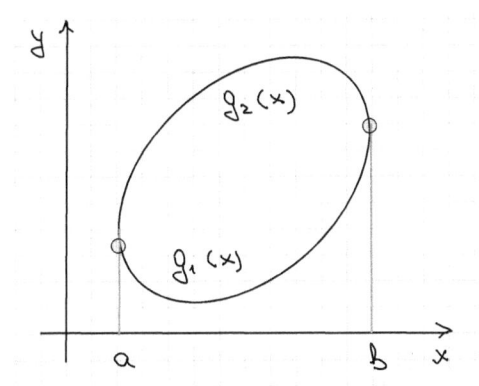

Fig. 6.11. A domain for double integration, where $g_1(x)$ and $g_2(x)$ describe boundaries of the area between points a and b.

the domain **no more than twice**, as in Figure 6.11. In this scenario, we can still apply the concept of iterated integrals to evaluate the double integral over the domain. The limits of integration will be determined based on the boundaries of the domain as defined by these parallel lines.

In the rectangular case (6.39), y is varied in the same interval $[c, d]$ for any value of $x = x_0$. Now, this interval is determined by x_0 as $[g_1(x_0), g_2(x_0)]$ so that

$$I_1(x) = \int_{g_1(x)}^{g_2(x)} f(x,y)dy.$$

Therefore, we can write

$$I = \int_a^b dx \int_{g_1(x)}^{g_2(x)} f(x,y)dy. \tag{6.40}$$

Note that the boundaries at $x = a$ and $x = b$ are either points or vertical straight lines.

If we want to change the order of integration, then we need to consider functions $h_1(y)$ and $h_2(y)$ bounding the area between points c and d (see Figure 6.12).

Then, the double integral I is defined as

$$I = \int_c^d dy \int_{h_1(y)}^{h_2(y)} f(x,y)dx. \tag{6.41}$$

It is evident that the order of integration does not alter the value of the integral. Regardless of whether we integrate first with respect to x or y, the result remains the same. This property is known as the independence of the order of integration, meaning that

$$\int_a^b dx \int_{g_1(x)}^{g_2(x)} f(x,y) dy = \int_c^d dy \int_{h_1(y)}^{h_2(y)} f(x,y) dx. \tag{6.42}$$

In practical work, it is often easier to perform the double integral in one order rather than the other. The choice of order of integration order depends on the specific function, the nature of the region, and the simplifications that can be achieved by adopting a particular order. By carefully selecting the integration order, we can often simplify the calculations and arrive at a more manageable expression for the double integral.

Note that when drawing a line parallel to the Ox axis, it is important that there are no more than two intersections with the boundary defined by $h_1(y)$ and $h_2(y)$. If there are more than two intersections, it indicates that the domain of integration needs to be split into sub-domains. Each sub-domain should have no more than two intersections with lines parallel to the Ox axis (or Oy axis). This allows us to properly define the integration limits and ensure accurate calculations.

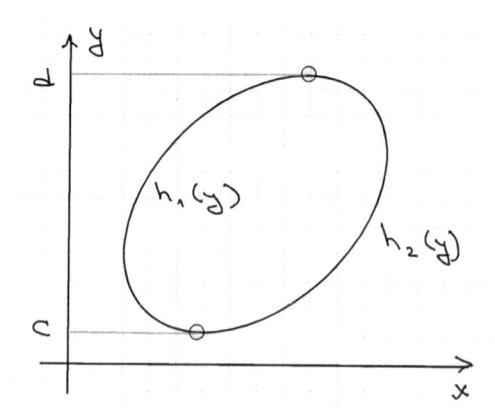

Fig. 6.12. A domain for double integration, where $h_1(y)$ and $h_2(y)$ describe boundaries of the area between points c and d.

Example 6.8. Consider an integral
$$I = \int_0^1 dx \int_{x^2}^{2x} f(x,y)dy,$$
where the area of integration shown in Figure 6.13. By changing the order of integration, we need to divide the integration interval over y into two sub-intervals because there is no single function describing the upper boundary of integration. Then, we have
$$I = \int_0^1 dy \int_{y/2}^{\sqrt{y}} f(x,y)dx + \int_1^2 dy \int_{y/2}^1 f(x,y)dx.$$
Depending on a shape of the area, we may need to divide the area of integration into more than two sub-domains. ∎

Attention: Very often, it is stated that a regular definite integral describes the area under a function. However, in our considerations, it is a double integral with $f(x,y) = 1$ that represents the calculation of an area. This may seem contradictory at first glance. Indeed, the area of a figure with arbitrary shape is evaluated as
$$\sigma = \int_a^b dx \int_{g_1(x)}^{g_2(x)} dy. \tag{6.43}$$
It seems that this has little in common with
$$\sigma = \int_a^b f(x)dx. \tag{6.44}$$
To resolve this, we need to understand that the interpretation of the definite integral as an area depends on the context. When dealing

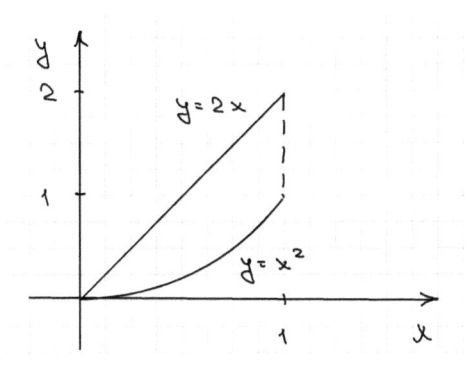

Fig. 6.13. Dividing the domain of integration into sub-domains.

with a regular definite integral of a single variable, it represents the signed area between the curve and the x-axis. On the other hand, when working with a double integral where $f(x, y) = 1$, the integral evaluates the area of a region or figure with an arbitrary shape in the xy plane.

Assume that $g_1(x) = 0$, then (6.43) can be evaluated as

$$\sigma = \int_a^b dx \int_0^{g_2(x)} dy = \int_a^b g_2(x) dx \Rightarrow \int_a^b f(x) dx \quad \text{if } g_2(x) = f(x).$$

Thus, the statement that a regular integral (6.44) represents an area is only correct if we speak about a figure with a flat base going along Ox axis. Therefore, the interpretation of an integral as an area depends on the context and the specific integration being performed.

6.5.3. Green's theorem

Green's theorem establishes a relationship between a line integral around a simple closed curve in a plane and a double integral over the region enclosed by that curve.

Let us consider an area S enclosed by the contour $ABCD$ with BC described by a function $y_2(x)$, and AD by $y_1(x)$. Assume that a *continuous function* $P(x, y)$ with a *continuous derivative* $\partial P/\partial y$ is defined in the area. We want to evaluate a double integral

$$\iint_S \frac{\partial P}{\partial y} dx dy.$$

According to (6.42), we have

$$\iint_S \frac{\partial P}{\partial y} dx dy = \int_a^b dx \int_{y_1(x)}^{y_2(x)} \frac{\partial P}{\partial y} dy.$$

The inner integral is easily calculated using the anti-derivative function $P(x, y)$, namely

$$\int_{y_1(x)}^{y_2(x)} \frac{\partial P}{\partial y} dy = P(x, y)\big|_{y=y_1(x)}^{y=y_2(x)} = P(x, y_2(x)) - P(x, y_1(x)).$$

Then, we have

$$\iint_S \frac{\partial P}{\partial y} dx dy = \int_a^b P(x, y_2(x)) dx - \int_a^b P(x, y_1(x)) dx.$$

Each of the two above integrals can be represented as a line integral of the second order (6.20), namely

$$\int_a^b P(x, y_2(x))dx = \int_{BC} P(x,y)dx, \quad \int_a^b P(x, y_1(x))dx$$
$$= \int_{AD} P(x,y)dx,$$

and then

$$\iint_S \frac{\partial P}{\partial y} dxdy = \int_{BC} P(x,y)dx - \int_{AD} P(x,y)dx$$
$$= \int_{BC} P(x,y)dx + \int_{DA} P(x,y)dx.$$

Since we want to extend the line integration to cover the entire contour $ABCD$, we can add two integrals equal to zero to the right side:

$$\int_{AB} P(x,y)dx = 0, \quad \int_{CD} P(x,y)dx = 0.$$

(Do you see why these integrals are equal to zero?)

Collecting all the integrals we have

$$\iint_S \frac{\partial P}{\partial y} dxdy = \int_{AB} P(x,y)dx + \int_{BC} P(x,y)dx$$
$$+ \int_{CD} P(x,y)dx + \int_{DA} P(x,y)dx.$$

The right side of the above equation is the integral taken over the entire closed contour L that bounds the region S but in the negative direction. In accordance with the convention for linear integrals over a closed contour, we can finally rewrite the resulting formula as follows:

$$\iint_S \frac{\partial P}{\partial y} dxdy = -\int_L P(x,y)dx.$$

In a similar way for a function $Q(x,y)$ in the area in Figure 6.15, we can derive

$$\iint_S \frac{\partial Q}{\partial x} dxdy = \int_L Q(x,y)dy.$$

Since an area of a arbitrary shape can be decomposed into a finite number of areas of type 1 (Figure 6.14) and type 2 (Figure 6.15), then we can write for continuous functions $P(x,y)$, $Q(x,y)$ with continuous derivatives $\partial P/\partial y$, $\partial Q/\partial x$:

$$\int_L P(x,y)dx + Q(x,y)dy = \iint_S \left(\frac{\partial Q}{\partial x} - \frac{\partial P}{\partial y}\right) dxdy. \qquad (6.45)$$

This is Green's theorem connecting line integrals of the second kind to double integrals. Green's theorem provides a powerful tool for calculating line integrals by transforming them into double integrals, which can often be more easily evaluated.

From Green's theorem follows that if

$$\frac{\partial P}{\partial y} = \frac{\partial Q}{\partial x}, \qquad (6.46)$$

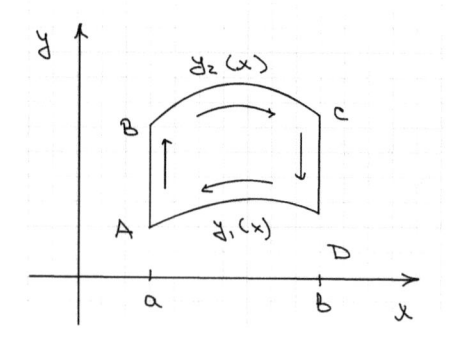

Fig. 6.14. Integrating the area $ABCD$ as a double integral.

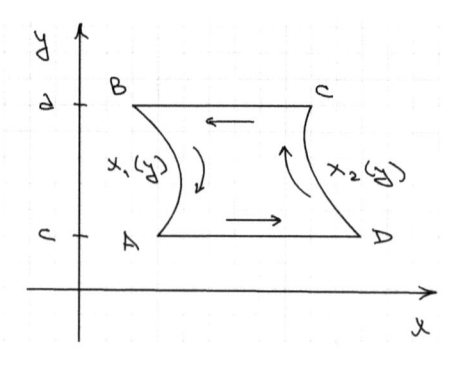

Fig. 6.15. Integrating the area $ABCD$ as a double integral.

then the second kind line integral over a closed contour turns to be zero. It is important to remember that Green's theorem is only valid for continuous functions with continuous partial derivatives $\partial P/\partial y$ and $\partial Q/\partial x$.

Example 6.9. Test Green's theorem for

$$P(x,y) = -\frac{y}{x^2+y^2}, \quad Q(x,y) = \frac{x}{x^2+y^2}.$$

In this case,

$$\frac{\partial P}{\partial y} = -\frac{x^2-y^2}{(x^2+y^2)^2}, \quad \frac{\partial Q}{\partial x} = \frac{y^2-x^2}{(x^2+y^2)^2}$$

or

$$\frac{x^2-y^2}{(x^2+y^2)^2} = \frac{\partial Q}{\partial x}.$$

Thus, we expect that a line integral

$$\int_L -\frac{y}{x^2+y^2}dx + \frac{x}{x^2+y^2}dy$$

over a closed contour should be zero. Let us evaluate it for a circular contour of radius one with its center at the origin. Using (6.20) with $x = \cos(t)$ and $y = \sin(t)$ with $0 \le t \le 2\pi$, we have for the line integral

$$\int_0^{2\pi} -\frac{\sin(t)}{\cos^2(t)+\sin^2(t)}(-\sin(t))dt + \frac{\cos(t)}{\cos^2(t)+\sin^2(t)}(\cos(t))dt$$

$$= \int_0^{2\pi} dt = 2\pi.$$

The line integral is NOT equal to zero! The point is that Green's theorem was derived under the assumption that the considered functions and their derivatives are continuous, and here this condition at the origin is violated. So, the moral of this story is to pay attention! ■

6.5.4. Applications: Mass, center of mass, and moment of inertia

Multiple integration (or multi-dimensional integration) has so many applications that it is not even feasible to cover it in a single book. In this chapter, we concentrate on a few of the most common applications in general physics courses, namely evaluating mass, center of mass, and moment of rotational inertial for various shapes.

First, let's recall the definitions for the center of mass and the moment of inertia for a set of particles.

The center of mass along x is defined as

$$x_{\text{cm}} = \frac{m_1 x_1 + m_2 x_2 + \cdots + m_n x_n}{m_1 + m_2 + \cdots + m_n} = \frac{1}{M} \sum_{i=1}^{n} m_i x_i.$$

If the particles are distributed in three dimensions, the center of mass must be identified by three coordinates, namely

$$x_{\text{cm}} = \frac{1}{M} \sum_{i=1}^{n} m_i x_i, \quad y_{\text{cm}} = \frac{1}{M} \sum_{i=1}^{n} m_i y_i, \quad z_{\text{cm}} = \frac{1}{M} \sum_{i=1}^{n} m_i z_i.$$

For the moment of inertia, we have

$$I = \sum_{i=1}^{n} m_i r_i^2,$$

where r_i are distances from the axis of rotation. It is convenient to represent the moment of inertia by using the parallel-axis theorem also known as the Huygens–Steiner theorem. The theorem relates the moment of inertia of a rigid body about an axis parallel to an axis through its center of mass to the moment of inertia about the center of mass itself. In two dimensions, this theorem can be stated as follows:

$$I_p = I_{\text{cm}} + M d^2, \tag{6.47}$$

where I_p is the moment of inertia about an axis parallel to and at a distance d from the axis through the center of mass, I_{cm} is the moment of inertia about the axis through the center of mass, and M is the mass of the object.

For rigid bodies, we replace the mass on ρdV, where ρ is the density and dV is the area of integration. We also replace the sums on integrals.

Line Integrals and Multiple Integrals 205

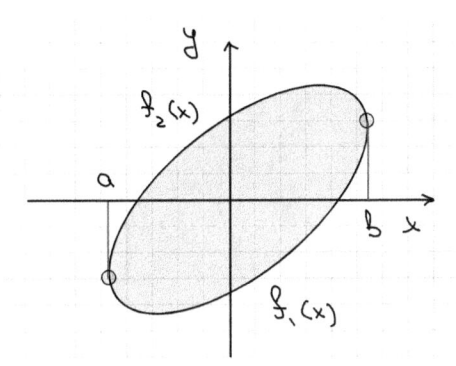

Fig. 6.16. A thin sheet of material of an arbitrary shape with the boundaries defined by $f_1(x)$ and $f_2(x)$.

6.5.5. *2D case*

For a thin sheet of material, as in Figure 6.16, with a surface density $\sigma(x, y)$, the definitions of mass, center of mass and rotational inertia are similar to one in 1D case:

Mass:
$$M = \int_a^b dx \int_{f_1(x)}^{f_2(x)} \sigma(x,y) dy. \qquad (6.48)$$

Center of mass:
$$x_{\text{cm}} = \frac{1}{M} \int_a^b x\, dx \int_{f_1(x)}^{f_2(x)} \sigma(x,y) dy,$$

$$y_{\text{cm}} = \frac{1}{M} \int_a^b dx \int_{f_1(x)}^{f_2(x)} y\sigma(x,y) dy. \qquad (6.49)$$

Rotational inertias around the Cartesian axes:
$$I_x = \int_a^b dx \int_{f_1(x)}^{f_2(x)} y^2 \sigma(x,y) dy,$$

$$I_y = \int_a^b x^2 dx \int_{f_1(x)}^{f_2(x)} \sigma(x,y) dy,$$

$$I_z = \int_a^b dx \int_{f_1(x)}^{f_2(x)} ((x-x_0)^2 + (y-y_0)^2)\sigma(x,y) dy. \qquad (6.50)$$

Note that for I_z we consider rotation around an axis parallel to z and passing through a point with coordinates x_0, y_0 in xy-plane.

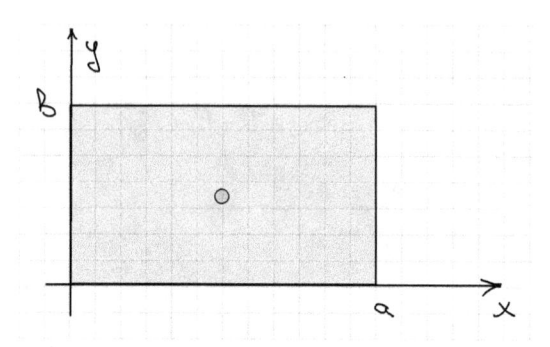

Fig. 6.17. A rectangular shape.

Example 6.10. In this example, we will consider rather a simple case — a rectangular thin sheet of material with constant density $\sigma(x, y) = \sigma_0$ (Figure 6.17). For this shape, we integrate over x from 0 to a, and for y from 0 to b. Then, we have the results:

Mass:
$$M = \int_0^a dx \int_0^b \sigma_0 dy = \sigma_0 \int_0^a dx \int_0^b dy = \sigma_0 ab.$$

Center of mass:
$$x_{\text{cm}} = \frac{1}{M} \int_0^a x dx \int_0^b \sigma_0 dy = \frac{1}{M} \sigma_0 \frac{a^2}{2} b = \frac{1}{\sigma_0 ab} \sigma_0 \frac{a^2}{2} b = \frac{a}{2},$$
$$y_{\text{cm}} = \frac{1}{M} \int_0^a dx \int_0^b y \sigma_0 dy = \frac{1}{M} \sigma_0 a \frac{b^2}{2} = \frac{1}{\sigma_0 ab} \sigma_0 a \frac{b^2}{2} = \frac{b}{2}.$$

Rotational inertias:
$$I_y = \int_0^a x^2 dx \int_0^b \sigma_0 dy = \sigma_0 \int_0^a x^2 dx \int_0^b dy$$
$$= \sigma_0 \frac{a^3}{3} b = \sigma_0 ab \frac{a^2}{3} = \frac{1}{3} M a^2.$$

Thus, we derived rotational inertia for a thin rectangular plate about axis going along edge attached to y. For z, we choose the point of rotation as the center of the plate, namely $(a/2, b/2)$. Then,

$$I_z = \int_0^a dx \int_0^b \left[\left(x-\frac{a}{2}\right)^2 + \left(y-\frac{b}{2}\right)^2\right]\sigma_0 dy$$

$$= \sigma_0 \int_0^a \left(x-\frac{a}{2}\right)^2 dx \int_0^b dy + \sigma_0 \int_0^a dx \int_0^b \left(y-\frac{b}{2}\right)^2 dy$$

$$= \sigma_0 \frac{a^3}{12}b + \sigma_0 \frac{b^3}{12}a = \sigma_0 ab \frac{1}{12}(a^2+b^2) = \frac{1}{12}M(a^2+b^2)$$

is another well-known result from general physics courses.[2] ∎

6.5.6. 3D case

The generalization of equations ((6.48)–(6.50)) for 3D case is rather straightforward. Let the volume density be $\rho(x, y, z)$, then

Mass:

$$M = \int_a^b dx \int_{f_1(x)}^{f_2(x)} dy \int_{g_1(x,y)}^{g_2(x,y)} \rho(x,y,z) dz. \qquad (6.51)$$

Center of mass:

$$x_{cm} = \frac{1}{M} \int_a^b x dx \int_{f_1(x)}^{f_2(x)} dy \int_{g_1(x,y)}^{g_2(x,y)} \rho(x,y,z) dz,$$

$$y_{cm} = \frac{1}{M} \int_a^b dx \int_{f_{,1}(x)}^{f_2(x)} y dy \int_{g_1(x,y)}^{g_2(x,y)} \rho(x,y,z) dz,$$

$$z_{cm} = \frac{1}{M} \int_a^b dx \int_{f_1(x)}^{f_2(x)} dy \int_{g_1(x,y)}^{g_2(x,y)} z\rho(x,y,z) dz. \qquad (6.52)$$

---

[2]H. Young and R. Freedman, *University Physics with Modern Physics*, 15th edn. Pearson, 2019.

Rotational inertias around the Cartesian axes:

$$I_x = \int_a^b dx \int_{f_1(x)}^{f_2(x)} dy \int_{g_1(x,y)}^{g_2(x,y)} (y^2 + z^2)\rho(x,y,z)dz,$$

$$I_y = \int_a^b dx \int_{f_1(x)}^{f_2(x)} dy \int_{g_1(x,y)}^{g_2(x,y)} (x^2 + z^2)\rho(x,y,z)dz, \quad (6.53)$$

$$I_z = \int_a^b dx \int_{f_1(x)}^{f_2(x)} dy \int_{g_1(x,y)}^{g_2(x,y)} (x^2 + y^2)\rho(x,y,z)dz.$$

Evaluation of integrals for 2D and 3D cases can often be a complex task. However, by changing variables and utilizing a suitable coordinate system that reflects the symmetry of the object, the process can be significantly simplified. Choosing appropriate coordinate transformations can help in reducing the integral to a more manageable form and exploiting the symmetries present in the problem. This technique can greatly ease the evaluation of integrals and enhance the efficiency of the calculations.

6.6. Change of variables in multiple integration

Change of variable in an integral, particularly from a function of a single variable $f(x)$, is considered one of the most effective methods for analytic integration. It is also a powerful technique for evaluating multiple integrals efficiently.

6.6.1. Jacobian

To prove that a change of variables in multiple integrals can be facilitated by using determinants, specifically the Jacobian determinant composed of partial derivatives, several lemmas and theorems are involved. In the case of a double integral, the Jacobian determinant can be expressed as follows:

$$\iint_R f(x,y)dxdy = \iint_{R'} g(u,v) J(x,y;u,v) dudv, \quad (6.54)$$

where

$$J(x,y;u,v) = \left|\frac{\partial(x,y)}{\partial(u,v)}\right| = \begin{vmatrix} \dfrac{\partial x}{\partial u} & \dfrac{\partial x}{\partial v} \\ \dfrac{\partial y}{\partial u} & \dfrac{\partial y}{\partial v} \end{vmatrix} = \frac{\partial x}{\partial u}\frac{\partial y}{\partial v} - \frac{\partial x}{\partial v}\frac{\partial y}{\partial u}. \quad (6.55)$$

Practically, the area element $dxdy$ should be replaced on $J(x,y;u,v)dudv$:

$$dxdy = \begin{vmatrix} \dfrac{\partial x}{\partial u} & \dfrac{\partial x}{\partial v} \\ \dfrac{\partial y}{\partial u} & \dfrac{\partial y}{\partial v} \end{vmatrix} dudv.$$

For triple integrals (from x,y,z to u,v,w), the Jacobian takes the form

$$J = \begin{vmatrix} \dfrac{\partial x}{\partial u} & \dfrac{\partial x}{\partial v} & \dfrac{\partial x}{\partial w} \\ \dfrac{\partial y}{\partial u} & \dfrac{\partial y}{\partial v} & \dfrac{\partial y}{\partial w} \\ \dfrac{\partial z}{\partial u} & \dfrac{\partial z}{\partial v} & \dfrac{\partial z}{\partial w} \end{vmatrix}. \quad (6.56)$$

Indeed, evaluating the Jacobian determinant for a given set of new variables is typically a straightforward process. However, determining the appropriate domains of integration in the new variables can be quite challenging. This is a common source of errors in calculations.

On the other hand, in physics, variable changes often involve switching between coordinate systems where the domains of integration are well-defined. In practice, physicists frequently employ commonly used coordinate systems such as Cartesian, cylindrical, and spherical coordinates. The choice of a coordinate system (and hence the variables) is guided by the symmetry of the problem at hand. This wise selection of a coordinate system can lead to significant simplifications in multiple integration, reducing the problem to a more manageable form and making the calculations more tractable.

Now, we are going to evaluate the Jacobian for transformation between Cartesian and **polar coordinates**: $x = r\cos\varphi$ and $y = r\sin\varphi$ (one of the simplest examples):

$$J(x,y;r,\varphi) = \begin{vmatrix} \frac{\partial x}{\partial r} & \frac{\partial x}{\partial \varphi} \\ \frac{\partial y}{\partial r} & \frac{\partial y}{\partial \varphi} \end{vmatrix} = \begin{vmatrix} \cos\varphi & -r\sin\varphi \\ \sin\varphi & r\cos\varphi \end{vmatrix} = r\cos^2\varphi + r\sin^2\varphi = r$$

(6.57)

or $dxdy = rdrd\varphi$.

For the **cylindrical coordinates**, $J(x,y,z;r,\theta,z) = r$, and we have

$$x = r\cos\theta,\ y = r\sin\theta,\ z = z,\ dxdydz = rdrd\theta dz. \quad (6.58)$$

For **spherical coordinates**, $J(x,y,z;r,\theta,\phi) = r^2\sin\phi$,

$$x = r\cos\theta\sin\phi,\ y = r\sin\theta\sin\phi,$$
$$z = r\cos\phi,\ dxdydz = r^2\sin\phi\, drd\theta dz. \quad (6.59)$$

Example 6.11. Evaluate mass, center of mass and rotational inertia about y axis for the following figure (Figure 6.18). Assume that the surface density is constant $\rho(x,y) = \rho_0$. Note that the equation of a circle is $x^2 + y^2 = R^2$.

It is clear that the symmetry of the problem calls for using polar coordinates. However, we want to compare work done using both the Cartesian and polar coordinate systems.

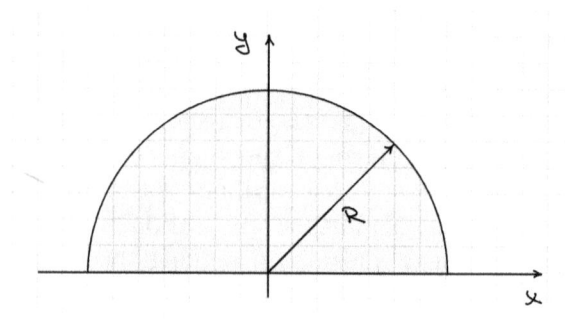

Fig. 6.18. A semicircular plate.

In Cartesian coordinates,

$$M = \int_a^b dx \int_{f_1(x)}^{f_2(x)} \rho(x,y) dy = \rho_0 \int_{-R}^{R} dx \int_0^{\sqrt{R^2-x^2}} dy$$

$$= \rho_0 \int_{-R}^{R} \sqrt{R^2 - x^2} dx = \frac{1}{2}\rho_0 \pi R^2,$$

$$x_{\text{cm}} = \frac{1}{M} \int_a^b x dx \int_{f_1(x)}^{f_2(x)} \rho(x,y) dy = \frac{1}{M}\rho_0 \int_{-R}^{R} x dx \int_0^{\sqrt{R^2-x^2}} dy$$

$$= \frac{1}{M}\rho_0 \int_{-R}^{R} x\sqrt{R^2 - x^2} dx = 0,$$

$$y_{\text{cm}} = \frac{1}{M} \int_a^b dx \int_{f_1(x)}^{f_2(x)} y\rho(x,y) dy = \frac{1}{M}\rho_0 \int_{-R}^{R} dx \int_0^{\sqrt{R^2-x^2}} y dy$$

$$= \frac{1}{2M}\rho_0 \int_{-R}^{R} (R^2 - x^2) dx = \frac{1}{2M}\rho_0 \frac{4}{3} R^3.$$

Since $M = \frac{1}{2}\rho_0 \pi R^2$, then

$$y_{\text{cm}} = \frac{4}{3\pi} R.$$

While we can continue our work in the Cartesian coordinate system, it is now opportune to explore the convenience offered by the polar coordinate system, given the symmetry of the object. We already evaluated the Jacobian for transformation from Cartesian to polar coordinates as $J(x, y; r, \varphi) = r$ (see 6.57). Therefore, $dxdy = rdrd\varphi$ and

$$M = \int_a^b dx \int_{f_1(x)}^{f_2(x)} \rho(x,y) dy = \rho_0 \int_0^R r dr \int_0^\pi d\varphi = \frac{1}{2}\rho_0 \pi R^2,$$

$$x_{\text{cm}} = \frac{1}{M}\rho_0 \int_0^R r dr \int_0^\pi r \cos\varphi d\varphi = \frac{1}{M}\rho_0 \int_0^R r^2 dr \int_0^\pi \cos\varphi d\varphi = 0,$$

$$y_{\text{cm}} = \frac{1}{M}\rho_0 \int_0^R r dr \int_0^\pi r \sin\varphi d\varphi = \frac{1}{M}\rho_0 \int_0^R r^2 dr \int_0^\pi \sin\varphi d\varphi$$

$$= \frac{1}{M}\rho_0 \frac{R^3}{3} 2 = \frac{4}{3\pi} R.$$

It is clear that we deal with easier integrals when using polar coordinates because we can separate the variables.

We can make one more step and evaluate the rotational inertia:

$$I_y = \int_a^b x^2 dx \int_{f_1(x)}^{f_2(x)} \rho(x,y) dy = \rho_0 \int_{-R}^{R} x^2 dx \int_0^{\sqrt{R^2-x^2}} dy$$

$$= \rho_0 \int_0^R r^2 r dr \int_0^\pi \cos^2 \varphi d\varphi = \frac{1}{8} \rho_0 \pi R^4.$$

Since $M = \frac{1}{2}\rho_0 \pi R^2$, then

$$I_y = \frac{1}{8}\rho_0 \pi R^4 = \frac{1}{4} M R^2.$$

As one can see, using polar coordinates makes it easier to perform the integration. ∎

Example 6.12. Let's consider one more example, namely we want to find a volume of a sphere of radius R. In Cartesian coordinates, we get

$$V = \int_{x_1}^{x_2} dx \int_{f_1(x)}^{f_2(x)} dy \int_{g_1(x,y)}^{g_2(x,y)} dz.$$

Since the equation of the sphere is $x^2 + y^2 + z^2 = R^2$, then

$$V = \int_{-R}^{R} dx \int_{-\sqrt{R^2-x^2}}^{\sqrt{R^2-x^2}} dy \int_{-\sqrt{R^2-x^2-y^2}}^{\sqrt{R^2-x^2-y^2}} dz.$$

This is not a trivial multiple integral, though it can be evaluated. Transferring to spherical coordinates we have $dxdydz = r^2 \sin\theta dr d\theta d\varphi$. Thus, the volume in spherical coordinates is a product of three simple integrals:

$$V = \int_0^R r^2 dr \int_0^\pi \sin\theta d\theta \int_0^{2\pi} d\varphi = \frac{R^3}{3} \cdot 2 \cdot 2\pi = \frac{4}{3}\pi R^3. \quad \blacksquare$$

Line Integrals and Multiple Integrals 213

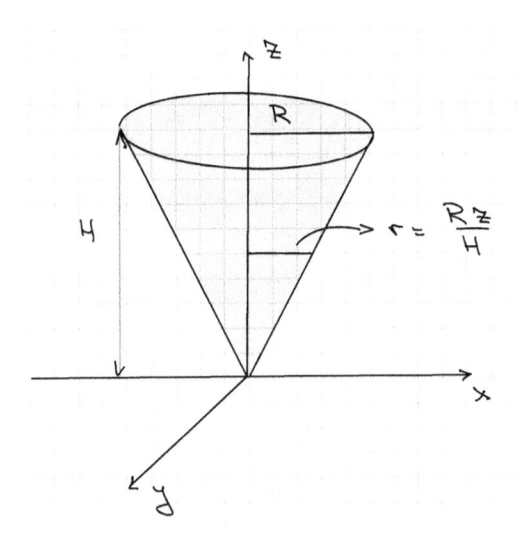

Fig. 6.19. A solid cone.

Example 6.13. Evaluate the mass, center of mass, and rotational inertia of a uniform solid cone with the given parameters: radius R, vertical size H, and uniform density $\rho(x, y, z) = \rho_0$ (Figure 6.19). In this example, it is convenient to work within a coordinate system that mirrors the symmetry inherent in the problem, specifically in cylindrical coordinates, where $x = r\cos\theta$, $y = r\sin\theta$, and $z = z$. We note that the angle θ is the angle in the xy-plane (measured counterclockwise from the x-axis). Since $dV = rdrd\theta dz$ and $z = r\frac{H}{R}$, then

$$M = \int_0^R rdr \int_0^{2\pi} d\theta \int_{r\frac{H}{R}}^H dz \rho(x,y,z) = \rho_0 \int_0^R rdr \int_0^{2\pi} d\theta \left(H - r\frac{H}{R}\right)$$

$$= 2\pi\rho_0 \int_0^R H\left(r - \frac{r^2}{R}\right) dr = \frac{1}{3}\pi\rho_0 H R^2.$$

From the symmetry of the problem, we expect that $x_{\text{cm}} = 0$ and $y_{\text{cm}} = 0$:

$$z_{\text{cm}} = \frac{1}{M}\rho_0 \int_0^R rdr \int_0^{2\pi} d\theta \int_{r\frac{H}{R}}^H zdz = \frac{1}{M} 2\pi\rho_0 \int_0^R r\left(H^2 - r^2\frac{H^2}{R^2}\right)\frac{1}{2} dr$$

$$= \frac{1}{2M} 2\pi\rho_0 H^2 \int_0^R \left(r - \frac{r^3}{R^2}\right) dr = \frac{1}{4}\pi\rho_0 H^2 R^2 = \frac{3}{4}H.$$

Next, we evaluate the rotational inertia I_z:

$$I_z = \rho_0 \int_0^R rdr \int_0^{2\pi} d\theta \int_{r\frac{H}{R}}^H r^2 dz = 2\pi\rho_0 \int_0^R r^3 \left(H - H\frac{r}{R}\right) dr$$

$$= \frac{1}{10}\pi\rho_0 H R^4.$$

Since $M = \frac{1}{3}\pi\rho_0 H R^2$, then we obtain

$$I_z = \frac{3}{10} M R^2.$$

For I_x, we integrate

$$I_x = \rho_0 \int_0^R rdr \int_0^{2\pi} d\theta \int_{r\frac{H}{R}}^H (y^2 + z^2) dz$$

$$= \rho_0 \int_0^R rdr \int_0^{2\pi} d\theta \int_{r\frac{H}{R}}^H (r^2 \sin^2\theta + z^2) dz.$$

The result of the integration is

$$I_x = \frac{3}{20} M R^2 + \frac{3}{5} M H^2. \qquad \blacksquare$$

6.7. Exercises

(1) Line integrals of the first kind:
   (a) Evaluate $\int_L (x^2 + y^2) dl$, where L is a line connecting points (a,a) and (b,b) with $b > a$.
   (b) Evaluate $\int_L y \, dl$, where L is described by $y^2 = 2ax$, connecting $(0,0)$ and $(2,2)$.

(2) Line integrals of the second kind:
   (a) Evaluate the integral $\int_L (x^2 - 2xy) dx + (y^2 - 2xy) dy$, where L is a parabola $y = x^2$ with $-1 \le x \le 1$.
   (b) Evaluate the integral $\int_L (x - y^2) dx + 2xy \, dy$ if one of the following lines is taken from a to c as the integration path, as shown in Figure 6.20:
      (a) from a to c as a straight line $y = x$,
      (b) $ad + dc$,
      (c) $ab + bc$.

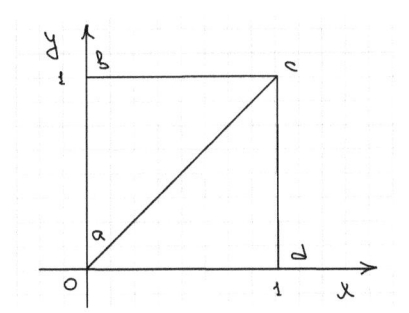

Fig. 6.20. Integration along various paths L.

(c) Evaluate the integral $\int_L (y^2 + 2xy)dx + (2xy + x^2)dy$ along the same three paths as in the problem before.
(d) Evaluate the integral $\int_L (x^2 + 2xy)dy$, where L means the top half of the ellipse $x^2/a^2 + y^2/b^2 = 1$ oriented in the counterclockwise direction.
(e) Evaluate the integral $\int_L y^2 dx - x^2 dy$, where L is a circle of unit radius with the center (a) at the origin and (b) at the point $(1, 1)$.

(3) Find areas:

(a) Area of the astroid: $x = a\cos^3(t)$, $y = a\sin^3 t$, where $0 \leq t \leq 2\pi$.
(b) Area of the folium of Descartes $x^3 + y^3 - 3axy = 0$, or given as a parametric equation $x = 3at/(1+t^3)$, $y = 3at^2/(1+t^3)$ with $0 \leq t \leq \infty$.

(4) Application of line integrals:

(a) Evaluate work done by force $\vec{F}(x, y) = \hat{i}x + \hat{j}y$ along path defined by the parametric equations: $x(t) = \cos(t)$, $y(t) = \sin(t)$.
(b) Find the mass of a wire following $y = \ln x$ between points a and b if linear density of the wire at each point is $\rho = x^2$.
(c) A rod of length L has a uniform linear density $\rho(x) = \rho_0$. Find the mass M, center of mass x_{cm}, and rotational inertia about an axis perpendicular to the rod at $x = 0$ and $x = L/2$.
(d) A rod of length L has a linear density that varies as $\rho(x) = \rho_0 \exp(-\alpha x)$. Find the mass M, center of mass x_{cm}, and rotational inertia about an axis perpendicular to the rod

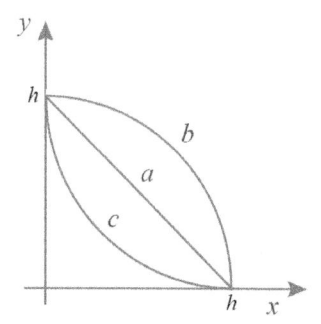

Fig. 6.21. A box sliding along various paths.

and passing through the heavier end. Show also that in the limit $\alpha \to 0$ the results are identical to the results from the problem before with $\rho(x) = \rho_0$.

(e) A wire has a semicircular shape in (x, y) plane $y > 0$ with $x^2 + y^2 = R^2$. Find mass M, center of mass x_{cm}, and rotational inertias I_x, I_y, and I_z of a wire of a uniform density ρ_0.

(f) A box is released from rest at the point (0, h) and slides down a frictionless track along one of three paths, as shown in Figure 6.21: (a) a straight line $y = h - x$, (b) a part of a circular shape $x = h \sin \varphi$, $y = h \cos \varphi$ with ($0 \leq \varphi \leq \pi/2$, and (c) another part of a circular shape $x = h(1 - \cos \varphi)$, $y = h(1 - \sin \varphi)$ with ($0 \leq \varphi \leq \pi/2$. Find the time it takes for the box to slide to the ground level.

Hint: From conservation of energy, the ball's speed v at any point on the track is given by $\sqrt{2gy}$, where g is the free-fall acceleration and y is the vertical position relative to the ground. Recall that at in any short interval the time it takes to travel a distance ds is $dt = ds/v$.

(g) Find the gravitational force exerted by a homogeneous semicircle of radius R with linear density $\rho = 1$ on a unit mass located at the origin.

(h) Find the gravitational force exerted by an infinite homogeneous straight line (with linear density $\rho = 1$) on a point of a unit mass ($m = 1$) located at a distance h from the straight line.

(5) Evaluate the double integrals over the rectangular areas:

(a)
$$\int_0^1 dx \int_0^2 2xy^2\,dy$$

(b)
$$\int_0^\pi dx \int_0^1 x\sin(xy)\,dy$$

(c)
$$\int_0^1 dx \int_0^1 \frac{x^2}{1+y^2}\,dy$$

(d)
$$\int_1^2 dy \int_3^4 \frac{1}{(x+y)^2}\,dx.$$

(6) Evaluate the double integrals and sketch the area:

(a)
$$\int_0^4 dx \int_0^{x/2} y\,dy$$

(b)
$$\int_0^1 dx \int_x^{e^x} y\,dy$$

(c)
$$\int_{y=0}^2 \int_{x=y/2}^1 (x+y)\,dx\,dy.$$

(d) Evaluate
$$S = \iint_D (x^2+y)\,dx\,dy,$$
where the area of integration D is bounded by two parabolas $y = x^2$ and $y^2 = x$.

(e) Evaluate
$$S = \iint_D \frac{x^2}{y^2} dx dy,$$
where the area of integration D is bounded by lines $x = 2$, $y = x$ and a hyperbola $y = 1/x$.

(7) Evaluate the integrals from Exercise 6 but change the order of integration.

(8) Evaluate the given double integrals in polar coordinates:

(a)
$$I = \frac{2}{\pi} \int_0^1 dx \int_0^{\sqrt{1-x^2}} \exp(x^2 + y^2) dy.$$

(b)
$$I = \int_0^1 dx \int_0^{\sqrt{1-x^2}} e^{-x^2-y^2} dy.$$

(9) Evaluate the integral
$$\int_{x=0}^{1/2} \int_{y=x}^{1-x} \left(\frac{x-y}{x+y}\right)^2 dx dy$$
by making the change of variables as
$$x = \frac{1}{2}(r-s), \quad y = \frac{1}{2}(r+s).$$

(10) A triangular lamina has vertices $(0,0)$, $(0,a)$, and $(a,0)$ and a uniform surface density σ_0. Find by integration its mass, the center of mass, and the rotational inertias I_y and I_z (about $x = 0, y = 0$).

(11) For a sphere of radius R and constant volume density ρ_0, find the moment of inertia around a diameter (you may choose the diameter along z-axis).

Answers and solutions

(1) First-kind line integrals:
   (a) $\frac{2\sqrt{2}}{3}[(b^3 - a^3)]$.
   (b) $\frac{1}{3a}[(a^2 + 4)^{3/2} - a^3]$.

(2) Second-kind line integrals:
   (a) $-\frac{14}{15}$.
   (b) (a) $\frac{5}{6}$, (b) $\frac{3}{2}$, (c) $1\frac{1}{2}$.
   (c) 2 for the all three paths.
   (d) $\frac{4}{3}ab^2$.
   (e) (a) 0, (b) -4π.

(3) Areas:
   (a) $\frac{3}{8}\pi a^2$.
   (b) $\frac{3}{2}a^2$.

(4) Applications of line integrals:
   (a) $W = \pi$.
   (b) $\frac{1}{3}[(1 + b^2)^{3/2} - (1 + a^2)^{3/2}]$.
   (c) $M = \rho_0 L$, $x_{cm} = L/2$, $I_0 = ML^2/3$, and $I_{L/2} = ML^2/12$.
   (d) $M = \rho_0(1 - e^{-\alpha L})/\alpha$, $x_{cm} = 1/\alpha - L/(e^{-\alpha L} - 1)$, and
   $I = \rho_0[e^{-\alpha L}(-\alpha L(\alpha L + 2) - 2) + 2]/\alpha^3$.
   (e) $M = \pi \rho_0 R$, $x_{cm} = 0$, $y_{cm} = 2R/\pi$, $I_x = MR^2/2$.
   (f) (a) $t = 2\sqrt{h/g}$, (b) $t = \infty$, and (c) $t = 1.85\sqrt{h/g}$.
   (g) $F_y = \frac{2G}{R}$, $F_x = 0$ where G is gravitational constant.
   (h) $F_y = -\frac{2G}{h}$, $F_x = 0$.

(5) Double integrals over rectangular areas:
   (a) 8/3.
   (b) π.
   (c) $\pi/12$.
   (d) $\ln(25/24)$.

(6) Double integrals:
   (a) 8/3.
   (b) $(3e^2 - 5)/12 \approx 1.4306$.

(c) $4/3$.
(d) $S = \int_0^1 dy \int_{y^2}^{\sqrt{y}} (x^2 + y) = 33/140$.
(e) $S = \int_1^2 dx \int_{1/x}^{x} \frac{x^2}{y^2} dy = 9/4$.

(8) Integral in polar coordinates:
   (a) $\frac{1}{\pi}(e - 1)$.
   (b) $\int_0^1 re^{-r^2} dr \int_0^{\pi/2} d\theta = \pi(1 - e^{-1})/4$.

(9) *Hint*: r and s limits are: r from 0 to 1, and s from 0 to r and the integral is $1/12$.

(10) $M = \sigma_0 a^2/2$, $x_{\text{cm}} = a/3$, $y_{\text{cm}} = a/3$, $I_y = Ma^2/5$, $I_z = Ma^2/3$.

Chapter 7

Fourier Series and Transforms

7.1. Function space and basis vectors

We routinely utilize basis or fundamental vectors to represent vectors in a vector space. For instance, in a three-dimensional space, a set of three linearly independent vectors forms what is known as a basis. Then, any vector $\vec{r}$ in that space can be expressed as a linear combination of the basis vectors. Choosing appropriate basis vectors that align with the coordinate system frequently yields practical benefits.

In the Cartesian coordinate system, we employ three mutually perpendicular unit vectors $\hat{e}_x, \hat{e}_y$, and $\hat{e}_z$ as the basis. This set of basis vectors is known as an orthonormal basis because they are both orthogonal to one another and normalized. It allows any vector to be conveniently expressed as a linear combination of these three base vectors as

$$\vec{r} = x\hat{e}_x + y\hat{e}_y + z\hat{e}_z,$$

where $\hat{e}_x$, $\hat{e}_y$, and $\hat{e}_z$ are unit vectors and x, y, and z are corresponding vector components of the vector $\vec{r}$. In various coordinate systems, there exists a wide range of basis sets for vectors. Selecting the appropriate basis within the vector space, one that reflects the symmetry of the problem, can significantly streamline the solution process.

In an n-dimensional space, you need precisely n linearly independent vectors to form a basis. However, what happens when we encounter a space with an infinite number of dimensions? In such instances, the elements of this infinite-dimensional space are *functions*, which can be either real or complex. For the purposes of this chapter, we focus on working with real functions.

7.1.1. Orthogonal set of base function

We define a scalar product[1] of two functions on an interval $[a, b]$ as

$$(f, g) = \int_a^b f(x)g(x)dx \qquad (7.1)$$

and the norm $|f|$ of a function $f(x)$ as

$$|f| = (f, f)^{1/2} = \left(\int_a^b f^2(x)dx\right)^{1/2}. \qquad (7.2)$$

If the norm $|f| < \infty$ exists,[2] then all such functions form an infinite-dimensional linear Hilbert space on the interval $[a, b]$. Two elements of the Hilbert space are called orthogonal if the scalar product of these elements is equal to zero:

$$(f, g) = \int_a^b f(x)g(x)dx = 0. \qquad (7.3)$$

Let us consider in such infinite-dimensional space a set of functions

$$f_1(x), f_2(x), f_3(x), \ldots f_n(x), \ldots. \qquad (7.4)$$

The set (7.4) is called an *orthonormal system* or orthonormal set of functions if all elements of the set are pairwise orthogonal and have a norm equal to one:

$$(f_i, f_j) = \int_a^b f_i(x)f_j(x)dx = \delta_{i,j}.$$

A distinctive feature of Hilbert space is its infinite dimensionality. It makes it difficult to check for the completeness of the orthogonal system of functions comparing to Euclidean space with finite

---

[1]Earlier in Chapter 3, we define a scalar product of two vectors as $\vec{a} \cdot \vec{b} = a_x b_x + a_y b_y + a_z b_z$.
[2]As is true for all piecewise continuous functions.

dimensions. Thus, a system of k pairwise orthogonal non-zero vectors in an n-dimensional Euclidean space is complete if $k = n$, and it is incomplete if $k < n$. However, in an infinite-dimensional space, an orthogonal system consisting of an infinite number of functions does not have to be complete. The question of completeness is considered later in this chapter.

Example 7.1. A classical example of an orthogonal set of functions on the interval $[-\pi, \pi]$ is a trigonometric set, or the Fourier set

$$1, \cos x, \sin x, \cos 2x, \sin 2x, \ldots, \cos nx, \sin nx, \ldots \qquad (7.5)$$

The orthogonality is based on the orthogonality property of the $\sin nx$ and $\cos mx$ functions for all $n, m > 0$:

$$\int_{-\pi}^{\pi} \sin(nx) \cos(mx) dx = 0,$$

$$\int_{-\pi}^{\pi} \sin(nx) \sin(mx) dx = \begin{cases} 0 & (n \neq m), \\ \pi & (n = m), \end{cases}$$

$$\int_{-\pi}^{\pi} \cos(nx) \cos(mx) dx = \begin{cases} 0 & (n \neq m), \\ \pi & (n = m). \end{cases}$$

And obviously, for $n = m = 0$, we have

$$\int_{-\pi}^{\pi} \sin(nx) \sin(mx) dx = 0, \quad \int_{-\pi}^{\pi} \cos(nx) \cos(mx) dx = 2\pi.$$

Note that the set (7.5) is orthogonal not only on the interval $[-\pi, \pi]$ but on any interval of length 2π.

This set can be easily made orthonormal by dividing each term on $1/\sqrt{\pi}$ (the first term we divide by $1/\sqrt{2\pi}$). Later, we work more closely with sets of trigonometric functions. ∎

Example 7.2. It is possible to create a set of orthogonal polynomials on $[-1, 1]$ from

$$1, x, x^2, x^3, \ldots x^n, \ldots.$$

One can see that the first two functions are orthogonal:

$$\int_{-1}^{1} 1 \cdot x dx = \left. \frac{x^2}{2} \right|_{-1}^{1} = 0,$$

and we can set $P_0(x) = 1$ and $P_1(x) = x$. However, the third function is not orthogonal to the first one. Therefore, for $P_2(x)$, we consider a linear combination of the first three functions as $P_2(x) = ax^2 + bx + c$. We can find the unknown coefficients form the orthogonality condition:

$$\int_{-1}^{1} 1 \cdot (ax^2 + bx + c)dx = 0, \quad \int_{-1}^{1} x \cdot (ax^2 + bx + c)dx = 0.$$

Evaluating the integrals and solving for the coefficients gives

$$a = -3c, b = 0, \quad P_2(x) = c(-3x^2 + 1).$$

Choosing the coefficients c so that $P_2(1) = 1$ we get $c = -1/2$ and

$$P_2(x) = \frac{3}{2}x^2 - \frac{1}{2}.$$

For finding $P_3(x)$, we combine the first three functions into $P_3(x) = ax^3 + bx^2 + cx + d$. By making it orthogonal to $P_0(x)$, $P_1(x)$, and $P_2(x)$, we find that

$$P_3(x) = \frac{5}{2}x^2 - \frac{3}{2}.$$

In the same manner, we can construct $P_4(x)$, $P_5(x)$, and so on. These polynomials, orthogonal to each other on the interval $[-1, 1]$, are well-known Legendre polynomials. Note that we can also derive the same results by applying the Gram–Schmidt orthogonalization process[3] to the set $1, x, x^2, x^3, \ldots$. This is a very tedious approach, and there are better ways to generate Legendre polynomials, for example, using the three-term recurrence relation[4] that allows us to compute Legendre polynomials for increasing degrees using the values of lower-degree Legendre polynomials:

$$(n+1)P_{n+1}(x) = (2n+1)xP_n(x) - nP_{n-1}(x).$$

To initiate the recurrence relation, we typically start with $P_0(x) = 1$ and $P_1(x) = x$. Legendre polynomials can be written in compact

---

[3]D. C. Lay, *Linear Algebra and Its Applications*, 5th edn. Pearson, 2015.
[4]M. Abramowitz and I. A. Stegun, *Handbook of Mathematical Functions*. Dover Publications, 1964.

form as

$$P_n(x) = \frac{1}{2^n n!} \frac{d^n}{dx^n}(x^2-1)^n \quad (n = 0, 1, 2, \ldots).$$

From the Legendre polynomials, we can create an orthonormal set of functions as

$$\phi_n(x) = \sqrt{\frac{2n+1}{2}} P_n(x) \quad (n = 0, 1, 2, \ldots).$$

A similar orthogonalization process can be done for any linearly independent system of functions on any interval if the integrals of the norms are finite in the considered interval. ∎

7.2. Generalized Fourier series

Generalized Fourier[5] series refers to a mathematical technique employed to represent periodic or non-periodic functions as an indefinite series of orthogonal functions. It is based on the properties of a complete set of orthogonal functions. A set is considered complete if it spans the entire space, meaning that any function in the space can be approximated arbitrarily closely by a finite linear combination of

---

[5] Jean-Baptiste Joseph Fourier (1768–1830) was a French mathematician and physicist renowned for his groundbreaking work in the fields of heat conduction and mathematical analysis. Born in Auxerre, France, Fourier's mathematical talents were evident early on, leading him to a distinguished academic career. He is best known for introducing Fourier series, a revolutionary mathematical tool that enables the representation of complex periodic functions as a sum of simpler trigonometric functions. This concept revolutionized the study of heat transfer and wave propagation, finding applications in diverse fields, such as physics, engineering, signal processing, and even art.

Fourier's profound insights extended beyond series expansions. He made significant contributions to the study of heat conduction, formulating Fourier's law, which describes how heat flows through a material. His analytical techniques allowed scientists to approach complex problems in heat distribution and diffusion with newfound clarity. Fourier's innovative thinking and ability to bridge the gap between theory and practical applications transformed the way researchers approached mathematical analysis and physical phenomena. His legacy endures through the widespread use of Fourier analysis and his lasting influence on the fields of mathematics, physics, and engineering.

functions from the set. Given a function $f(x)$ and a complete set of orthonormal basis functions $\varphi_k(x)$, the generalized Fourier series representation of $f(x)$ is written as

$$f(x) = \sum_{k=1}^{\infty} f_k \varphi_k(x), \qquad (7.6)$$

where f_k coefficients are determined as the scalar product of two functions (7.1):

$$f_k = (f, \varphi_k) = \int_a^b f(x) \varphi_k(x) dx. \qquad (7.7)$$

If a set of functions $\varphi_k(x)$ is orthogonal but not orthonormal, then to achieve this the coefficients are defined as

$$f_k = \frac{(f, \varphi_k)}{(\varphi_k, \varphi_k)} = \frac{\int_a^b f(x) \varphi_k(x) dx}{\int_a^b \varphi_k^2(x) dx}. \qquad (7.8)$$

Most common examples of basis functions $\varphi_k(x)$ are sines and cosines, orthogonal polynomials (Legendre, Chebyshev, Hermite, and Laguerre polynomials), Bessel functions, and spherical harmonics.

The objective of generalized Fourier series is to find the best approximation for $f(x)$ using a *finite* number of terms in the series. Such an approximation provides a powerful tool for interpolation and for solving ordinary and partial differential equations. For instance, in quantum mechanics, the generalized Fourier series is used to represent quantum states and wavefunctions in terms of basis functions, such as the eigenfunctions of quantum operators.

A finite sum of a generalized Fourier series

$$S_n = \sum_{k=1}^{n} f_k \varphi_k(x) \qquad (7.9)$$

is called a partial sum of the Fourier series (7.9). This partial sum has a remarkable and practical property. Let us consider a linear combination of the first n elements of the orthonormal system but with arbitrary coefficients c_k:

$$\sum_{k=1}^{n} c_k \varphi_k(x).$$

We want to find a deviation of the sum above from the function $f(x)$:

$$\left|\sum_{k=1}^{n} c_k \varphi_k(x) - f\right|^2 = \left(\sum_{k=1}^{n} c_k \varphi_k(x) - f, \sum_{k=1}^{n} c_k \varphi_k(x) - f\right)$$

$$= \sum_{k=1}^{n} c_k^2 (\varphi_k(x), \varphi_k(x)) - 2\sum_{k=1}^{n} c_k(f, \varphi_k) + (f, f)$$

$$= \sum_{k=1}^{n} c_k^2 - 2\sum_{k=1}^{n} c_k f_k + |f|^2$$

$$= \sum_{k=1}^{n} (c_k - f_k)^2 - \sum_{k=1}^{n} f_k^2 + |f|^2.$$

As one can see, the smallest deviation of the partial sum takes place when $c_k = f_k$. Thus, the generalized Fourier series provides the best approximation of a function $f(x)$ by a series. And for such series, we can write Bessel identity

$$\left|\sum_{k=1}^{n} f_k \varphi_k(x) - f\right|^2 = |f|^2 - \sum_{k=1}^{n} f_k^2.$$

From Bessel identity follows Bessel inequality

$$\sum_{k=1}^{n} f_k^2 < |f|^2. \tag{7.10}$$

Both Bessel identity and inequality play an important role for analyzing Fourier series.

We are interested in conditions when the Bessel inequality transforms into an exact equality as $n \to \infty$. It is true for *complete* orthonormal sets of functions. An orthonormal set $\varphi_k(x)$ is called a complete set on an interval $[a, b]$ if for any function $f(x)$ and any positive ϵ there exists such an integer $N(x, \epsilon)$ that for $n > N$ the norm

$$\left|\sum_{k=1}^{N} f_k \varphi_k(x) - f\right|^2 < \epsilon.$$

Thus, if an orthogonal set is complete, then for every function $f(x)$, the Bessel inequality reduces to the exact identity

$$\sum_{k=1}^{\infty} f_k^2 = |f|^2, \tag{7.11}$$

also called *Parseval's identity*.

Another property of a complete set of functions is that there exists no non-zero function that is orthogonal to every function in the set, thus a set of functions is *closed*. Practically, a set of orthonormal functions in Hilbert space is complete if and only if it is closed. The completeness of an orthonormal set of functions in Hilbert space is well discussed elsewhere.[6]

Weierstrass's theorem proves that if $f(x)$ is continuous on the closed interval $[a, b]$, then there exists a sequence of polynomials $P_n(x)$ such that

$$\lim_{n \to \infty} P_n(x) = f(x)$$

uniformly on $[a, b]$.[7] Weierstrass's theorem together with the consideration above lead to a very important special case of complete and closed functions, namely Legendre polynomials, that are orthogonal on the interval $[-1, 1]$. Thus, we can represent a function $f(x)$ that is continuous on the interval $[-1, 1]$ as

$$f(x) = \sum_{k=1}^{\infty} a_k P_k(x),$$

where $P_k(x)$ are Legendre polynomials and a_k are the Fourier coefficients (7.8).

---

[6]See, for example, F. W. Byron and R. W. Fuller, *Mathematics of Classical and Quantum Physics*, 1992, p. 217.

[7]Note that the Weierstrass theorem for approximation of a function is generally stronger than Taylor's theorem for expansion in power series. More details can be found in mathematical literature.

7.3. Trigonometric Fourier series

The most important complete set of functions is the set of the trigonometric functions on the interval $[-\pi, \pi]$:

$$1, \cos x, \sin x, \cos 2x, \sin 2x, \ldots, \cos nx, \sin nx, \ldots. \qquad (7.12)$$

The Fourier series for a function $f(x)$ on the interval $[-\pi, \pi]$ is defined as

$$f(x) = \frac{a_0}{2} + \sum_{k=1}^{\infty}(a_k \cos kx + b_k \sin kx), \qquad (7.13)$$

where the Fourier coefficients a_k and b_k are computed, according to (7.9), as

$$a_k = \frac{1}{\pi}\int_{-\pi}^{\pi} f(x) \cos kx\, dx \quad (k=0,1,2,\ldots), \qquad (7.14)$$

$$b_k = \frac{1}{\pi}\int_{-\pi}^{\pi} f(x) \sin kx\, dx \quad (k=1,2,\ldots). \qquad (7.15)$$

The process of expanding a given function $f(x)$ into a Fourier series is commonly known as Fourier analysis, also referred to as harmonic analysis.

The Bessel inequality (7.10) takes the form

$$\frac{a_0^2}{2} + \sum_{k=1}^{\infty}(a_k^2 + b_k^2) \leq \frac{1}{\pi}\int_{-\pi}^{\pi} f^2(x)\, dx. \qquad (7.16)$$

Note that the Fourier coefficients above were first derived by Euler.

The above formulas are valid for integration over any interval of length 2π. In general, if a function $f(x)$, defined for all real values of x, has a period T, i.e., $f(x+T) = f(x)$, then the integral of $f(x)$ over any interval of length T has a definite value, independent of the initial point of the interval, i.e., the value of integral

$$\int_c^{c+T} f(x)\, dx \qquad (7.17)$$

is independent of c.

Example 7.3. Show that (7.17) is correct.
Indeed, let us represent c as $c = kT + h$, where k is some integer number and $0 \leq h \leq T$. Then,

$$\int_c^{c+T} f(x)dx = \int_{kT+h}^{(k+1)T+h} f(x)dx$$

$$= \int_{kT+h}^{(k+1)T} f(x)dx + \int_{(k+1)T}^{(k+1)T+h} f(x)dx.$$

We introduce new variables: $u_1 = x - kT$ in the first integral and $u_2 = x - (k+1)T$ in the second integral. Thus, we obtain

$$\int_c^{c+T} f(x)dx = \int_h^T f(u_1 + kT)du_1 + \int_0^h f(u_2 + (k+1)T)du_2.$$

Since the function $f(x)$ is periodic with T, then

$$\int_c^{c+T} f(x)dx = \int_h^T f(x)dx + \int_0^h f(x)dx = \int_0^T f(x)dx.$$

∎

7.3.1. Fourier coefficients by direct integration

Earlier, we defined the Fourier coefficients a_k and b_k (7.14) and (7.15) for the trigonometric Fourier series

$$f(x) = \frac{a_0}{2} + \sum_{k=1}^{\infty}(a_k \cos kx + b_k \sin kx) \tag{7.18}$$

by using the generalized Fourier series. For functions satisfying the Dirichlet conditions (see the following), we can calculate the Fourier coefficients by direct integration. Integrating both parts of (7.21) over the interval $[-\pi, \pi]$ and replacing the integral from the sum by the sum of integrals, we get

$$\int_{-\pi}^{\pi} f(x)dx = \int_{-\pi}^{\pi} \frac{a_0}{2}dx + \sum_{k=1}^{\infty}\left(a_k \int_{-\pi}^{\pi} \cos(kx)dx + b_k \int_{-\pi}^{\pi} \sin(kx)dx\right).$$

Using
$$\int_{-\pi}^{\pi} \cos(kx)dx = 0, \quad \int_{-\pi}^{\pi} \sin(kx)dx = 0 \quad (k=0,1,2,\ldots),$$
we obtain
$$\int_{-\pi}^{\pi} f(x)dx = \frac{a_0}{2} 2\pi = a_0 \pi,$$
and then the coefficient a_0 is
$$a_0 = \frac{1}{\pi} \int_{-\pi}^{\pi} f(x)dx.$$

Now, we can find the other coefficients. Multiplying both parts of (7.21) by $\cos nx$ and integrating as before, we get
$$\int_{-\pi}^{\pi} f(x)\cos(nx)dx = \frac{a_0}{2} \int_{-\pi}^{\pi} \cos(nx)dx$$
$$+ \sum_{k=1}^{\infty} \left(a_k \int_{-\pi}^{\pi} \cos(kx)\cos(nx)dx \right.$$
$$\left. + b_k \int_{-\pi}^{\pi} \sin(kx)\cos(nx)dx \right).$$

According to the properties of orthogonality of the trigonometric functions (see Example 7.1), all integrals will be equal to zero except the one
$$\int_{-\pi}^{\pi} \cos(kx)\cos(nx)dx = \pi \quad \text{when } k=n.$$

Then,
$$\int_{-\pi}^{\pi} f(x)\cos(nx)dx = a_n \pi$$
or
$$a_n = \frac{1}{\pi} \int_{-\pi}^{\pi} f(x)\cos(nx)dx \quad (n=0,1,2,\ldots). \tag{7.19}$$

In the same way, by multiplying both parts of (7.21) by $\sin nx$ and integrating, we can get

$$b_n = \frac{1}{\pi} \int_{-\pi}^{\pi} f(x) \sin(nx) dx \quad (n = 1, 2, \ldots). \tag{7.20}$$

Thus, we obtain the same Fourier coefficients as defined from the generalized Fourier series.

If the given function $f(x)$ is either even or odd, then the Fourier series can be Recall that if $f(x)$ is an even function in the interval $(-a, a)$, i.e., $f(-x) = f(x)$, then

$$\int_{-a}^{a} f(x) dx = 2 \int_{0}^{a} f(x) dx,$$

and if $f(x)$ is an odd function, i.e., $f(-x) = -f(x)$, then

$$\int_{-a}^{a} f(x) dx = 0.$$

Keeping in mind that $\cos(kx)$ are even functions and $\sin(kx)$ are odd functions, we can easily find that for functions $f(x)$ even on $[-\pi, \pi]$ we have all $b_k = 0$, and for odd functions we have all $a_k = 0$. When calculating the Fourier coefficients in such cases, we change the limits of integration from $[-\pi, \pi]$ to $[0, \pi]$ and multiply all the non-zero coefficients by 2.

7.3.2. Dirichlet's conditions

The above calculations of the Fourier coefficients a_k and b_k provide a helpful guide, but they should be acknowledged as lacking rigor. Several assumptions were made during the process. First, we assumed that the given function $f(x)$ could be expanded into series trigonometric Fourier series. Additionally, we replaced the integral of an infinite sum with the sum of integrals of individual terms, known as integrating the series term by term. However, it is important to note that this approach is not always applicable and may introduce errors.[8] A more rigorous approach to the problem will take the

---

[8]If the terms of the series $u_1(x) + u_2(x) + \cdots + u_n(x) + \cdots$ are continuous in the interval (a, b), and the series converges uniformly, then it can be integrated term by term between arbitrary limits α, β lying in the interval (a, b), namely

$$\int_{\alpha}^{\beta} \sum_{n=1}^{\infty} u_n(x) dx = \sum_{n=1}^{\infty} \int_{\alpha}^{\beta} u_n(x) dx.$$

following form: Consider the function $f(x)$ defined in a given interval $-\pi, \pi$. Then, calculate the coefficients a_k and b_k by (7.19) and (7.20) and substitute it into the series

$$f(x) = \frac{a_0}{2} + \sum_{k=1}^{\infty}(a_k \cos kx + b_k \sin kx). \qquad (7.21)$$

The main question at hand is to determine whether the series obtained in this manner converges within the specified interval. Furthermore, if the series does converge, we seek to ascertain whether its sum is equal to $f(x)$ for all values of x in the given interval. Positive answers to both questions impose restrictive assumptions about the function $f(x)$.

First, we assume that the function $f(x)$ is piecewise continuous in the interval $[-\pi, \pi]$ and has a piecewise continuous derivatives, i.e., the function $f(x)$ has a finite number of points of discontinuities in the interval. We further assume that all these discontinuities have the following property: If $x = a$ is a point of discontinuity, then there exist finite limits for $f(x)$ from the right $f(a+0)$ and from the left $f(a-0)$. Such points of discontinuity are known as discontinuities of the first kind.

Second, we assume that the total interval $[-\pi, \pi]$ can be divided into a finite number of sub-intervals such that $f(x)$ varies monotonically in each. The above conditions are known as *Dirichlet conditions*.

The following Dirichlet theorem is a fundamental one in the theory of Fourier series: If a function $f(x)$ satisfies the Dirichlet conditions in the interval $[-\pi, \pi]$, then its Fourier series

(a) converges uniformly to $f(x)$ in every closed subinterval of $[-\pi, \pi]$ in which $f(x)$ is continuous,
(b) is equal to

$$\frac{f(x+0) + f(x-0)}{2}$$

at points of discontinuity x of $f(x)$, i.e., its Fourier series converges to the arithmetic mean of the left- and right-hand limits of the function, and
(c) is equal to

$$\frac{f(-\pi+0) + f(\pi-0)}{2}$$

at the ends of the interval. The last part (part c)) needs some clarification. The terms of the Fourier series (7.5) are periodic functions with the period 2π:

$$\cos n(x + 2\pi) = \cos(nx + 2\pi n) = \cos nx,$$
$$\sin n(x + 2\pi) = \sin(nx + 2\pi n) = \sin nx.$$

Therefore, if the series converges in the interval $[-\pi, \pi]$, then it converges for all values of x having the same values, with the period 2π, as in the interval $[-\pi, \pi]$. Thus, if we use the Fourier series outside the interval $[-\pi, \pi]$, then we must assume that the function $f(x)$ is a periodic one with a period 2π. However, a function $f(x)$ can be periodic, but at the ends of the interval $x = \pm\pi$, we may have $f(-\pi) \neq f(\pi)$. Therefore, for the end points, we are interested in the limits $f(-\pi + 0)$ and $f(\pi - 0)$ that can be different, but the Fourier series must be the same for both $x = -\pi$ and $x = \pi$ since the trigonometric functions are periodic.

Parseval's identity for the Fourier series of a function satisfying Dirichlet's condition takes the form

$$\frac{a_0^2}{2} + \sum_{k=1}^{\infty}(a_k^2 + b_k^2) = \frac{1}{\pi}\int_{-\pi}^{\pi} f^2(x)dx. \qquad (7.22)$$

For a curious reader: The Dirichlet theorem is based on the analysis of a partial sum of the Fourier series at some point x_0, namely

$$S_n(x_0) = \frac{a_0}{2} + \sum_{k=1}^{n}(a_k \cos kx_0 + b_k \sin kx_0).$$

Substituting a_k and b_k from (7.14) and (7.15) in the above equation, and keeping in mind that $\cos kx_0$ and $\sin kx_0$ are constant numbers, we can write

$$S_n(x_0) = \frac{1}{2}\int_{-\pi}^{\pi} f(x)dx$$
$$+ \sum_{k=1}^{n}\frac{1}{\pi}\int_{-\pi}^{\pi} f(x)[\cos kx \cos kx_0 + \sin kx \sin kx_0]dx.$$

After some steps involving using trigonometric identities and changing variables, we can write (it is a good exercise to try and verify this) the partial sum as Dirichlet's integral:

$$S_n(x_0) = \frac{1}{\pi} \int_0^\pi [f(x_0+u) + f(x_0-u)] \frac{\sin\left(n+\frac{1}{2}\right)u}{2\sin\frac{u}{2}} du. \quad (7.23)$$

7.3.3. Case of non-periodic function

Here is an important remark regarding the Dirichlet theorem. The terms of series (7.21) are periodic functions with a period of 2π. Thus, if we are using the Fourier series outside the interval $(-\pi, \pi)$, we must assume that the function $f(x)$ is periodically continued outside this interval with a period of 2π. This periodic continuation ensures that the Fourier series representation remains valid and accurately approximates the function over the extended domain. From this point of view, the ends of the interval $x = \pm\pi$ will be points of discontinuity for the function if $f(-\pi+0) \neq f(\pi-0)$. Figure 7.1 illustrates a function $f(x) = x$ that is continuous within the interval $-\pi, \pi$. However, when extended periodically, it results in discontinuities due to the mismatch of values at the ends of the interval.

Example 7.4. We expand $f(x) = x$ in Fourier series within the interval $-\pi \leq x \leq \pi$. All products $x\cos(kx)$ are odd functions of x, therefore all a_k coefficients are equal to zero. Then, for b_k coefficients, we have

$$b_k = \frac{2}{\pi} \int_0^\pi x\sin(kx)dx$$

$$= \frac{2}{\pi}\left(-\frac{x\cos kx}{k}\bigg|_{x=0}^{x=\pi} + \frac{1}{k}\int_0^\pi \cos(kx)dx\right) = \frac{2(-1)^{k-1}}{k}.$$

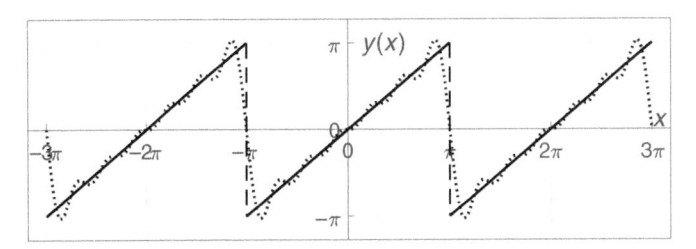

Fig. 7.1. Extension of $f(x) = x$ as a periodic function with a period of 2π (solid line), the Fourier series with six terms (dotted line).

The resulting Fourier series is presented in Figure 7.1. As we see, at points $x = \pm\pi$, we have the first kind of discontinuity. Thus, Dirichlet's theorem gives in this case for the Fourier series

$$2\left(\frac{\sin x}{1} - \frac{\sin 2x}{2} + \cdots + \frac{(-1)^{k-1}\sin kx}{k} + \cdots\right)$$
$$= \begin{cases} x & \text{for } -\pi < x < \pi \\ 0 & \text{for } x = \pm\pi \end{cases}.$$

■

Example 7.5. In this example, we want to find the Fourier series of $f(x) = x^2$ for $-\pi \le x \le \pi$. In this case, all $x^2 \sin kx$ are odd functions and all the b_k coefficients are equal to zero. For a_k coefficients, we have

$$a_0 = \frac{2}{\pi}\int_0^\pi x^2 dx = \frac{2\pi^2}{3},$$

$$a_k = \frac{2}{\pi}\int_0^\pi x^2 \cos(kx)dx = \frac{2}{\pi}\left(\left.\frac{x^2 \sin kx}{k}\right|_{x=0}^{x=\pi} - \frac{2}{k}\int_0^\pi x\sin(kx)dx\right)$$

$$= \frac{4}{\pi k}\left(\left.\frac{x\cos kx}{k}\right|_{x=0}^{x=\pi} - \frac{1}{k}\int_0^\pi \cos(kx)dx\right) = (-1)^k \frac{4}{k^2}.$$

From Figure 7.2, we can see that in this case the Fourier series does not have discontinuities, and we have

$$x^2 = \frac{\pi^2}{3} + 4\sum_{k=1}^\infty (-1)^k \frac{\cos kx}{k^2} \quad (-\pi \le x \le \pi).$$

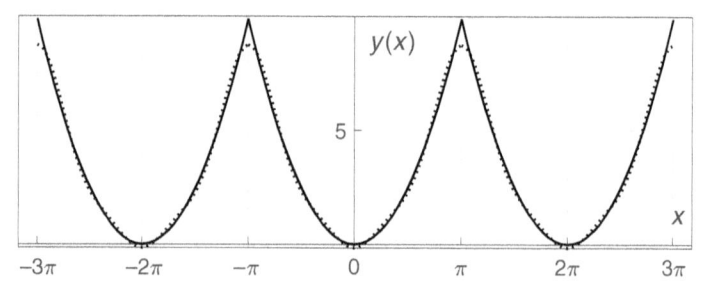

Fig. 7.2. Extension of $f(x) = x^2$ as a periodic function with a period of 2π (solid line), the Fourier series with just three terms (dotted line).

For $x = 0$, the above expansion gives

$$\frac{\pi^2}{12} = 1 - \frac{1}{4} + \frac{1}{9} - \frac{1}{16} + \cdots + (-1)^{k-1}\frac{1}{k^2} + \cdots.$$

If we introduce

$$s_1 = 1 + \frac{1}{4} + \frac{1}{9} + \frac{1}{16} + \cdots$$

and

$$s_2 = 1 + \frac{1}{9} + \frac{1}{25} + \frac{1}{49} + \cdots,$$

then we have

$$s_1 = s_2 + \frac{1}{4} + \frac{1}{16} + \frac{1}{36} + \cdots = s_2 + \frac{1}{4}s_1, \quad s_2 = \frac{3}{4}s_1.$$

Then, from the Fourier expansion for x^2 at $x = 0$, we have

$$1 - \frac{1}{4} + \frac{1}{9} - \frac{1}{16} + \cdots = s_2 - \frac{1}{4}s_1 = \frac{1}{2}s_1 = \frac{\pi^2}{12},$$

or, we have found the exact sum of the $1/n^2$ terms, known as the Basel problem or Euler's Basel problem:

$$s_1 = 1 + \frac{1}{4} + \frac{1}{9} + \frac{1}{16} + \cdots + \frac{1}{n^2} + \cdots = \sum_{n=1}^{\infty} \frac{1}{n^2} = \frac{\pi^2}{6}. \blacksquare$$

Example 7.6. It is instructive to consider a function given by a couple of formulas, for example,

$$f(x) = \begin{cases} c_1, & -\pi < x < 0 \\ c_2, & 0 < x < \pi \end{cases}.$$

Then,

$$a_0 = \frac{1}{\pi}\int_{-\pi}^{\pi} f(f)dx = \frac{1}{\pi}\left[\int_{-\pi}^{0} c_1 dx + \int_{0}^{\pi} c_2 dx\right] = c_1 + c_2,$$

$$a_k = \frac{1}{\pi}\int_{-\pi}^{\pi} f(f)\cos(kx)dx$$

$$= \left[\int_{-\pi}^{0} c_1 \cos(kx)dx + \int_{0}^{\pi} c_1 \cos(kx)dx\right] = 0,$$

$$b_k = \frac{1}{\pi}\int_{-\pi}^{\pi} f(f)\sin(kx)dx = \left[\int_{-\pi}^{0} c_1 \sin(kx)dx + \int_{0}^{\pi} c_2 \sin(kx)dx\right]$$

$$= (c_1 - c_2)\frac{(-1)^k - 1}{k\pi}.$$

Then, for the Fourier series, we have

$$f(x) = \frac{1}{2} + \frac{2}{\pi}\left(\frac{\sin x}{1} + \frac{\sin 3x}{3} + \frac{\sin 5x}{5} + \cdots\right)$$

$$= \begin{cases} c_1 & -\pi < x < 0, \\ c_2 & x < x < \pi, \\ (c_1 + c_2)/2 & x = 0, \pm\pi. \end{cases}$$

Thus, a function defined by several formulas can be represented as a single series. Figure 7.3 shows the step function extended -3π to 3π interval together with the Fourier series with three terms. The trigonometric series turns out to be a universal tool for "gluing" functions together by erasing the line between functions that allow a single analytic representation in the entire domain of definition and functions defined using several analytic expressions. Fourier's discovery of this fact played a large role in the history of mathematics, since it led to a significant expansion of the concept of functions. ■

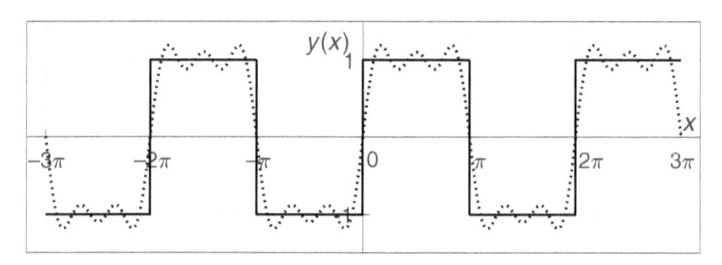

Fig. 7.3. A step function (solid line) and its Fourier series representation with three terms (dotted line).

7.3.4. Fourier series in $[0, \pi]$ interval

We can easily check that the system of cosine functions

$$1, \cos x, \cos 2x, \ldots, \cos nx, \ldots$$

is orthogonal on the interval $0 \leq x \leq \pi$. Then, for this interval, we can write Fourier series as cosine Fourier series:

$$f(x) = \frac{a_0}{2} + \sum_{k=1}^{\infty} a_k \cos(kx), \quad a_k = \frac{2}{\pi} \int_0^{\pi} f(x) \cos(kx) dx. \quad (7.24)$$

We can also note that the system of sine functions

$$\sin x, \sin 2x, \ldots, \sin nx, \ldots$$

is a set of orthogonal functions on the same $0 \leq x \leq \pi$ interval. Then, the sine Fourier series takes the form

$$f(x) = \sum_{k=1}^{\infty} b_k \sin(kx), \quad b_k = \frac{2}{\pi} \int_0^{\pi} f(x) \sin(kx) dx. \quad (7.25)$$

Such series can be applied to an arbitrary function $f(x)$ on $[0, \pi]$ interval. Such functions can be expanded in $[0, \pi]$ either in a cosine or sine series. Both series have a sum inside the interval equal to $f(x)$ or to the arithmetic mean at points of discontinuity. Outside $[0, \pi]$, however, they represent quite different functions: The cosine series gives a function derived from $f(x)$ by even continuation in the neighboring interval $[-\pi, 0]$, followed by periodic continuation with period 2π outside the interval $[-\pi, \pi]$. The sine series gives the function obtained by odd continuation to the $[-\pi, 0]$, followed by periodic continuation with period 2π outside $[-\pi, \pi]$. Thus, in the cosine series,

$$f(-0) = f(+0), \quad f(-\pi + 0) = f(\pi - 0),$$

while for the sine series,

$$f(-0) = -f(+0), \quad f(-\pi + 0) = -f(\pi - 0).$$

We illustrate this in the following example for $f(x) = x$ and $f(x) = x^2$.

Example 7.7. Expand (a) in a Fourier cosine series, the function $f(x) = x$, and (b) in a Fourier sine series, the function $f(x) = x^2$ on $[0, \pi]$ interval.

(a) For $f(x) = x$, we have

$$x = \frac{a_0}{2} + \sum_{k=1}^{\infty} a_k \cos(kx),$$

where

$$a_0 = \frac{2}{\pi} \int_0^\pi x\, dx = \pi,$$

$$a_k = \frac{2}{\pi} \int_0^\pi x \cos(kx)\, dx = \frac{2}{k^2 \pi}[(-1)^k - 1],$$

then

$$x = \frac{\pi}{2} - \frac{4}{\pi}\left(\frac{\cos x}{1^2} + \frac{\cos 3x}{3^2} + \cdots + \frac{\cos(2k+1)x}{(2k+1)^2} + \cdots\right) \quad 0 < x < \pi.$$

For the $[-\pi, 0]$ interval, the series is a symmetrical continuation so that for the whole $[-\pi, \pi]$ interval we have

$$|x| = \frac{\pi}{2} - \frac{4}{\pi}\left(\frac{\cos x}{1^2} + \frac{\cos 3x}{3^2} + \frac{\cos 5x}{5^2} + \cdots\right) \quad -\pi < x < \pi.$$

Outside this interval it gives the function obtained by periodic continuation of $|x|$ from $[-\pi, \pi]$ (see Figure 7.4).

(b) The Fourier sine series coefficients for $f(x) = x^2$ on $[0, \pi]$ take the form

$$b_k = \frac{2}{\pi} \int_0^\pi x^2 \sin(kx)\, dx = \frac{2(-1)^{k-1}\pi}{k} + \frac{4[(-1)^k - 1]}{\pi k^2},$$

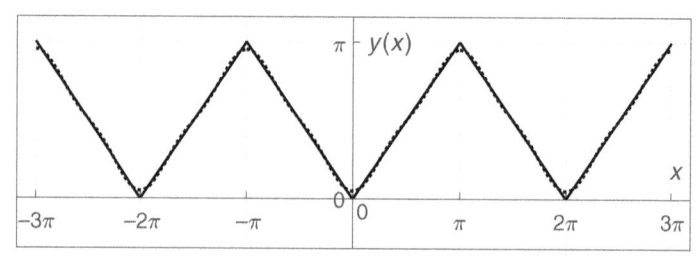

Fig. 7.4. Expansion of $f(x) = x$ on $(0, \pi)$ in $\cos(kx)$ series. The series expansion (dotted line) has been calculated with just four terms.

with

$$x^2 = 2\pi \left[\frac{\sin x}{1} - \frac{\sin 2x}{2} + \frac{\sin 3x}{3} - \cdots \right]$$
$$- \frac{8}{\pi} \left[\frac{\sin x}{1^2} + \frac{\sin 3x}{3^2} + \frac{\sin 5x}{5^2} + \cdots \right],$$

for $0 < x < \pi$, as shown in Figure 7.5. ∎

7.3.5. *Fourier series in $[-L, L]$ interval*

So far we have considered functions defined in the interval $[-\pi, \pi]$. If these functions are periodic of period 2π, then the Fourier expansions are as good everywhere as within the interval $[-\pi, \pi]$. It is rather natural to extend the Fourier series expansions to the interval $[-L, L]$ and functions of period $2L$. In this case, we replace the variable x on $\pi x/L$. Note that the condition of periodicity cannot be removed if the series is to be convergent outside the basic interval. Thus, for the Fourier series on the $[-L, L]$ interval, we have

$$f(x) = \frac{a_0}{2} + \sum_{k=1}^{\infty} \left(a_k \cos \frac{k\pi}{L} x + b_k \sin \frac{k\pi}{L} x \right), \qquad (7.26)$$

with

$$a_k = \frac{1}{L} \int_{-L}^{L} f(x) \cos \frac{k\pi}{L} x \, dx \quad (k = 0, 1, 2, \ldots), \qquad (7.27)$$

$$b_k = \frac{1}{L} \int_{-L}^{L} f(x) \sin \frac{k\pi}{L} x \, dx \quad (k = 1, 2, \ldots). \qquad (7.28)$$

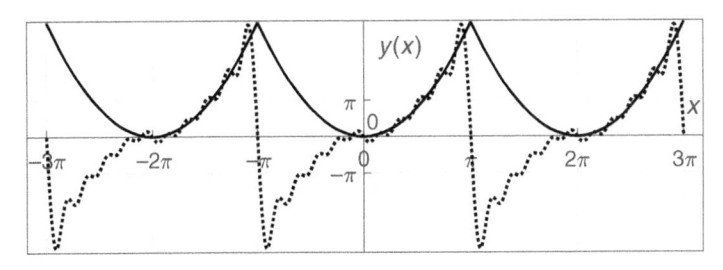

Fig. 7.5. Expansion of $f(x) = x^2$ on $(0, \pi)$ in $\sin(kx)$ series. The series expansion (dotted line) has been calculated with 10 terms.

The same can be done for the cosine and sine series for the $[0, L]$ interval. Then, for the cosine Fourier series, we have

$$\frac{a_0}{2} + \sum_{k=1}^{\infty} a_k \cos \frac{k\pi}{L} x, \quad a_k = \frac{2}{L} \int_0^L f(x) \cos \frac{k\pi}{L} x \, dx,$$

and for the sine series,

$$\sum_{k=1}^{\infty} b_k \sin \frac{k\pi}{L} x, \quad b_k = \frac{2}{L} \int_0^L f(x) \sin \frac{k\pi}{L} x \, dx.$$

Note that the interval $(-L, L)$ can be substituted with any interval $(c, c + 2L)$ of length $2L$, as we have previously indicated for an interval of length 2π. In this scenario, the sum of the series (7.26) yields $f(x)$ in the interval $(c, c+2L)$, and while computing the coefficients using formulas (7.27) and (7.28), the integration interval $(-L, L)$ should be replaced with $(c, c + 2L)$.

Example 7.8. Let's consider a piecewise-defined function:

$$f(x) = \begin{cases} -1 & \text{for } -L \leq x < 0, \\ 1 & \text{for } 0 \leq x < L. \end{cases}$$

The period of this function is $2L$, and we want to find the Fourier series expansion in terms of sines and cosines. By calculating the coefficients with (7.27) and (7.28), we obtain that all $a_k = 0$ (since the function $f(x)$ is odd), thus we have only sine terms in the expansion

$$f(x) = \frac{4}{\pi} \left[\sin(\pi x) + \frac{1}{3} \sin(3\pi x) + \frac{1}{5} \sin(5\pi x) + \frac{1}{7} \sin(7\pi x) \dots \right].$$

This example demonstrates how to find the Fourier series expansion for a given piecewise function on the interval $[-L, L]$. ∎

7.3.6. Complex forms of Fourier

The Fourier expansion can be written in complex form by expressing both $\sin kx$ and $\cos kx$ using complex exponentials:

$$\sin kx = \frac{e^{ikx} - e^{-ikx}}{2i}, \quad \cos kx = \frac{e^{ikx} + e^{-ikx}}{2}.$$

Substituting the above equation into the Fourier series (7.13) gives

$$f(x) = \sum_{k=-\infty}^{\infty} c_k e^{ikx} \tag{7.29}$$

with

$$c_k = \frac{1}{2\pi} \int_{-\pi}^{\pi} f(x) e^{-ikx} dx. \tag{7.30}$$

There are easier ways to derive the complex form of the Fourier series, i.e., by finding the coefficients c_k directly for the complex form.

Note that the coefficients c_k can be expressed in a_k and b_k coefficients as

$$c_k = \begin{cases} (a_k - ib_k)/2 & (k > 0), \\ (a_k + ib_k)/2 & (k < 0), \\ a_0/2 & (k = 0). \end{cases}$$

For real functions, the Fourier sum must be real, which brings the following properties for the coefficients: The coefficient c_0 is real, $c_{-k} = c_k^*$. Furthermore, for even functions, all c_k are real, and for odd functions, $c_0 = 0$ and all c_k are imaginary.

For $-L < x < L$ interval, the complex form of the Fourier series is

$$f(x) = \sum_{k=-\infty}^{\infty} c_k e^{ik\pi x/L} \quad (-L \leq x \leq L), \tag{7.31}$$

with

$$c_k = \frac{1}{2L} \int_{-L}^{L} f(x) e^{-ik\pi x/L} dx. \tag{7.32}$$

7.3.7. Applications of Fourier series

Fourier series has numerous applications in almost every field of science and engineering. Here are some examples: series solutions for ordinary and partial differential equations (especially with periodic boundary conditions). In Chapter 9, Section 9.6, we apply Fourier series expansion to solving second-order ordinary differential equations. Fourier series are helpful in representing wavefunctions in quantum mechanics as superpositions of sinusoidal functions. In heat transfer analysis, Fourier's law of heat conduction is based on the principles of Fourier series. Here are more examples: audio and visual signal processing, MRI and CT scans in medical imaging, data and image compression, circuit analysis in electrical engineering, and vibration analysis in mechanical engineering.

7.4. Fourier transform

The Fourier series provide an invaluable way to expand a function within an interval of its periodicity $[-L, L]$. However, many functions are not periodic. By expanding the interval to $[-\infty, \infty]$, i.e., considering $L \to \infty$, we can expand the Fourier series expansion to non-periodic functions.

Let a function $f(x)$ satisfy Dirichlet's condition, and also there exists a finite integral

$$\int_{-\infty}^{\infty} |f(x)| dx = C.$$

For functions satisfying Dirichlet's condition within $[-L, L]$, we have

$$f(x) = \frac{a_0}{2} + \sum_{k=1}^{\infty} \left(a_k \cos \frac{k\pi x}{L} + b_k \sin \frac{k\pi x}{L} \right),$$

where

$$a_k = \frac{1}{L} \int_{-L}^{L} f(x) \cos \frac{k\pi x}{L} dx, \quad b_k = \frac{1}{L} \int_{-L}^{L} f(x) \sin \frac{k\pi x}{L} dx.$$

Combining all together, we can write the Fourier series as

$$f(x) = \frac{1}{2L} \int_{-L}^{L} f(t) dt + \frac{1}{L} \sum_{k=1}^{\infty} \int_{-L}^{L} f(t) \cos \frac{k\pi(t-x)}{L} dt.$$

For $L \to \infty$, the first term in equation above tends to zero since

$$\left| \frac{1}{2L} \int_{-L}^{L} f(t) dt \right| \leq \frac{1}{2L} \int_{-L}^{L} |f(t)| dt = \frac{C}{2L} \to 0.$$

Introducing a new variable ω that takes positive values

$$\omega_1 = \frac{\pi}{L}, \; \omega_2 = \frac{2\pi}{L}, \ldots, \omega_k = \frac{k\pi}{L}, \ldots$$

with $\delta\omega = \pi/L$, we can rewrite the sum above as

$$\frac{1}{\pi} \sum_{(\omega)} \Delta\omega \int_{-L}^{L} f(t) \cos\omega(t-x) dt.$$

For $L \to \infty$, the above formula can be rewritten as

$$\frac{1}{\pi} \int_0^\infty d\omega \int_{-\infty}^{\infty} f(t) \cos\omega(t-x) dt.$$

Then, we arrive at the Fourier formula, also called Fourier theorem, namely

$$f(x) = \frac{1}{\pi} \int_0^\infty d\omega \int_{-\infty}^{\infty} f(t) \cos\omega(t-x) dt. \tag{7.33}$$

At the discontinuity points (if any), we replace $f(x)$ with

$$\frac{f(x+0) + f(x-0)}{2}.$$

The Fourier formula can we written as

$$f(x) = \frac{1}{\pi} \int_0^\infty d\omega \left(\int_{-\infty}^{\infty} f(t) \cos(\omega t) dt + \int_{-\infty}^{\infty} f(t) \sin(\omega t) dt \right). \tag{7.34}$$

Introducing two new functions

$$g_c(\omega) = \frac{1}{\pi} \int_{-\infty}^{\infty} f(t) \cos(\omega t) dt, \quad g_s(\omega) = \frac{1}{\pi} \int_{-\infty}^{\infty} f(t) \sin(\omega t) dt,$$

we can write (7.33) as

$$f(x) = \int_0^\infty [g_c(\omega) \cos(\omega x) + g_s(\omega) \sin(\omega x)] d\omega.$$

This formula gives us an expansion of $f(x)$ in the infinite interval $(-\infty, \infty)$ in terms of harmonic oscillations with frequencies varying continuously from 0 to ∞. The functions $g_c(\omega)$ and $g_s(\omega)$ provide the amplitude distribution as functions of the frequency. Note that for even functions $f(x)$, we have $g_s(\omega) = 0$, and for odd functions, $g_c(\omega) = 0$. For a finite interval $(-L, L)$, we have frequencies $\omega_k = k\pi/L$ that form an arithmetic series.

The Fourier formula (7.34) is often written in the complex form as

$$f(x) = \frac{1}{2\pi} \int_{-\infty}^{\infty} d\omega \int_{-\infty}^{\infty} f(t) e^{i\omega(t-x)} dt. \tag{7.35}$$

It is easy to show the equivalence of (7.33) and (7.35) using

$$e^{i\omega(t-x)} = \cos \omega(t-x) + i \sin \omega(t-x).$$

It is more common to write the complex form of Fourier formula as

$$f(x) = \frac{1}{2\pi} \int_{-\infty}^{\infty} e^{-i\omega x} d\omega \int_{-\infty}^{\infty} f(t) e^{i\omega t} dt. \tag{7.36}$$

Introducing the Fourier transform as

$$F(\omega) = \frac{1}{\sqrt{2\pi}} \int_{-\infty}^{\infty} f(t) e^{i\omega t} dt, \tag{7.37}$$

we can write for the function $f(x)$

$$f(x) = \frac{1}{\sqrt{2\pi}} \int_{-\infty}^{\infty} F(\omega) e^{-i\omega x} d\omega. \tag{7.38}$$

The two functions $f(x)$ and $F(\omega)$ are called a Fourier transform pair. From the representation (7.38), we can understand the meaning of $F(\omega)$. In this representation, for a small interval of frequencies from ω to $\omega + d\omega$, we have $[F(\omega)d\omega]e^{-i\omega x}$, where $F(\omega)d\omega$ corresponds to an amplitude of oscillations with frequency ω. This means that $F(\omega)$ can be regarded as the "amplitude density" corresponding to a small frequency interval and calculated per unit length of this interval. Therefore, the function $F(\omega)$ is also called the spectral density of the function $f(x)$.

For even functions, the Fourier transform is conveniently written as Fourier cosine transform:

$$F_c(\omega) = \sqrt{\frac{2}{\pi}} \int_0^{\infty} f(t) \cos(\omega t) dt, \quad f(x) = \sqrt{\frac{2}{\pi}} \int_0^{\infty} F_c(\omega) \cos(\omega x) d\omega. \tag{7.39}$$

And for odd functions, we have a sine Fourier transform:

$$F_s(\omega) = \sqrt{\frac{2}{\pi}} \int_0^\infty f(t)\sin(\omega t)dt, \quad f(x) = \sqrt{\frac{2}{\pi}} \int_0^\infty F_s(\omega)\sin(\omega x)d\omega. \tag{7.40}$$

Example 7.9. In this example, we explore the Fourier transform for a Gaussian function:

$$f(x) = Ce^{-ax^2}, \tag{7.41}$$

where C is a normalization and $a > 0$ is a constant. For normalization, we use

$$\int_{-\infty}^\infty f(x)dx = 1,$$

then $C = \sqrt{a/\pi}$. It is good to note that the Gaussian function is widely used in physics and engineering due to its mathematical properties and its connection to probability distributions. Here are several examples. In quantum mechanics, the probability density of finding a particle in space is often modeled by a squared modulus of a complex-valued Gaussian function. Gaussian beams are commonly used in optics to describe the intensity profile of laser beams. Gaussian filters are frequently employed in signal processing for smoothing and noise reduction. In Magnetic Resonance Imaging (MRI), the spatial distribution of magnetic resonance signals is often modeled using a Gaussian function. The Gaussian function is the probability density function of the normal distribution, which appears in various statistical and probabilistic models. Gaussian functions are employed in image processing for blurring and sharpening images.

Now, we proceed with calculating the Fourier transform of the Gaussian function:

$$F(\omega) = \frac{1}{\sqrt{2\pi}} \int_{-\infty}^\infty Ce^{-ax^2}e^{i\omega x}dx.$$

The above integral can be evaluated analytically by using the transformation:

$$-ax^2 + i\omega x = -a\left(x^2 - \frac{i\omega x}{a}\right) = -a\left(x - \frac{i\omega}{2a}\right)^2 - \frac{\omega^2}{4a}.$$

Introducing a new variable

$$z = \sqrt{a}\left(x - \frac{i\omega}{2a}\right),$$

we have

$$F(\omega) = \frac{C}{\sqrt{2\pi}} \int_{-\infty}^{\infty} e^{-a\left(x - \frac{i\omega}{2a}\right)^2 - \frac{\omega^2}{4a}} dx = \frac{C}{\sqrt{2\pi a}} e^{-\frac{\omega^2}{4a}} \int_{L} e^{-z^2} dz,$$

where the path L is a straight line in a complex plane z that is parallel to the real axis x and passing through $z = -i\omega/(2\sqrt{a})$. This integral can be evaluated using Cauchy's theorem (see Chapter 12) as

$$\int_{L} e^{-z^2} dz = \int_{-\infty}^{\infty} e^{-x^2} dx = \sqrt{\pi}.$$

Then, the Fourier transform is

$$F(\omega) = \frac{C}{\sqrt{2a}} e^{-\frac{\omega^2}{4a}} = D e^{-\frac{\omega^2}{4a}}, \qquad (7.42)$$

where D is a constant.[9] As we can see in Figure 7.6, the ω dependence of the Fourier transform is similar to the original function $f(x)$ dependence on x. By comparing equations (7.41) and (7.42), we can see that the smaller the a, the wider the original function $\exp(-ax^2)$, and the narrower the Fourier transform $\exp(-\omega^2/4a)$. We can define the width of the "bell-like" Gaussian functions (7.41) and (7.42) as the width of the function at a level where its height decreases by e times from the maximum value. Then, the width of (7.41) is $\Delta x = 2/\sqrt{a}$, and for (7.42), we have $\Delta\omega = 4/\sqrt{a}$. Thus, if we change the parameter a, then one of the functions becomes narrower, and the other becomes as much wider so that

$$\Delta x \Delta \omega = 8.$$

This result has fundamental importance since it is directly related to the "uncertainty principle". From the uncertainty principle follows that if a system oscillates with a variable frequency, then it makes no sense to talk about the value of the frequency at a given moment of time. The time interval during which this frequency is determined

---

[9] For normalization $\int_{-\infty}^{\infty} F(\omega) d\omega = 1$, we set $D = 1/2\sqrt{a\pi}$.

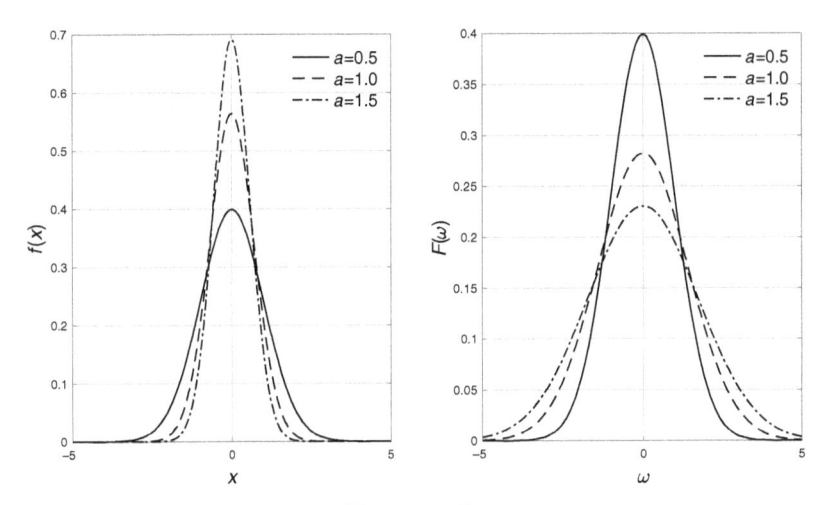

Fig. 7.6. The function $f(x) = \sqrt{\frac{a}{\pi}} \exp(-ax^2)$ (on the left) and its Fourier transform $F(\omega) = \frac{1}{2\sqrt{a\pi}} \exp\left(-\frac{\omega^2}{4a}\right)$ (on the right) for various values of a. Both functions are normalized.

cannot be taken much less than the oscillation period. In particular, in quantum mechanics, Heisenberg's uncertainty principle states that there is a fundamental limit to how precisely we can simultaneously know two complementary properties of a particle, such as its position and momentum. Mathematically, it can be expressed as follows for position and momentum:

$$\Delta x \cdot \Delta p \geq \frac{\hbar}{2},$$

and for energy and time as

$$\Delta E \cdot \Delta t \geq \frac{\hbar}{2}. \qquad \blacksquare$$

Example 7.10. Find the Fourier transform of $f(x) = e^{-a|x|}$ for $a > 0$.

The Fourier transform is

$$F(\omega) = \frac{1}{\sqrt{2\pi}} \int_{-\infty}^{\infty} e^{-a|x|} e^{i\omega x} dx$$

$$= \frac{1}{\sqrt{2\pi}} \left(\int_{-\infty}^{0} e^{ax} e^{i\omega x} dx + \int_{0}^{\infty} e^{-ax} e^{i\omega x} dx \right).$$

Evaluating the two integrals gives

$$F(\omega) = \frac{1}{\sqrt{2\pi}} \left(\frac{1}{a+i\omega} + \frac{1}{a-i\omega} \right) = \sqrt{\frac{2}{\pi}} \frac{a}{a^2+\omega^2}.$$

It is interesting to explore the Fourier transform in the limit $a \to 0$, i.e., when $f(x) = 1$. In Chapter 10 (see Section 10.1), it is shown that for $a = 1/k$ with $k \to \infty$

$$\lim_{k\to\infty} \frac{1}{\pi} \frac{k}{1+(kx)^2} \to \delta(x).$$

Or the Fourier transform of $f(x) = 1$ gives the delta function $\delta(x)$ as

$$F(\omega) = \frac{1}{\sqrt{2\pi}} \int_{-\infty}^{\infty} e^{i\omega x} dx = \sqrt{2\pi} \delta(\omega).$$

As we can see from Figure 7.7, the wider the function $e^{-a|x|}$, the narrower the Fourier transform $F(\omega)$. It is interesting to explore if the uncertainty principle holds for the function $f(x) = exp(-a|x|)$ and its Fourier transform.

In the limit $a \to 0$, the Fourier transform $\lim_{a\to 0} F(\omega) \to \sqrt{2\pi}\delta(\omega)$. ∎

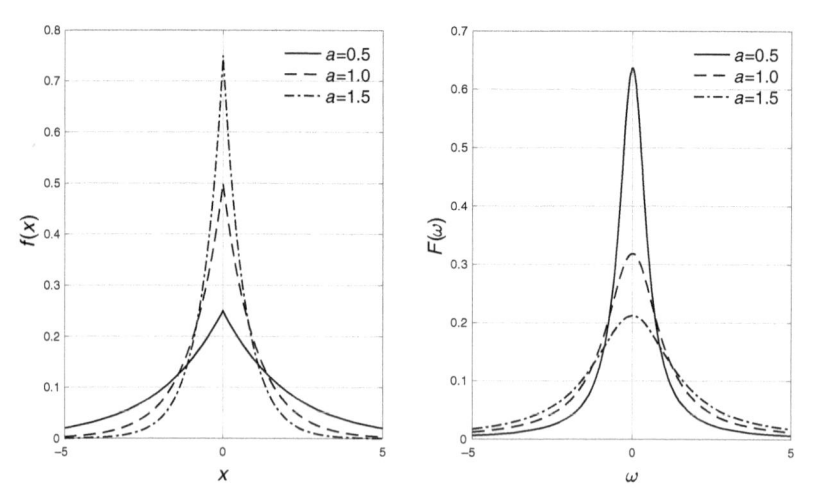

Fig. 7.7. The function $f(x) = \frac{a}{2} \exp(-a|x|)$ (on the left) and its Fourier transform $F(\omega) = \frac{1}{\pi} \frac{a}{a^2+\omega^2}$ (on the right) for various values of a. Both functions are normalized.

7.4.1. *Properties of Fourier transform*

The Fourier transform possess a number of practical properties:

(1) The Fourier transform is a linear operator. If $f(x) = f_1(x) + f_2(x)$, then
$$F(\omega) = F_1(\omega) + F_2(\omega).$$

(2) The Fourier transform for the derivative of a function $f'(x)$ is
$$\frac{1}{\sqrt{2\pi}} \int_{-\infty}^{\infty} f'(t) e^{i\omega t} dt = i\omega F(\omega).$$

This property can be proved using integration by parts.

(3) The Fourier transform for $f(ax), a > 0$, is
$$\frac{1}{a} F\left(\frac{\omega}{a}\right).$$

(4) For a function shifted by a constant, i.e., $f(x-b)$, the Fourier transform is
$$e^{i\omega b} F(\omega).$$

7.4.2. *Applications of Fourier transform*

The Fourier transform is a powerful mathematical tool used in various fields to analyze signals and functions in terms of their frequency components. Here are some examples of applications of the Fourier transform. Fourier transform is widely used in spectral analysis for analyzing signals in terms of their frequency content. Fourier transform is used for image analysis and manipulation. It helps in tasks such as image compression, filtering, and enhancement by transforming images between the spatial and frequency domains. In quantum mechanics, the Fourier transform is used to move between position and momentum representations of wavefunctions. Fourier transform is used in medical imaging techniques like Magnetic Resonance Imaging (MRI) and Computed Tomography (CT) scans. It helps reconstruct images from the measured data in the frequency domain.

7.5. Exercises

(1) Derive the Fourier series expansion of $f(x) = e^{ax}$ in $-\pi < x < \pi$ if $a > 0$.
(2) Derive the Fourier series expansion of a triangular wave $f(x)$ in $-\pi < x < \pi$, where
$$f(x) = \begin{cases} \pi + x & -\pi < x < 0, \\ \pi - x & 0 < x < \pi. \end{cases}$$
(3) Derive the Fourier series expansion of a function $f(x)$ in $-\pi < x < \pi$, where
$$f(x) = \begin{cases} 0 & -\pi < x < 0, \\ x & 0 < x < \pi. \end{cases}$$
(4) Derive the Fourier series expansion of $f(x) = x \sin x$ in $-\pi < x < \pi$.
(5) Derive the Fourier series expansion of $f(x) = x \cos x$ in $-\pi < x < \pi$.
(6) Derive the cosine Fourier series expansion of $f(x) = \cos(ax)$ in $-\pi < x < \pi$.
(7) Derive the sine Fourier series expansion of $f(x) = \sin(ax)$ in $-\pi < x < \pi$.
(8) Derive the sine Fourier series expansion of $f(x) = x$ in $0 < x < \pi$.
(9) Derive the cosine Fourier series expansion of $f(x) = x^2$ in $0 < x < \pi$.
(10) Derive the cosine and sine Fourier series expansions of $f(x) = e^{ax}$ in $0 < x < \pi$.
(11) Derive the cosine and sine Fourier series expansions of $f(x) = \sin(ax)$ in $0 < x < \pi$. Treat separately cases whether a is non-integer or integer.
(12) Expand in the Fourier series $f(x) = x^2$ in the interval $-2 < x < 2$.
(13) Find the exponential Fourier transform of the function
$$f(x) = \begin{cases} 1 & |x| \le a, \\ 0 & |x| > a \end{cases}$$
and plot figures for $f(x)$ and $F(\omega)$.

(14) Derive the Fourier transform for $f(x) = e^{-a|x|}$ for $a > 0$.
(15) Find the Fourier transform of $f(x) = e^{-ax^2}$ with $a > 0$.

Answers and solutions

(1)
$$e^{ax} = \frac{2}{\pi}\sinh(a\pi)\left(\frac{1}{2a} + \sum_{k=1}^{\infty}\frac{(-1)^k}{a^2+k^2}[a\cos kx - k\sin kx]\right)$$
$$\times (-\pi < x < \pi).$$

(2)
$$\frac{\pi}{2} + \frac{4\cos x}{\pi\,1^2} + \frac{4\cos 3x}{\pi\,3^2} + \cdots = \frac{\pi}{2} + \frac{4}{\pi}\sum_{1(\text{odd}k)}^{\infty}\frac{\cos kx}{k^2} \quad (-\pi < x < \pi).$$

(3)
$$\frac{\pi}{4} - \frac{2}{\pi}\cos x + \sin x - \frac{\sin 2x}{2} - \frac{2\cos 3x}{9\pi} + \frac{\sin 3x}{3}$$
$$- \frac{\sin 4x}{4} \cdots \quad (-\pi < x < \pi).$$

(4)
$$x\sin x = 1 - \frac{1}{2}\cos x + 2\sum_{k=2}^{\infty}(-1)^k\frac{1}{k^2-1}\cos kx \quad (-\pi \le x \le \pi).$$

(5)
$$x\cos x = -\frac{1}{2}\sin x + 2\sum_{k=2}^{\infty}(-1)^k\frac{k}{k^2-1}\sin kx \quad (-\pi < x < \pi).$$

(6)
$$\cos ax = \frac{\sin a\pi}{a\pi} + \frac{2\sin a\pi}{\pi}\sum_{k=1}^{\infty}(-1)^k\frac{a\cos kx}{a^2-k^2} \quad (-\pi \le x \le \pi).$$

(7)
$$\sin ax = \frac{2\sin a\pi}{d\pi}\sum_{k=1}^{\infty}(-1)^k\frac{k\sin kx}{a^2-k^2} \quad (-\pi < x < \pi).$$

(8)
$$f(x) = \sum_{k=1}^{\infty} \frac{2}{n}(-1)^{n+1} \sin(nx) \quad (0 < x < \pi)$$

(9)
$$f(x) = \frac{\pi^2}{3} + \sum_{k=1}^{\infty} (-1)^k \frac{4}{k^2} \cos(kx) \quad (0 < x < \pi)$$

(10)
$$e^{ax} = \frac{e^{a\pi} - 1}{a\pi} + \frac{2a}{\pi} \sum_{k=1}^{m\infty} (-1)^k \frac{e^{a\pi} - 1}{a^2 + k^2} \cos kx \quad (0 \le x \le \pi).$$

(11)
$$\sin ax = \frac{1 - \cos a\pi}{\pi} \left(1 + 2a \sum_{k=1}^{\infty} \frac{\cos 2kx}{a^2 - (2k)^2} \right)$$
$$+ 2a \frac{1 + \cos a\pi}{\pi} \sum_{k=1}^{\infty} \frac{\cos(2k-1)x}{a^2 - (2k-1)^2}$$
$$\times a \text{ noninteger} \quad (0 < x < \pi).$$

$$\sin ax = \frac{4a}{\pi} \sum_{k=1}^{\infty} \frac{\cos(2k-1)x}{a^2 - (2k-1)^2} \quad a \text{ integer, even} \quad (0 < x < \pi).$$

$$\sin ax = \frac{4}{\pi} \left(1 + 2a \sum_{k=1}^{\infty} \frac{\cos 2kx}{(a-1)^2 - (2k)^2} \right)$$
$$\times a \text{ integer, odd} \quad (0 < x < \pi).$$

(12)
$$x^2 = \frac{4}{3} + 16 \sum_{k=1}^{\infty} \frac{(-1)^k}{k^2 \pi^2} \cos\left(\frac{k\pi x}{2}\right) \quad (-2 < x < 2).$$

(13)
$$F(\omega) = \sqrt{\frac{2}{\pi}} \frac{\sin a\omega}{\omega}.$$

(14)
$$F(\omega) = \sqrt{\frac{2}{\pi}} \frac{a}{a^2 + \omega^2}.$$

(15)
$$F(\omega) = \frac{1}{\sqrt{2a}} e^{-\frac{\omega^2}{4a}}.$$

Chapter 8

First-Order Ordinary Differential Equations

8.1. A note on classification of ordinary differential equations

A differential equation is an equation which involves one or more independent variables, one or more dependent variables, and derivatives of the dependent variables with respect to some or all of the independent variables. If there is just one independent variable, then the derivatives are all ordinary derivatives, and the equation is an *ordinary differential equation* (ODE). If there is more than one independent variable and partial derivatives appear in the equation, the equation is called a *partial differential equation* (PDE).

Let x be the independent variable and y a function of this variable. The general form of the ordinary differential equation becomes

$$F(x, y, y', y'', y''', \ldots, y^{(n)}) = 0.$$

The order of a differential equation is the order of the highest derivative which appears in the equation. For example, the general form of a second-order ordinary differential equation, in which y is the dependent variable and x is the independent variable, is

$$F(x, y, y', y'') = 0,$$

where the second derivative y'' must appear in this function. Lower-order derivatives may be missing.

If $F(x, y, y', y'')$ is a linear function of the variables y, y', y'' that is

$$F(x, y, y', y'') = a_2(x)y'' + a_1(x)y' + a_0(x)y + f(x) = 0,$$

then the differential equation is said to be linear.

If a function $y = \varphi(x)$ satisfies the differential equation, i.e., if the equation reduces to an identity on replacing y, y' and so on by $\varphi(x), \varphi'(x)$, and so on, the function $\varphi(x)$ is said to be a solution of the differential equation. Thus, when we are solving a differential equation, we are essentially integrating it.

8.1.1. *General and particular solutions*

In the simplest case, when a first-order differential equation has the form

$$y'(x) = f(x),$$

the total set of solutions is given by the formula

$$y(x) = \int f(x)dx + C,$$

where C is an arbitrary constant. We thus obtain, in this elementary case, a solution of the differential equation containing an arbitrary constant. A solution containing an arbitrary constant also occurs in the general case of a first-order differential equation; such a solution is referred to as the general solution of the equation

$$y = y(x, C).$$

Generally, a differential equation has an infinite number of solutions. In the case of the nth-order ordinary differential equation, the general solution can be written in the form

$$y = y(x, C_1, C_2, \ldots, C_n).$$

Clearly, to define any given physics problem, we must specify more than just the differential equation. When applying differential equations, we are usually not as interested in finding a family of solutions (the general solution) as we are in finding a solution that satisfies some additional requirement. In many physical problems, we need to find the particular solution that satisfies a condition

First-Order Ordinary Differential Equations

or a set of conditions. Thus, assigning definite values to constants $C_1, C_2, \ldots, C_n$, we obtain particular solutions.

Regarding ordinary differential equations, we say we have an *initial-value problem* if all the specified values of the solution and its derivatives are given at one point. Then, constants $C_1, C_2, \ldots, C_n$ can be found from the initial conditions

$$y(x = x_0) = y_0, \quad y'(x = x_0) = y'_0, \ldots y^{(n-1)}(x = x_0) = y_0^{(n-1)}.$$

These problems are most frequently encountered in dynamical problems where time is the independent variable and the data are given at some initial time, say $t = 0$. For this reason, when we are dealing with initial-value problems, we often use t as the independent variable.

We take as an example the motion along a straight line of a point-like mass m under the action of a force F, which depends on time t, on the position of the point, and on its velocity. If we take as axis OX the straight line along which the point moves, the force F can be considered as a given function of t, x, and dx/dt. By Newton's second law, the product of the mass of the particle and its acceleration must be equal to the force acting. This gives us the differential equation of motion:

$$m\frac{d^2x}{dt^2} = F\left(t, x, \frac{dx}{dt}\right).$$

Integration of this second-order equation determines the relationship between x and t, i.e., the motion of the particle under the action of the given force. In order to obtain a definite solution of the problem, we must also specify the initial conditions of the motion, i.e., the position of the particle and its velocity at some initial instant, say at $t = 0$:

$$x(t = 0) = x_0, \quad \frac{dx}{dt}(t = 0) = v_0.$$

If, on the other hand, data for some differential equation are given at more than one point, we say we have a *boundary value problem*. These problems occur most frequently when the independent variable is a space variable. For this reason, when we are dealing with boundary value problems, we shall generally use x as the dependent variable.

Note that we can always replace an nth-order differential equation on a system of n first-order differential equations with n unknown functions. For example, given the second-order ODE

$$F(x, y, y', y'') = 0,$$

then by introducing $y = y_1$, $y' = y_2$, we get a system of two first-order differential equations for two unknown functions y_1 and y_2.

8.2. First-order ordinary differential equations: Basic concepts

An ordinary first-order differential equation can be written as

$$F(x, y, y') = 0,$$

where F is a function of three arguments with y being the unknown function of x. The equation above is called unresolved with respect to the derivative. In a particular case, a large class of first-order differential equations can be written as

$$\frac{dy}{dx} = f(x, y). \tag{8.1}$$

Such a differential equation is called *resolved with respect to the derivative*. A solution of equation (8.1) is any differentiable function $y = y(x)$ that turns its differential equation into an identity.

As we noted before, an ordinary differential equation has infinite number of solutions called general solutions. We can write these as

$$y = \varphi(x, C),$$

where C is a constant of integration. In order to select one solution from the infinite set (or family) of solutions of a given equation (8.1), it is necessary to specify an additional condition. The simplest additional condition is that we specify an initial condition at some point in the domain. Analytically, this is written as

$$y|_{x=x_0} = y_0, \quad \text{or} \quad y(x_0) = y_0.$$

The differential equation (8.1) with this condition is called the initial problem or the Cauchy problem:

$$\frac{dy}{dx} = f(x, y), \quad y(x_0) = y_0.$$

First-Order Ordinary Differential Equations 259

We should note that there is no systematic technique that enables us to solve even such first-order ordinary differential equations. However, depending on the form of $f(x, y)$, we often can find closed-form solutions.

8.3. Separable-variable equations

If $f(x, y)$ can be written as $f(x, y) = g(x)h(y)$, where $g(x)$ is a function of x only and $h(y)$ is a function of y only, then the variables are said to be separable. Then, the equations above can be rewritten as

$$\frac{dy}{dx} = g(x)h(y). \tag{8.2}$$

Rearranging this equation so that the terms depending on x and on y appear on opposite sides (i.e., are separated), and integrating, we obtain the general solution $y(x)$:

$$\int \frac{dy}{h(y)} = \int g(x)dx + C. \tag{8.3}$$

Note that the justification for simply integrating both sides in equation (8.3) is based on the substitution rule for differentiation, namely

$$\int \frac{dy}{h(y)} = \int \frac{1}{h(y)} \frac{dy}{dx} dx = \int \frac{1}{h(y)} [g(x)h(y)]dx = \int g(x)dx.$$

Finding the general solution $y(x)$ from (8.3) depends of course on the ease with which the integrals in the above equation can be evaluated. A particular solution for a given condition $y = y_0$ at $x = x_0$ fixes the constant C.

Another way to find a particular solution for equation (8.2) is as follows. Let

$$H(y) = \int \frac{1}{h(y)} dy, \quad G(x) = \int g(x)dx.$$

Now, the general solution (8.3) can be written as

$$H(y) = G(x) + C.$$

For a given boundary condition $H(y_0) = G(x_0) + C$ or $C = H(y_0) - G(x_0)$,

$$H(y) = G(x) + H(y_0) - G(x_0) \quad \text{or} \quad H(y) - H(y_0) = G(x) - G(x_0).$$

According to the fundamental theorem of calculus (in some textbooks, it is called the second fundamental theorem of calculus),

$$H(y) - H(y_0) = \int_{y_0}^{y} \frac{dy}{h(y)},$$

$$G(x) - G(x_0) = \int_{x_0}^{x} g(x)dx.$$

Then, a particular solution of (8.2) can be found from

$$\int_{y_0}^{y} \frac{dy}{h(y)} = \int_{x_0}^{x} g(x)dx. \tag{8.4}$$

Thus, there is no need to look for integration constant, as in (8.3), because this information it is already embedded in the limits of integration.

Note that a differential equation of a form

$$\frac{dy}{dx} = f(ax + by + c),$$

where a, b, and c are constants, can be reduced to an equation with separable variables by substitution $u = ax + by + c$. Indeed, differentiating u with respect to x, we have $du/dx = a + b(dy/dx)$ and then we easily get

$$\frac{du}{dx} = bf(u) + a \quad \text{or} \quad \frac{du}{bf(u) + a} = dx.$$

In this case, we may obtain solutions of a type $u = const$ that are roots of the equation $bf(u) + a = 0$.

Example 8.1. Find general and particular solutions of the equations

$$\frac{dy}{dx} = 2xy, \quad y(0) = 2.$$

Rewriting the equation to separate variables and integrating both sides gives

$$\frac{dy}{y} = 2xdx, \quad \int \frac{dy}{y} = \int 2xdx,$$

then

$$\ln y = x^2 + C \quad \text{or} \quad y(x) = Ce^{x^2},$$

where C is a constant of integration which can be found from the initial condition $y(0) = C = 2$.

Note that we could find the particular solution immediately by using (8.4):

$$\int_2^y \frac{dy}{y} = \int_0^x 2x\,dx, \quad \ln(y) - \ln(x) = x^2, \quad y(x) = 2e^{x^2}. \qquad \blacksquare$$

8.4. First-order linear ordinary differential equations

If the function $f(x,y)$ in equation (8.1) can be represented as a sum of two terms $f(x,y) = -P(x)y + Q(x)$ (linear in y), then we can reduce the equation (8.1) to a linear form:

$$y' + P(x)y = Q(x). \tag{8.5}$$

We consider the initial-value problem based on this equation, which requires a solution continuous for a given interval $0 \le x \le b$ satisfying $y(x_0) = y_0$. For now, we assume that $P(x)$ and $Q(x)$ are continuous for $0 \le x \le b$.

Ordinary linear differential equations are called homogeneous if $Q(x)$ is equal to zero and they are considered non-homogeneous equations if $Q(x) \ne 0$.

8.4.1. First-order linear homogeneous ordinary differential equation

The homogeneous differential equation takes the form

$$\frac{dy}{dx} + P(x)y = 0. \tag{8.6}$$

Since

$$\frac{dy}{dx} = -P(x)y, \quad \text{then} \quad \frac{dy}{y} = -P(x)dx,$$

and we can integrate both sides as

$$\int \frac{dy}{y} = -\int P(x)dx + C,$$

$$\ln y = -\int P(x)dx + C \quad \text{or} \quad y = \exp\left(-\int P(x)dx + C\right).$$

Thus, a general solution of (8.6) is

$$y = C_1 \exp\left(-\int P(x)dx\right). \tag{8.7}$$

A particular solution for $y(x_0) = y_0$ at $x = x_0$ can be derived using (8.4) as follows:

$$\int_{y_0}^{y} \frac{dy}{y} = \int_{x_0}^{x} P(x)dx.$$

The integration can be easily performed and finally, the particular solution for the differential equation (8.6) is

$$y = y_0 \exp\left(-\int_{x_0}^{x} P(x)dx\right). \tag{8.8}$$

Attention: There are two ways to find a particular solution. You can start from the general solution (8.7) an then find the constant of integration C_1 or you can directly use (8.8). It is advisable to use (8.8) since in this case you immediately have the particular solution.

In a special case of when the coefficient $P(x)$ is a constant, $P(x) = p$, the solution (8.8) has a simple form:

$$y = y_0 e^{-p(x-x_0)}.$$

Note that the solution above describes both exponential decay and exponential growth problems.

Example 8.2. First-order homogeneous ODE

$$xy' - xy = y \quad y(1) = 1.$$

This equation can be rewritten as

$$y' - \left(1 + \frac{1}{x}\right)y = 0.$$

Now, we can use solution (8.8) with $P(x) = -(1 + 1/x)$. For the integral, we have

$$-\int_{1}^{x} -\left(1 + \frac{1}{x}\right) dx = (x + \ln x)|_{1}^{x} = x + \ln x - 1.$$

Then,
$$y(x) = e^{(x+\ln x - 1)} = xe^{x-1}.$$

Note that this equation can also be solved using separation of variables, namely rewriting the equation as

$$\frac{dy}{y} = \left(1 + \frac{1}{x}\right) dx. \qquad \blacksquare$$

8.4.2. First-order linear non-homogeneous ODE: Lagrange's method

Now, we consider equation (8.5) when the non-homogeneous term $Q(x) \neq 0$

$$\frac{dy}{dx} + P(x)y = Q(x).$$

We assume that the general solution of this equation can be written as $y(x) = u(x)y_1(x)$, where $y_1(x)$ is the solution

$$y_1(x) = \exp\left(-\int P(x)dx\right)$$

of the homogeneous equation (8.6), i.e.,

$$\frac{dy_1}{dx} + P(x)y_1 = 0.$$

Note that it was Lagrange who suggested this approach for solving first-order linear non-homogeneous ordinary differential equation.

Substituting $y(x) = u(x)y_1(x)$ into the non-homogeneous equation

$$\frac{du y_1}{dx} + P(x)uy_1 = Q(x),$$

we get

$$\frac{du}{dx}y_1 + u\frac{dy_1}{dx} + P(x)uy_1 = Q(x)$$

or

$$\frac{du}{dx}y_1 + u\left(\frac{dy_1}{dx} + P(x)y_1\right) = Q(x).$$

Since $(dy_1)/dx + P(x)y_1 = 0$, then we have differential equation for the unknown function $u(x)$

$$\frac{du}{dx} y_1 = Q(x) \quad \text{or} \quad du = \frac{Q(x)}{y_1} dx.$$

This is readily integrated to obtain

$$u(x) = \int \frac{Q(x)}{y_1} dx + C.$$

Thus, the general solution takes the form

$$y(x) = C y_1(x) + y_1(x) \int \frac{Q(x)}{y_1(x)} dx, \qquad (8.9)$$

where $y_1(x)$ is the solution of complementary homogeneous equations.

If we are interested in a particular solution of (8.9) with the initial condition at $x = x_0$, then we have

$$y(x) = \frac{y(x_0)}{y_1(x_0)} y_1(x) + y_1(x) \int_{x_0}^{x} \frac{Q(x)}{y_1(x)} dx. \qquad (8.10)$$

It is convenient to rewrite the above equation in terms of a dummy variable ξ as

$$y(x) = \frac{y_0}{y_1(x_0)} y_1(x) + y_1(x) \int_{x_0}^{x} \frac{Q(\xi)}{y_1(\xi)} d\xi,$$

In this case, one can see a clear difference between variables inside and outside the integrals.

The solution of the non-homogeneous equation takes a simple form when the coefficients $P(x)$ and $Q(x)$ are constants:

$$\frac{dy}{dx} + py = q.$$

Then, from (8.9) follows

$$y(x) = Ce^{-px} + \frac{q}{p}.$$

This solution can also be derived by noting that $u = B = \text{const}$ is a particular solution. Substituting it into the equation gives $0 + pB = q$ or $B = q/p$.

Example 8.3. The first-order non-homogeneous ODE
$$x^2 y' + 3xy = 1, \quad y(1) = 2.$$
The equation can be rewritten as
$$y' + \frac{3}{x}y = \frac{1}{x^2}, \quad \text{that corresponds to } P(x) = \frac{3}{x}, \quad Q(x) = \frac{1}{x^2}.$$
Then, the solution of the homogeneous equation is
$$y_1(x) = \exp\left(-\int \frac{3}{x}dx\right) = \exp(-3\ln x) = \frac{1}{x^3}$$
and the general solution of the homogeneous equation is
$$y(x) = C\frac{1}{x^3} + \frac{1}{x^3}\int \frac{1}{x^2}x^3 dx = C\frac{1}{x^3} + \frac{1}{2x}.$$
The constant of integration can be found from the initial conditions $y(1) = 2$, i.e., $y(1) = C + 1/2$, then $C = 3/2$ and the particular solution is
$$y(x) = \frac{3}{2x^3} + \frac{1}{2x}. \quad \blacksquare$$

8.5. Exact equations: Integrating factors

The first-order nonlinear ODE of form (8.1) can be rewritten as
$$A(x,y)dx + B(x,y)dy = 0, \tag{8.11}$$
where $f(x,y) = -A(x,y)/B(x,y)$. Note that either of $A(x,y)$ and $B(x,y)$ or both can be nonlinear functions of y.

8.5.1. *Exact equations*

If the form of the differential equation (8.11) is an exact differential, that is,
$$A(x,y)dx + B(x,y)dy = \frac{\partial F}{\partial x}dx + \frac{\partial F}{\partial y}dy = dF = 0$$
for some function $F(x,y)$ which has continuous derivatives, then from $dF(x,y) = 0$ we can get a solution of the differential equation.

Indeed, in the equation above, we have

$$\frac{\partial F}{\partial x} = A(x, y), \quad \frac{\partial F}{\partial y} = B(x, y).$$

Since for second-order partial derivatives

$$\frac{\partial^2 F}{\partial x \partial y} = \frac{\partial^2 F}{\partial y \partial x},$$

then we require

$$\frac{\partial A(x, y)}{\partial y} = \frac{\partial B(x, y)}{\partial x} \qquad (8.12)$$

for $A(x, y)dx + B(x, y)dy$ to be an exact differential. Then, the integration of $dF = 0$ gives the general solution of (8.11) as $F(x, y) = C$, where C is an arbitrary constant.

Note that the exact equation can also be solved as initial value problem satisfying $y(x_0) = y_0$, then $F(x_0, y_0) = 0$. More specifically, we can write the solution as a path integral (see Figure 8.1):

$$F(x, y) = \int_{x_0}^{x} A(x, y_0)dx + \int_{y_0}^{y} B(x, y)dy = 0. \qquad (8.13)$$

The terms correspond to integration along x from x_0, y_0 to x, y_0 with $dy = 0$ and then along y from x, y_0 to x, y with $dx = 0$.

Example 8.4. Let's consider equation

$$(x^2 + 2xy)dx + (x^2 - y^3)dy = 0, \quad y(0) = y_0 = 0.$$

Here,

$$\frac{\partial A(x, y)}{\partial y} = \frac{\partial (x^2 + 2x)}{\partial y} = 2x, \quad \frac{\partial B(x, y)}{\partial x} = \frac{\partial (x^2 - y^2)}{\partial x} = 2x$$

so that condition (8.12) is satisfied. Now, we integrate according to (8.13):

$$\int_0^x (x^2 + 2xy_0)dx + \int_0^y (x^2 - y^3)dy = \frac{x^3}{3} + x^2y - \frac{y^4}{4} = 0.$$

Thus, the solution of the equation is a parametric equation of both x and y. ∎

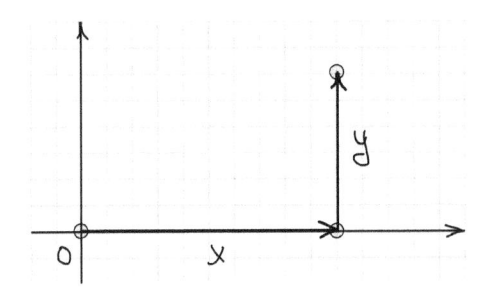

Fig. 8.1. Integration.

8.5.2. Reducible to exact form

If a differential equation can be written in the form

$$A(x,y)dx + B(x,y)dy = 0 \quad \text{but} \quad \frac{\partial A(x,y)}{\partial y} \neq \frac{\partial B(x,y)}{\partial x},$$

then we have an inexact equation. However, we can make such an equation exact by introducing an integrating factor. Multiplying the equation above by a function $\mu(x,y) \neq 0$

$$\mu(x,y)A(x,y)dx + \mu(x,y)B(x,y)dy = 0,$$

we try to find such $\mu(x,y)$ that

$$\frac{\partial}{\partial y}\mu(x,y)A(x,y) = \frac{\partial}{\partial x}\mu(x,y)B(x,y).$$

The function $\mu(x,y)$ is called the integrating factor. The equation above can be written as

$$\mu(x,y)\left(\frac{\partial A(x,y)}{\partial y} - \frac{\partial B(x,y)}{\partial x}\right) = B(x,y)\frac{\partial \mu(x,y)}{\partial x} - A(x,y)\frac{\partial \mu(x,y)}{\partial y}.$$

This is a partial differential equation. Its solution exists if A, B, $\partial A/\partial y, \partial B/\partial x$ are continuous in the domain of integration and $A^2(x,y) + B^2(x,y) \neq 0$. Solving the above equation is in general not simpler than solving the original equation (8.11). Methods for finding an integrating factor $\mu(x,y)$ are known only for certain very special types of differential equations. For example, if an integrating factor exists, that is, a function of either x or y alone, then we can find it. For example, if we assume that the integrating factor is a function of x alone, i.e., $\mu = \mu(x)$, then the equation above can be rearranged as

$$\frac{1}{\mu(x)} \frac{\partial \mu}{\partial x} = \frac{1}{B(x,y)} \left(\frac{\partial A(x,y)}{\partial y} - \frac{\partial B(x,y)}{\partial x} \right) = f(x)$$

where we require that $f(x)$ is a function of x only. Then, for the integrating factor $\mu(x)$, we get

$$\mu(x) = \exp\left[\int f(x) dx \right] \quad \text{where}$$

$$f(x) = \frac{1}{B(x,y)} \left(\frac{\partial A(x,y)}{\partial y} - \frac{\partial B(x,y)}{\partial x} \right).$$

Conversely, if $(1/B)(\partial A/\partial y - \partial B/\partial x)$ does not depend on y, then the $\mu(y)$ given by this formula is an integrating factor.

Similarly, if $\mu = \mu(y)$, then

$$\mu(y) = \exp\left[\int g(y) dy \right] \quad \text{where}$$

$$g(y) = \frac{1}{B(x,y)} \left(\frac{\partial A(x,y)}{\partial y} - \frac{\partial B(x,y)}{\partial x} \right).$$

8.6. Some first-order differential equations in physics

8.6.1. Cooling and heating

Under certain conditions, the rate of change of the temperature of a body, immersed in a medium whose temperature (kept constant) differs from it, is proportional to the difference in temperature between it and the medium. In mathematical symbols, this statement is written as

$$\frac{dT_B}{dt} = -k(T_B - T_M),$$

where $k > 0$ is a proportionality constant, T_B is the temperature of the body at any time t, and T_M is the constant temperature of the medium.

In solving problems in which proportionality constant k is not given, it is necessary to know another condition in addition to the initial condition. In the temperature problem, for example, we also need to know, in addition to the initial condition, the temperature of the body at some future time t. With these two sets of conditions, it will then be possible to determine the value of the proportionality constant k. Thus, we must use the above equation twice: once to find k and the second time to find the desired answer.

Example 8.5. The temperature in a room is 25°C. A thermometer which has been kept in it is placed outside. In 5 minutes, the thermometer reading is 15°C. Five minutes later, it is 10°C. Find the outdoor temperature.

The cooling equation can be solved using separation of variables. Then, according to (8.4), the particular solution can be derived from

$$\int_{T_0}^{T} \frac{dT}{T - T_M} = -k \int_{t_0}^{t} dt,$$

then

$$\ln(T - T_M)|_{T_0}^{T} = -kt|_{t_0}^{t}, \quad \text{or}$$
$$\ln(T - T_M) - \ln(T_0 - T_M) = -k(t - t_0),$$
$$T - T_M = (T_0 - T_M)e^{-k(t-t_0)}.$$

Finally, the temperature of a body at any given time t is

$$T = T_M + (T_0 - T_M)e^{-k(t-t_0)}.$$

Now, we are ready to find the outdoor temperature T_M. For the given conditions: If $T_0 = 25°$, then

$$15° = T_M + (25° - T_M)e^{-5k},$$
$$10° = T_M + (25° - T_M)e^{-10k}.$$

Thus, we have two equations with two unknowns T_M and k. Noting that $e^{-10k} = (e^{-5k})^2$ reduces the problem to simple algebra. From the first equation,

$$e^{-5k} = \frac{15 - T_M}{25 - T_M}, \quad (e^{-5k})^2 = e^{-10k} = \left(\frac{15 - T_M}{25 - T_M}\right)^2,$$

and from the second equation,
$$e^{-10k} = \frac{10 - T_M}{25 - T_M}.$$
Then,
$$\left(\frac{15 - T_M}{25 - T_M}\right)^2 = \frac{10 - T_M}{25 - T_M}.$$
Solving the above equation gives the temperature $T_M = 5°$. ∎

8.6.2. Decay and growth

As an example, the number of bacteria in a yeast culture grows at a rate which is proportional to the number present. If the population of a colony of yeast bacteria doubles in one hour, find the number of bacteria which will be present at the end of h hours. Then, in mathematical symbols, the first sentence of the problem states
$$\frac{dN}{dt} = \lambda N.$$
Note that for growth problems, $\lambda > 0$, and for decay problems, $\lambda < 0$.

Example 8.6. Suppose the rate at which bacteria in a culture grow is proportional to the number present at any time but the population is being reduced at a constant rate R by the removal of bacteria for experimental purposes. Write the differential equation for the number N of bacteria as a function of time t. Solve the differential equation if there are N_0 bacteria when $t = 0$.

Let αN be the rate of increase of bacteria and R be the removal rate, then
$$\frac{dN}{dt} = \alpha N - R,$$
this equation can be solved using separation of variables:
$$\frac{dN}{\alpha N - R} = dt, \quad \int_{N_0}^{N} \frac{dN}{\alpha N - R} = \int_{0}^{t} dt,$$
and using
$$\int \frac{dx}{ax - b} = \frac{1}{a} \ln(ax - b),$$

we obtain
$$t = \frac{1}{\alpha}(\ln(\alpha N - R) - \ln(\alpha N_0 - R)) = \frac{1}{\alpha}\ln\left(\frac{\alpha N - R}{\alpha N_0 - R}\right).$$

Solving for N gives
$$N = N_0 e^{\alpha t} - \left(\frac{R}{\alpha}\right)(e^{\alpha t} - 1).$$

In the case of $R = 0$, we have a regular exponential growth $N = N_0 e^{\alpha t}$. ∎

8.6.3. Motion along a straight line: Vertical, horizontal, and inclined

By Newton's second law, the rate of change of the momentum of a body $\vec{p} = m\vec{v}$ is proportional to the resultant external force $\vec{F}$ acting upon it. For one-dimensional motion (and assuming the mass is independent of time), in mathematical symbols, the second law says
$$m\frac{dv}{dt} = F.$$

The force F can be gravity, air resistance (linear or quadratic), friction, or a combination of more than one force.

Example 8.7. Upward vertical motion with linear drag force.
A projectile of mass m kg is thrown vertically upward from the ground with an initial velocity v_0 m/s. The projectile is the subject to linear air resistance so that the force of air resistance is given by $f = -bv$. Find its velocity and position as a function of time. For how long and how far will it rise?

Then, Newton's second law reads (remember this is one-dimensional motion, no vector notations, the direction is either $+$ or $-$)
$$m\frac{dv}{dt} = -mg - bv, \quad \text{or} \quad \frac{dv}{dt} + kv = -g \quad \text{where } k = \frac{b}{m}.$$

With the velocity upward ($v > 0$), the retarding force is downward as well as the force of gravity. This equation is a first order linear non-homogeneous ODE that can be solved by using a general method for $y' + P(x)y = Q(x)$ with the general solution (8.9). In our case, the independent variable is time t and then $P = k$ and $Q = -g$ (note that we also can use separation of variables for solving this ODE):

$$v_1(t) = \exp\left(-k \int dt\right) = \exp(-kt),$$

where $v_1(t)$ is the solution of the homogeneous equation, and then

$$v(t) = C \exp(-kt) + \exp(-kt) \int (-g) \exp(+kt) dt$$

$$= C \exp(-kt) - \frac{g}{k}.$$

It is convenient to use the definition of terminal velocity, namely $v_t = g/k = mg/b$. Now, we can find the constant of integration C from the initial condition

$$v(t=0) = v_0 = -v_t + C, \quad C = v_0 + v_t,$$

then finally,

$$v(t) = v_0 e^{-kt} + v_t(e^{-kt} - 1).$$

The vertical position as a function of time is

$$y(t) = \int v(t) dt + C_y,$$

$$y(t) = -v_t t - (v_0 + v_t)\frac{1}{k}e^{-kt} + C_y.$$

Setting $y_0 = 0$ at $t = 0$ we obtain

$$y(t=0) = 0 = -(v_0 + v_t)\frac{1}{k} + C_y, \quad C_y = (v_0 + v_t)\frac{1}{k}.$$

Then, we can write

$$y(t) = -v_t t + (v_0 + v_t)\frac{1}{k}(1 - e^{-kt}).$$

We can easily verify the solution for $v(t)$ and $y(t)$ in the limit $b \to 0$ by keeping first few terms in the Taylor series for $\exp(-bt/m)$.

It is also interesting to find the time for the object to reach its highest point and its position y_{max} at that point. It can be done by setting $v(t) = 0$ for that point, in particular

$$v(t) = 0 = v(t) = v_0 e^{-kt} + v_t(e^{-kt} - 1),$$

then

$$e^{-kt} = \frac{v_t}{v_0 + v_t} \quad \text{or} \quad t = -\frac{1}{k} \ln \frac{v_t}{v_0 + v_t}$$

$$= \frac{1}{k} \ln \left(1 + \frac{v_0}{v_t}\right) = \frac{m}{b} \ln \left(1 + \frac{v_0 b}{mg}\right).$$

It is also easy to show (using Taylor's series for $\ln x$) that in the limit $b \to 0$ we have $t \to v_0/g$.

The maximum vertical position can be found either by substitution the time we found into $y(t)$ or by integrating

$$h_{max} = \int_0^{t_{max}} v(t) dt,$$

$$h_{max} = \frac{m}{b} v_0 + \frac{m^2 g}{b^2} \ln \frac{mg}{mg + bv_0}. \quad \blacksquare$$

8.6.4. Electric circuits

Some electric circuits can be described with first-order differential equations with time-dependent solutions.

(a) **R–C circuit:** A series electric circuit contains a resistance R, a capacitance C, and a battery supplying a time-varying electromotive force (EMF) $\mathcal{E}(t)$. The charge q on the capacitor therefore obeys the equation

$$R\frac{dq}{dt} + \frac{q}{C} = \mathcal{E}(t).$$

Note that the electric current in such a circuit can be determined by using the definition of the current as $i = dq/dt$.

(b) **R–L circuit:** A circuit that includes both a resistor R and an inductor L and possibly an electromotive force is called an R–L circuit. The differential equation for the current in the circuit is

$$L\frac{di}{dt} + Ri = \mathcal{E}(t).$$

Both the equations, for the R–C and R–L circuits, are first-order non-homogeneous differential equations with constant coefficients and can be solved using the methods discussed earlier.

Example 8.8. We analyze charging a capacitor in an R–C circuit with the initial condition $q(t = 0) = 0$ and a constant EMF $\mathcal{E}(t) = \mathcal{E}_0$. The differential equation

$$R\frac{dq}{dt} + \frac{q}{C} = \mathcal{E}_0$$

can be solved either using Lagrange's method (8.9) or by separating variables after rewriting the original equation as

$$\frac{dq}{q - \mathcal{E}_0 C} = -\frac{1}{RC} dt.$$

Then, the general solution for the given initial condition is

$$q(t) = \mathcal{E}_0 C \left(1 - e^{-t/RC}\right).$$

The instantaneous current i is

$$i = \frac{dq}{dt} = \frac{\mathcal{E}_0}{R} e^{-t/RC}.$$ ■

8.6.5. Rocket motion

Consider a rocket of mass m, traveling vertically and ejecting spent fuel at the exhaust speed v_{ex} relative to the rocket. The rocket mass changes at a rate dm/dt. Then, the differential equation describing the rocket motion is

$$m\frac{dv}{dt} = -v_{\text{ex}}\frac{dm}{dt} + F_{\text{external}},$$

where $v_{\text{ex}} dm/dt$ is called the thrust of the rocket and F_{external} is an external force acting on the rocket (typically gravity, i.e., $F_{\text{external}} = -mg$).

Example 8.9. A rocket with initial mass m_0 and initial speed v_0 accelerates in free space (no external forces). Find its change in velocity as its mass changes from m_0 to m.

The differential equation of the rocket motion in this case takes a simple form

$$m\frac{dv}{dt} = -v_{ex}\frac{dm}{dt}$$

that can be rewritten as

$$dv = -v_{ex}\frac{dm}{m}.$$

Integrating both parts

$$\int_{v_0}^{v} dv = -v_{ex}\int_{m_0}^{m}\frac{dm}{m}$$

gives

$$v = v_0 + v_{ex}\ln\frac{m_0}{m}. \qquad \blacksquare$$

8.7. Exercises (Math)

Attention: It is always a good habit to verify your solution by substituting it into the original differential equation.

Find the general solution for each of the following differential equations:

(1) $2xy' + y = 2x^{5/2}$.
(2) $y' + y\cos x = \sin 2x$.
(3) $x^2 y' + 3xy = 1$.
(4) $(1 + e^x)y' + 2e^x y = (1 + e^x)e^x$.
(5) $y' = \cos x - y\tan x$.
(6) $y' = 2x(x^2 + y)$.
(7) $e^{-y}dx - (2y + xe^{-y})dy = 0$.
(8) $y'\cos x + y\sin x = 1$.

Find the solution of each of the following differential equations subject to the indicated condition:

(9) $y' + xy = x$, $y(0) = 2$.
(10) $y\,dy + (xy^2 - 8x)dx = 0$, $y(1) = 3$.

(11) $y' + 2xy^2 = 0$, $y(2) = 1$.
(12) $(x + 2y)y' = 1$, $y(0) = -1$.
(13) $y' = -y + e^x$, $y(1) = 1/e$.
(14) $y' + y = \cos x$, $y(0) = 3/2$.

8.8. Physics problems

(1) **Horizontal motion I:** A mass m has velocity v_0 at time $t = 0$ and moves along the x axis in a medium where the drag force is $f = -bv$ so that the equation of motion is

$$m\frac{dv}{dt} = -bv.$$

Find its velocity in terms of the time t and the other given parameters. At what time (if any) will it come to rest? Find the position as a function of time. What is the limit $x(t)$ as $t \to \infty$?

(2) **Horizontal motion II:** Consider the same problem but with the quadratic drag force $f = -cv^2$.

(3) **Vertical downward motion I:** An object of mass m falls from rest under gravity subject to an air resistance proportional to its velocity:

$$m\frac{dv}{dt} = mg - bv,$$

where b is a positive constant and the y axis is directed down. Find its velocity and the distance it falls as functions of time. Find the limiting value of v as t tends to infinity; this limit is called the terminal velocity.

(4) **Vertical downward motion II:** Consider the same problem but with the quadratic drag force $f = -cv^2$.

(5) **Vertical upward motion with quadratic drag:** Consider an object that is thrown vertically up with initial speed v_0 in a quadratic medium:

$$m\frac{dv}{dt} = -mg - cv^2.$$

Find the object's' velocity $v(t)$ and position $y(t)$ as functions of time assuming $v(0) = v_0$ and $y(0) = 0$. Find the time for the object to reach its highest point. Find its position y_{max} at that point in time. Show that in the limit $c \to 0$ your results correspond to motion in vacuum.

(6) **Temperature change I:** A bucket of cold water with temperature $5°$ is brought from the outside and left in a room at temperature $25°$. If the temperature of the water is $10°$ after 20 min, what will its temperature be in 40 minutes? Note that all the temperatures are given in degrees Celsius.

(7) **Temperature change II*:** A slab of metal with temperature T_0 is placed in an oven. The temperature in the oven steadily increases from T_0 to T_1 within time Δt, i.e., $T_{oven} = T_0 + t(T_1 - T_0)/\Delta t$. Find and solve the differential equation describing the heating of the metal in the oven.

(8) **Fluid flowing from a tank:** A differential equation describing the height of a fluid level in a cylindrical tank of radius R losing fluid through a circular hole in the bottom with a radius r is

$$R^2 \frac{dh}{dt} = -r^2 \sqrt{2gh},$$

where g is free-fall acceleration. How much time will it take for the fluid to be completely drained from the tank if initially the tank was filled to a height of H.

(9) **R–L circuit:** Find electric current as a function of time in an R–L circuit if the emf changes with time as $\mathcal{E}(t) = \mathcal{E}_0 \cos \omega t$:

$$L \frac{di}{dt} + Ri = \mathcal{E}_0 \cos \omega t.$$

Assume the initial condition is $i(t=0) = i_0$.

(10) **Rocket motion I:** Consider a rocket taking off vertically (from rest) in a gravitational field g. The differential equation for rocket motion can be written as

$$m \frac{dv}{dt} = -v_{ex} \frac{dm}{dt} - mg,$$

where m is the mass of the rocket and v_{ex} is the speed of the exhaust (relative to the rocket). Find its speed v as a function of time and as a function of mass m left, assuming that the rocket ejects mass at a constant rate

$$m(t) = m_0 - \frac{dm}{dt}t.$$

(11) **Motion with variable mass***: A drop of water moves in a gravitational field g. Due to the condensation, the mass of the drop increases as $dm = Avdt$, where A is a constant and v is the velocity of the drop. Find the speed of the drop as a function of its mass if at the initial moment of time the drop was at rest with mass m_0.

(12) **Vertical motion from the earth****: An object is shot straight up from the surface of the earth with an initial velocity v_0. The force of gravity on the object is $F = -GmM/r^2$, where r is the distance form the center of the earth. Assuming that there is no air resistance, find the velocity v of the object as a function of the distance, how high the object will rise, and the time t as a function of r.

Answers and solutions

(1) $y(x) = \frac{x^{5/2}}{3} + C\frac{1}{\sqrt{x}}$.

(2) $y(x) = 2\sin(x) - 2 + Ce^{-\sin(x)}$.

(3) $y(x) = \frac{1}{2x} + C\frac{1}{x^3}$.

(4) $y(x) = \frac{e^x}{(e^x+1)^2}\left(1 + e^x + \frac{1}{3}e^{2x}\right) + C\frac{1}{(e^x+1)^2}$.

(5) $y(x) = \cos(x)(x + C)$.

(6) $y(x) = Ce^{x^2} - x^2 - 1$.

(7) $xe^{-y(x)} - y^2(x) = C$.

(8) $y(x) = \sin(x) + C\cos(x)$.

(9) $y(x) = e^{-x^2/2} + 1$.

(10) $y^2(x) = 8 + e^{(1-x^2)}$.

(11) $y(x) = \frac{1}{x^2-3}$.

(12) $y(x) = -\frac{x}{2} - 1$.

(13) $y(x) = \frac{1}{2}e^{-x}\left(e^{2x} + 2 - e^2\right)$.

(14) $y(x) = \frac{1}{2}\left(2e^{-x} + \sin(x) + \cos(x)\right)$.

Chapter 9

Second-Order Linear Differential Equations

Very many problems in physics are formulated in terms of second-order ordinary differential equations. For example, one of the fundamental laws of physics, Newton's second law, is a second-order ordinary differential equation. In the case of one-dimensional motion, we have

$$m\frac{d^2x}{dt^2} = F_{\text{net}}\left(t, x, \frac{dx}{dt}\right),$$

where the net force may depend on time, position, and velocity of an object. Having initial conditions for both the position $x(t_0) = x_0$ and velocity $x'(t_0) = x'_0$ we can find the solution of the equation at any moment in time. Such a problem is called initial value problem or Cauchy problem.

A large class of equations, of great interest for physics, is that of linear second-order differential equations:

$$a(x)y'' + b(x)y' + c(x)y = d(x),$$

where $a(x)$, $b(x)$, $c(x)$, and $d(x)$ are analytic within the region (of variable x) under consideration. Here we should note that there are no general methods for analytic solutions of second-order linear differential equations. However, there is a large class of equations, namely those with constant coefficients (when a, b, and c are constants), that can be solved analytically. Besides, there are many well-studied special functions that are solutions of some more complicated forms of the second-order linear differential equations.

It is common to divide the equation by $a(x)$ assuming that $a(x) \neq 0$ (points x_0 where $a(x_0) = 0$ must be given special consideration). Thus,

$$y'' + p(x)y' + q(x)y = f(x), \tag{9.1}$$

where the new notations should be obvious. A differential equation is called *homogeneous* if $f(x) = 0$ and *non-homogeneous* if $f(x) \neq 0$. Here we note that there is no general analytic solution for equation (9.1).

In the following, we first consider the general properties of homogeneous and non-homogeneous differential equations. Such general properties do not give immediate solutions but provide a systematic way to find solutions.

9.1. General solution of the homogeneous equation

We start our consideration of general properties with homogeneous equations of the form

$$y'' + p(x)y' + q(x)y = 0. \tag{9.2}$$

It follows from the linearity of equation (9.2) with respect to the function $y(x)$ and its derivatives that if $y = y_1(x)$ is a solution of the equation, then obviously $y(x) = Cy_1(x)$, where C is any constant, is also a solution. Similarly, if $y_1(x)$ and $y_2(x)$ are solutions, then

$$y(x) = C_1 y_1(x) + C_2 y_2(x) \tag{9.3}$$

is also a solution, with arbitrary constants C_1 and C_2 provided that $y_1(x)$ and $y_2(x)$ are linearly independent. This can be easily verified by direct substitution of solution (9.3) into equation (9.2). Thus, further solutions of the linear homogeneous equation can be obtained by multiplying existing solutions by arbitrary constants and adding. When we refer in the following to a solution of equation (9.2), it is assumed to differ from the trivial solution $y = 0$.

9.1.1. Wronskian and linear independence

Two solutions $y_1(x)$ and $y_2(x)$ of equation (9.2) are said to be linearly independent if no identity with respect to x exists of the form

$$C_1 y_1(x) + C_2 y_2(x) = 0,$$

where C_1 and C_2 are non-zero constant coefficients. In other words, the linear independence of y_1 and y_2 implies that the ratio y_2/y_1 is not a constant, i.e., that the derivative of the ratio

$$\frac{d}{dx}\left(\frac{y_2}{y_1}\right) = \frac{y_1 y_2' - y_2 y_1'}{y_1^2} \neq 0,$$

i.e., not identically zero for any value of x. We introduce into the discussion the expression

$$W(y_1, y_2) = y_1 y_2' - y_2 y_1', \qquad (9.4)$$

called the Wronskian of the solutions y_1 and y_2. If $W \neq 0$, then we have linear independence, and if $W = 0$, then the two solutions are linearly dependent. Note that Wronskian is a function of x.

First, we show that two solutions are linearly independent (for any value of x) if and only if $W(y_1, y_2)$ differs from zero, i.e., when $W_0 \neq 0$. Differentiating the Wronskian with respect to x

$$\frac{d}{dx}W(y_1, y_2) = y_1' y_2' + y_1 y_2'' - y_2' y_1' - y_2 y_1'' = y_1 y_2'' - y_2 y_1''.$$

Since y_1 and y_2 are solutions of equation (9.2), we can write

$$y_1'' + p(x) y_1' + q(x) y_1 = 0; \quad y_2'' + p(x) y_2' + q(x) y_2 = 0.$$

We multiply the first equation by $-y_2$ and the second by y_1 and add

$$y_1 y_2'' - y_2 y_1'' + p(x)(y_1 y_2' - y_2 y_1') = 0$$

so that

$$\frac{d}{dx}W(y_1, y_2) + p(x) W(y_1, y_2) = 0.$$

This is a linear homogeneous equation for $W(y_1, y_2)$ with the solution

$$W(y_1, y_2) = W_0 \exp\left(-\int_{x_0}^{x} p(t) dt\right).$$

It follows from this formula that $W(y_1, y_2)$ is either identically zero, if the constant W_0 is zero, or is non-zero for all values of x, since the exponential function does not vanish. We assume here that $p(x)$ is continuous. Then,

$$\frac{d}{dx}\left(\frac{y_2}{y_1}\right) = \frac{y_1 y_2' - y_2 y_1'}{y_1^2} = \frac{W(y_1, y_2)}{y_1^2} = \frac{W_0}{y_1^2} \exp\left(-\int_{x_0}^{x} p(t)dt\right) \tag{9.5}$$

and hence it follows that two solutions y_1 and y_2 of equation (9.2) are linearly independent (for any value of x) if, and only if, $W(y_1, y_2)$ differs from zero, i.e., when $W_0 \neq 0$.

We now show that if y_1 and y_2 are linearly independent solutions of equation (9.2), then (9.3) gives us, with suitable choice of constants C_1 and C_2, the solution of (9.2) satisfying any previously assigned initial conditions:

$$y(x_0) = y_0, \quad y'(x_0) = y_0'.$$

Let $y_{10}, y_{20}, y_{10}', y_{20}'$ denote the values of y_1 and y_2 and their derivatives for $x = x_0$. To satisfy the initial conditions above, we have to determine the C_1 and C_2 in (9.3) from the system of equations:

$$C_1 y_{10} + C_2 y_{20} = y_0, \quad C_1 y_{10}' + C_2 y_{20}' = y_0'.$$

It follows from the linear independence of y_1 and y_2 that

$$W_0 = y_{10} y_{20}' - y_{20} y_{10}' \neq 0$$

so that the system written gives us fully defined values for C_1 and C_2, which proves our assertion.

But by the existence and uniqueness theorem, every solution of equation (9.2) is fully defined by its initial conditions, and we can therefore state the following proposition: If y_1 and y_2 are two linearly independent solutions of equation (9.2), all the solutions of the equation are given by (9.3). Then, the problem of integrating (9.2) thus reduces to finding two linearly independent solutions.

Integrating (9.5) we can get *Liouville's formula*

$$y_2(x) = W_0(x) y_1(x) \int \frac{1}{y_1^2(x)} \exp\left(-\int_{x_0}^{x} p(t)dt\right) dx. \tag{9.6}$$

Thus if we have one particular solution for the homogeneous equation (9.2), the second solution can be derived using Liouville's formula, and we can set the constant W_0 simply to one.

9.1.2. Reduction of order

The Liouville formula can also be derived using the reduction of order method. The method constructs a second solution $y_2(x)$ of a homogeneous differential equation (9.2) when we already know one non-trivial solution $y_1(x)$ of the equation. Suppose that $y_1(x)$ is a non-trivial solution of (9.2),

$$y_1'' + p(x)y_1' + q(x)y_1 = 0.$$

Let us assume that the second solution can be written as

$$y_2(x) = u(x)y_1(x). \tag{9.7}$$

The two solutions y_1 and y_2 are linearly independent since the ratio of the two solutions is not a constant. The derivatives of y_2 are

$$y_2' = u'y_1 + uy_1', \quad y_2'' = u''y_1 + u'y_1' + u'y_1' + uy_1''.$$

Substituting y_2, y_2', and y_2'' in (9.2), we have

$$y_2'' + p(x)y_2' + q(x)y_2 = u''y_1 + u'y_1' + u'y_1' + uy_1'' \\ + (u'y_1 + uy_1')p(x) + q(x)u(x)y_1(x).$$

We can rewrite it as

$$u[y_1'' + p(x)y_1' + q(x)y_1] + y_1 u'' + (2y_1' + p(x)y_1)u' = 0.$$

Since y_1 is a solution, then the quantity in the first parenthesis is zero and the above equation is reduced to

$$y_1 u'' + (2y_1' + p(x)y_1)u' = 0.$$

This linear differential equation can be solved by making the substitution $u(x)' = v(x)$. Thus, the above equation is reduced to first-order linear homogeneous differential equation:

$$y_1 v' + (2y_1' + p(x)y_1)v = 0.$$

Using (8.6), we can write the solution for $v(x)$ as

$$v(x) = C \exp\left(-\int \left[2\frac{y_1'}{y_1} + p(x)\right] dx\right)$$

$$= C \exp\left(-2 \int \frac{y_1'}{y_1} dx\right) \exp\left(-\int p(x) dx\right).$$

Using for the derivative
$$\frac{d}{dx}\ln y_1(x) = \frac{y_1'}{y_1},$$
we can write
$$\exp\left(-2\int \frac{y_1'}{y_1}dx\right) = \exp(-2\ln y_1(x)) = \frac{1}{y_1^2}.$$
And since $y_2(x) = y_1(x)u(x) = y_1(x)\int v(x)dx$, we finally have the second solution of equation (9.2):
$$y_2(x) = y_1(x)\int \frac{1}{y_1^2(x)}\exp\left(-\int p(t)dt\right)dx. \qquad (9.8)$$

9.2. General solution of the non-homogeneous equation

Now, we consider a non-homogeneous equation where $p(x), q(x)$, and $f(x)$ are continuous in an interval $a < x < b$, and $f(x) \neq 0$:
$$y'' + p(x)y' + q(x)y = f(x). \qquad (9.9)$$
Let u be a particular solution of the above equation so that substituting u in the above equation makes the identity
$$u'' + p(x)u' + q(x)u = f(x).$$
We assume that the general solution of (9.9) can be written as a sum of this particular solution $u(x)$ and an unknown function $h(x)$, i.e.,
$$y(x) = h(x) + u(x).$$
Substituting $y(x) = h(x) + u(x)$ into (9.9), we get
$$[h'' + p(x)h' + q(x)h] + [u'' + p(x)u' + q(x)u] = f(x)$$
or
$$h'' + p(x)h' + q(x)h = 0.$$
As we see, this last equation is the homogeneous equation (9.2) corresponding to equation (9.9). If y_1 and y_2 are two linearly independent solutions of this equation, then we have the formula
$$y(x) = C_1 y_1(x) + C_2 y_2(x) + u(x) \qquad (9.10)$$
as a general solution of (9.9), where C_1 and C_2 are arbitrary constants. The property can be stated thus, the general solution of a

non-homogeneous linear equation of the second order is equal to the sum of the general solution of the corresponding homogeneous equation and any solution of the non-homogeneous equation.

9.2.1. Lagrange's method of variation of parameters

While equation (9.10) provides a general form of a solution for (9.9), the function $u(x)$ is still to be found. The method of variation of parameters is a general and infallible technique for determining a particular solution $u(x)$. The method is known as Lagrange's method of varying the arbitrary constants or variation of parameters.

Let y_1 and y_2 be two linearly independent solutions of the homogeneous equations with a general solution $C_1y_1 + C_2y_2$ (when $f(x) = 0$). We shall seek a solution of (9.9) in the same form as a solution of a homogeneous equation, except for taking C_1 and C_2 as functions of x instead of as constants:

$$u(x) = v_1(x)y_1(x) + v_2(x)y_2(x). \tag{9.11}$$

Evaluating first and second derivatives and substituting them in equation (9.9) yields

$$[v_1'' y_1 + 2v_1' y_1' + v_1 y_1'' + v_2'' y_2 + 2v_2' y_2' + v_2 y_2'']$$
$$+ p(x)[v_1' y_1 + v_1 y_1' + v_2' y_2 + v_2 y_2'] + q(x)[v_1 y_1 + v_2 y_2] = f(x).$$

Rearranging the terms in the equation we get

$$v_1[y_1'' + p(x)y_1' + q(x)y_1] + v_2[y_2'' + p(x)y_2' + q(x)y_2] + v_1'' y_1$$
$$+ v_1' y_1' + v_2'' y_2 + v_2' y_2' + p(x)(v_1' y_1 + v_2' y_2) + v_1' y_1' + v_2' y_2' = f(x).$$

The two terms in the first line are exactly zero since y_1 and y_2 are solutions of the complementary homogeneous equation. The terms in the second line can be rearranged as

$$\frac{d}{dx}[v_1' y_1 + v_2' y_2] + p(x)[v_1' y_1 + v_2' y_2] + v_1' y_1' + v_2' y_2' = f(x). \tag{9.12}$$

Since we have two functions, v_1 and v_2, to find, we need to generate two equations based on the equation above. The structure of equation (9.12) provides a hint. If we subject the unknown functions to the

condition that $v_1'(x)y_1(x) + v_2'(x)y_2(x) = 0$, then we have a system with two equations:
$$\begin{aligned} v_1'(x)y_1(x) + v_2'(x)y_2(x) &= 0, \\ v_1'(x)y_1'(x) + v_2'(x)y_2'(x) &= f(x), \end{aligned} \quad (9.13)$$
for determining $v_1'(x)$ and $v_2'(x)$.

Now, we get our attention to solving system (9.13). From the linear independence of solutions y_1 and y_2,
$$W(y_1, y_2) = y_1 y_2' - y_2 y_1' \neq 0,$$
follows that system (9.13) fully defines $v_1'(x)$ and $v_2'(x)$, namely
$$v_1' = -\frac{y_2 f}{W}, \quad v_2' = \frac{y_1 f}{W}.$$
We find $v_1(x)$ and $v_2(x)$ by carrying out the integration, then a particular solution $u(x)$ takes the form
$$u(x) = -y_1(x) \int \frac{y_2(x) f(x)}{W(y_1(x), y_2(x))} dx + y_2(x) \int \frac{y_1(x) f(x))}{W(y_1(x), y_2(x))} dx. \quad (9.14)$$
When computing the indefinite integrals, we do not need to introduce any constants. This is because for the general solution
$$\begin{aligned} y &= a_1 y_1 + a_2 y_2 + (v_1 + b_1) y_1 + (v_2 + b_2) y_2 \\ &= C_1 y_1 + C_2 y_2 + v_1 y_1 + v_2 y_2, \end{aligned}$$
where $C_1 = a_1 + b_1$ and $C_2 = a_2 + b_2$. Therefore, non-homogeneous linear differential equations are only slightly more complicated than homogeneous ones.

Summary: For finding a general solution of the non-homogeneous equation (9.9), first we need to find solutions y_1 and y_2 of the corresponding homogeneous equation. Then, a particular solution u can be found by straightforward integration (9.14), and finally, (9.10) gives general solution as
$$\begin{aligned} y(x) = C_1 y_1(x) + C_2 y_2(x) &- y_1(x) \\ \times \int \frac{y_2(x) f(x)}{W(y_1, y_2)} dx &+ y_2(x) \int \frac{y_1(x) f(x)}{W(y_1, y_2)} dx, \end{aligned} \quad (9.15)$$
where the constant of integrations C_1 and C_2 can be found from given initial or boundary conditions.

9.3. Homogeneous equations with constant coefficients

The sections above still leave open the question of how to find linearly independent solutions y_1 and y_2 for the homogeneous equation. Here, we consider a special but very common case of homogeneous equations with constant coefficients:

$$y''(x) + py'(x) + qy(x) = 0, \qquad (9.16)$$

where p and q are given numbers. Euler suggested looking for partial solutions of this equation in the form

$$y(x) = e^{\lambda x},$$

where λ is an unknown constant that can be real or complex. Then, we are going to use the property

$$(e^{\lambda x})' = \lambda e^{\lambda x}.$$

With $y = e^{\lambda x}$, $y' = \lambda e^{\lambda x}$, and $y'' = \lambda^2 e^{\lambda x}$, equation (9.16) reads

$$\lambda^2 e^{\lambda x} + p\lambda e^{\lambda x} + q e^{\lambda x} = 0$$

or

$$e^{\lambda x}(\lambda^2 + p\lambda + q) = 0.$$

Since the exponential function $e^{\lambda x}$ is never zero on $(-\infty, +\infty)$, then (9.16) will be satisfied if and only if λ is a root of the quadratic equation

$$\lambda^2 + p\lambda + q = 0; \qquad (9.17)$$

this latter equation is called the characteristic equation of the linear homogeneous equation with constant coefficients (9.16). As we know from algebra, a quadratic equation may have two distinct real roots, two equal roots, or complex roots only.

9.3.1. *Two distinct real roots*

If the quadratic equation has two distinct real roots $\lambda = \lambda_1$ and $\lambda = \lambda_2$, then

$$\lambda_{1,2} = \frac{-p \pm \sqrt{p^2 - 4q}}{2},$$

and we have two linearly independent solutions of the equation:
$$y_1(x) = e^{\lambda_1 x}, \quad y_2(x) = e^{\lambda_2 x}.$$

The linear independence follows easily from the fact that the ratio $e^{\lambda_1 x}/e^{\lambda_2 x} = e^{(\lambda_1 - \lambda_2)x}$ is not a constant. Thus, the general solution of (9.16) is

$$y(x) = C_1 e^{\lambda_1 x} + C_2 e^{\lambda_2 x}. \tag{9.18}$$

For given initial conditions $y(x_0 = 0) = y_0$ and $y'(x_0 = 0) = y'_0$, the two coefficients C_1 and C_2 provide a unique solution. The coefficients can be easily found from

$$y_0 = C_1 + C_2,$$
$$y'_0 = C_1 \lambda_1 + C_2 \lambda_2.$$

The determinant of this system of linear non-homogeneous equations (for two unknowns C_1 and C_2) is

$$\det \begin{vmatrix} C_1 & C_2 \\ C_1 \lambda_1 & C_2 \lambda_2 \end{vmatrix} = C_1 C_2 \lambda_2 - C_1 C_2 \lambda_1$$
$$= C_1 C_2 (\lambda_2 - \lambda_1) \neq 0 \quad (\lambda_1 \neq \lambda_2).$$

Solving the system of linear equations we get unique solutions for C_1 and C_2, then the solution of the differential equation is

$$y(x) = \frac{\lambda_2 y_0 - y'_0}{\lambda_2 - \lambda_1} e^{\lambda_1 x} - \frac{\lambda_1 y_0 - y'_0}{\lambda_2 - \lambda_1} e^{\lambda_2 x}.$$

9.3.2. *Two equal roots*

We now take the case when (9.16) has equal roots, i.e., when $p^2 - 4q = 0$, and the single root of the equation being given by

$$\lambda_1 = \lambda_2 = \lambda = -\frac{p}{2}.$$

Since the method described has led us only to the one solution

$$y_1(x) = e^{\lambda x} = e^{-\frac{p}{2}x},$$

the other solution remains to be found; this is done by using the reduction of order method (9.7), i.e., by looking for the second

solution in the form
$$y_2(x) = u(x)e^{-\frac{p}{2}x}.$$
Since for two equal roots we have $p^2 = 4q$, or the original equation (9.16) reads as
$$y''(x) + py'(x) + \frac{p^2}{4}y(x) = 0.$$
Substituting y_2 into the above equation gives
$$\left(u(x)e^{-\frac{p}{2}x}\right)'' + p\left(u(x)e^{-\frac{p}{2}x}\right)' + \frac{p^2}{4}u(x)e^{-\frac{p}{2}x} = 0.$$
After differentiating, we get
$$e^{-\frac{p}{2}x}\left(u'' - \frac{p}{2}u' - \frac{p}{2}u' + u\frac{p^2}{4} + pu' - u\frac{p^2}{2} + u\frac{p^2}{4}\right) = 0, \quad \text{or}$$
$$e^{-\frac{p}{2}x}u'' = 0.$$
Thus, the above equation is true when $u'' = 0$ or $u' = C$ and $u = Cx$. (Do you see why a second constant of integration is unnecessary?) Finally, for equal roots, the general solution is
$$y(x) = (C_1 + C_2 x)e^{\lambda x}. \tag{9.19}$$
The integration constants can be found, when needed, from initial conditions.

9.3.3. Complex roots

When the characteristic equation (9.17) has complex roots, they must be complex conjugates, i.e., if $\lambda_1 = \alpha + i\beta$, then $\lambda_2 = \alpha - i\beta$ (do you see why?) and the two linear independent solutions are
$$y_1(x) = e^{(\alpha+i\beta)x} = e^{\alpha x}(\cos\beta x + i\sin\beta x),$$
$$y_2(x) = e^{(\alpha-i\beta)x} = e^{\alpha x}(\cos\beta x - i\sin\beta x).$$
Thus, the general solution is
$$y(x) = C_1 e^{\lambda_1 x} + C_2 e^{\lambda_2 x}$$
$$= (C_1 + C_2)e^{\alpha x}\cos\beta x + i(C_1 - C_2)e^{\alpha x}\sin\beta x$$
which in general is a complex function. If the equation itself has only real coefficients, we may want to express the solution in real

numbers. There are a couple of ways to construct a general solution as a real function. We may show that Re y_1 and Im y_1 are also solutions of (9.16). On the other hand, there is an easier way, i.e., we use that a sum (or a difference) of two solutions is also a solution of a homogeneous differential equation. We obtain further solutions by taking the linear combinations of the above y_1 and y_2 solutions:

$$\frac{1}{2}(y_1 + y_2) = e^{\alpha x} \cos \beta x, \quad \frac{1}{2i}(y_1 - y_2) = e^{\alpha x} \sin \beta x.$$

These two solutions are also linearly independent so that in the case when equation (9.16) has complex roots $\alpha \pm \beta$, the general solution of the equation is

$$y(x) = e^{\alpha x}(C_1 \cos \beta x + C_2 \sin \beta x). \tag{9.20}$$

The past equation can be rewritten as

$$y(x) = Ce^{\alpha x}(\cos(\beta x + \varphi)), \tag{9.21}$$

where C and φ are arbitrary constants that can be defined by initial conditions.

Assume that the initial conditions are given as $y(x = 0) = y_0$ and $y'(x = 0) = y'_0$, then

$$y_0 = C \cos \varphi \quad \text{and}$$

$$y'_0 = -C\beta \sin \varphi.$$

We can easily find φ from the above equation as

$$\varphi = \arctan\left(-\frac{y'_0}{\beta y_0}\right).$$

Finding C will require a couple more steps. Using

$$\cos \varphi = \frac{y_0}{C}, \quad \sin \varphi = -\frac{y'_0}{C\beta},$$

and since $\cos^2 \varphi + \sin^2 \varphi = 1$, then

$$\frac{y_0^2}{C^2} + \frac{y_0'^2}{C^2 \beta^2} = 1,$$

and finally,

$$C = \left(y_0^2 + \frac{y_0'^2}{\beta^2}\right)^{1/2}.$$

In the special case when $y_0' = 0$, this reduces to
$$C = y_0, \quad \varphi = 0.$$

Example 9.1. Find a solution of the homogeneous equation satisfying the given conditions:
$$y'' + 6y' + 9y = 0, \quad y(0) = 1, \, y'(0) = 1.$$
The characteristic equation for the above differential equation is
$$\lambda^2 + 6\lambda + 9 = 0.$$
This equation has two equal roots, $\lambda_{1,2} = -3$. Then, the general solution for the differential equation, according to (9.19), is
$$y(x) = (C_1 + C_2 x)e^{-3x}.$$
Using the given conditions, we can easily find the first coefficient $y(0) = C_1 = 1$. To find the second coefficient, we differentiate the general solution
$$y'(x) = C_2 e^{-3x} - 3(C_1 + C_2 x)e^{-3x} = (-3C_1 + C_2 - 3C_2 x)e^{-3x},$$
$$y'(0) = -3C_1 + C_2 = 1, \quad C_2 = 4.$$
Thus, the solution for given conditions is
$$y(x) = (1 + 4x)e^{-3x}. \quad \blacksquare$$

9.4. Non-homogeneous equations with constant coefficients

A non-homogeneous linear equation with constant coefficients has a form
$$y'' + py' + qy = f(x), \tag{9.22}$$
where p and q are given real numbers and $f(x)$ is a given function of x.

According to the discussion in Section 9.2, a general solution of linear non-homogeneous equation $y'' + p(x)y' + q(x)y = f(x)$ (see equation (9.9)) can be written as
$$y(x) = C_1 y_1(x) + C_2 y_2(x) + u(x),$$

where y_1 and y_2 are linear independent solutions for a corresponding homogeneous equation $y''+p(x)y'+q(x)y = 0$ and $u(x)$ is a particular solution for the non-homogeneous equation.

In the previous section, we have learned how to find linear independent solutions y_1 and y_2 for a homogeneous equation with constant coefficients. Therefore, we now can concentrate on finding a particular solution $u(x)$ for equation (9.22).

9.4.1. Lagrange's method: Variation of constants

Lagrange's method of variation of parameters (see Section 9.2.1) has provided a path to find $u(x)$ when y_1 and y_2 are known (equation 9.14), namely

$$u(x) = -y_1(x) \int \frac{y_2(x) f(x)}{W(y_1(x), y_2(x))} dx + y_2(x) \int \frac{y_1(x) f(x)}{W(y_1(x), y_2(x))} dx,$$

with

$$W(y_1, y_2) = y_1 y_2' - y_2 y_1'.$$

Thus, finding a particular solution is reduced to a routine integration.

Note that in case of a differential equation with constant coefficients, the particular solution above can be simplified by using explicit solutions for the corresponding homogeneous equations. For example, a homogeneous equation with two distinct roots has solutions as $y_1(x) = \exp(\lambda_1 x)$ and $y_2(x) = \exp(\lambda_2 x)$. Substituting these solutions into the solution for the particular solution $u(x)$, we get

$$u(x) = \frac{1}{\lambda_2 - \lambda_1} \left(-e^{\lambda_1 x} \int e^{-\lambda_1 x} f(x) dx + e^{\lambda_2 x} \int e^{-\lambda_2 x} f(x) dx \right). \tag{9.23}$$

Example 9.2. Find a solution of the non-homogeneous differential equation for the given conditions:

$$y'' - 5y' + 6y = 2e^x + 6x - 5, \quad y(0) = 0, \; y'(0) = 1.$$

The complementary homogeneous differential equation

$$y'' - 5y' + 6y = 0$$

has the following characteristic equation:
$$\lambda^2 - 5\lambda + 6 = 0$$

with two distinct roots, $\lambda_1 = 3$ and $\lambda_2 = 2$. Thus, the linearly independent solutions of the homogeneous equation are $y_1(x) = e^{3x}$ and $y_2(x) = e^{2x}$. The Wronskian for the two solutions y_1 and y_2 is $W(y_1, y_2) = -e^{5x}$. The corresponding integrals can be easily evaluated:

$$\int \frac{y_2(x)f(x)}{W(y_1(x), y_2(x))} dx = -\int e^{-5x} e^{2x} (2e^x + 6x - 5) dx$$
$$= e^{-3x}(e^x + 2x - 1),$$

$$\int \frac{y_1(x)f(x)}{W(y_1(x), y_2(x))} dx = -\int e^{-5x} e^{3x} (2e^x + 6x - 5) dx$$
$$= e^{-2x}(3x + 2e^x - 1),$$

then
$$u(x) = -e^{3x} e^{-3x}(e^x + 2x - 1) + e^{2x} e^{-2x}(3x + 2e^x - 1) = e^x + x,$$

and then the general solution of the non-homogeneous equation is
$$y(x) = C_1 e^{3x} + C_2 e^{2x} + e^x + x.$$

Now, we can find the coefficients C_1 and C_2 satisfying the given conditions:
$$y(0) = C_1 + C_2 + 1 = 0,$$
$$y'(0) = 3C_1 + 2C_2 + 2 = 1.$$

Solving the system of linear equations, we have $C_1 = 1$, $C_2 = -2$, and finally,
$$y(x) = e^{3x} - 2e^{2x} + e^x + x. \qquad \blacksquare$$

9.4.2. Euler's method: Undetermined coefficients

With special forms $f(x)$ of the right-hand side of equation (9.22), a particular solution $u(x)$ can be found without the integration involved in Lagrange's method of variation of the arbitrary parameters. Let the right-hand side $f(x)$ of the non-homogeneous equation be a sum of two terms:

$$y'' + py' + qy = f_1(x) + f_2(x),$$

and let $u_1(x)$ and $u_2(x)$ be particular solutions of the non-homogeneous equations whose right-hand sides are respectively $f_1(x)$ and $f_2(x)$, i.e.,

$$u_1'' + pu_1' + qu_1 = f_1(x), \quad u_2'' + pu_2' + qu_2 = f_2(x).$$

We obtain on adding

$$(u_1 + u_2)'' + p(u_1 + u_2)' + q(u_1 + u_2) = f_1(x) + f_2(x)$$

so that $u_1 + u_2$ is a particular solution of the above equation. Thus, if the function $f(x)$ can be (or already is) written as a sum of two or more terms, then we need to find a particular solution corresponding to every term in $f(x)$, and the overall particular solution is the sum of all u_i.

The method of undetermined coefficients works very well only if $f(x)$ consist of terms, each of which has a *finite number of linearly independent derivatives*. This restriction implies that $f(x)$ can only contain terms such as $a, x^n, e^{ax}, \sin(ax), \cos(ax)$, and a combination of such terms, where a is a constant and n is a positive integer. For most common cases, we have a finite set that includes

$$x^n : x^n, x^{n-1}, x^{n-2}, \ldots, 1;$$
$$e^{ax} : e^{ax};$$
$$\sin(ax) : \sin(ax), \cos(ax);$$
$$\cos(ax) : \cos(ax), \sin(ax).$$

For example, for a non-homogeneous term, such as $f(x) = 4x^2 + 6e^x - 2\sin(x)$, successive differentiation gives $f'(x) = 8x + 6e^x - 2\cos(x)$, $f''(x) = 8 + 6e^x + 2\sin(x)$, $f'''(x) = 6e^x + 2\cos(x)$. As one can see, the third derivative does not bring new linearly independent terms. Thus, only the set $x^2, x, 1, e^x, \sin(x), \cos(x)$ is the set of linearly independent functions.

To find $u(x)$ using the method of undetermined coefficients, we must compare the terms of $f(x)$ in with solutions y_1 and y_2 of the complementary homogeneous equation. We may have three different possibilities. Each possibility must be treated separately.

Case 1: There is no single term in $f(x)$ that is the same as a term in the solution of the homogeneous equation. In this case, a particular solution $u(x)$ is be a linear combination of the terms in $f(x)$ and all its linearly independent derivatives.

Example 9.3. Let's find a general solution of the non-homogeneous differential equation from Example 9.2:

$$y'' - 5y' + 6y = 2e^x + 6x - 5.$$

The solutions of the homogeneous equations are $y_1(x) = e^{3x}$ and $y_2(x) = e^{2x}$. For $f(x) = 2e^x + 6x - 5$, the successive differentiation gives $f'(x) = 2e^x + 6$ with no new linearly independent terms. Hence, the trial solution for $u(x)$ is a linear combination

$$u(x) = Ae^x + Bx + C,$$

with $u'(x) = Ae^x + b$ and $u''(x) = Ae^x$. Substituting the trial solution into the original equation reads

$$Ae^x - 5(Ae^x + B) + 6(Ae^x + bx + C) = 2e^x + 6x - 5.$$

For the equality to hold at any x, we have

$$Ae^x - 5Ae^x + 6Ae^x = 2e^x; \quad A = 1,$$
$$6Bx = 6x; \quad B = 1,$$
$$-5B + 6C = -5; \quad C = 0,$$

and thus the particular solution is $u(x) = e^x + x$, the same as in Example 9.2. The general solution of the equation is therefore

$$y(x) = C_1 e^{3x} + C_2 e^{2x} + e^x + x. \quad \blacksquare$$

Case 2: $f(x)$ contains a term which (ignoring constant coefficients) is x^n times a term in the homogeneous solution. In this case, a particular solution will be a linear combination of $x^{n+1}u(x)$ and all its linearly independent derivatives (ignoring constant coefficients). If in addition $f(x)$ contains terms which belong to Possibility 1, then the proper terms called for by this case must be included in $u(x)$. For example, for the equation $y'' - 3y' + 2y = 2x^2 + 3e^{2x}$, the solutions of the complementary homogeneous equation are $y_1 = e^x$ and $y_2 = e^{2x}$. Then, we see that $f(x)$ has the term $y_2 = e^{2x}$. Then, after successive differentiation of $f(x)$ the trial solution is

$$u(x) = Ax^2 + Bx + C + Dxe^{2x};$$

in the last term, we multiplied e^{2x} by x^{0+1}.

Case 3: If the homogeneous solution has the form $(C_1+C_2x)e^{\lambda x}$ (i.e., $\lambda_1 = \lambda_2 = \lambda$) AND there is a term in $f(x)$ that is x^n times a term in the homogeneous solution, then a particular solution will be a linear combination of $x^{n+2}u(x)$ and all its linearly independent derivatives. If in addition $f(x)$ contains terms which belong to Possibilities 1 and 2, then the proper terms called for by these cases must also be added to the proper solution.

9.5. Series solutions

Most nonlinear as well as linear equations with variable coefficients are not integrated in elementary functions, such as algebraic, trigonometric, and logarithmic functions. However, very many problems in mathematical physics are expressed in terms of such second-order equations. A popular method of solving such equations is finding solutions in the form of a series. This approach works for both homogeneous and non-homogeneous linear differential equations. In the following, we mostly concentrate on homogeneous second-order linear ODEs:

$$y'' + p(x)y' + q(x)y = 0. \tag{9.24}$$

For now, we treat the coefficients $p(x)$ and $q(x)$ as analytic functions of x.

Series solutions for non-homogeneous equations can be found in the same manner as for homogeneous ODEs.

9.5.1. Analytic coefficients

If the functions $p(x)$ and $q(x)$ can be written as Taylor series[1] about a point $x = x_0$ and the series are convergent for $|x - x_0| < R$, then the solution for (9.24) can be represented as Taylor series

$$y(x) = y(x_0) + y'(x_0)(x - x_0) + \frac{y''(x_0)}{2}(x - x_0)^2$$

$$+ \cdots + \frac{y^{(k)}(x_0)}{k!}(x - x_0)^k + \cdots \qquad (9.25)$$

that is also convergent for any $|x - x_0| < R$. In the following, we consider for simplicity $x_0 = 0$. Note that in vary many practical applications Taylor series for $p(x)$ and $q(x)$ are convergent everywhere on the real line.

For the initial value problem, $y(0)$ and $y'(0)$ are given by the initial conditions. Then, we may find a particular solution by *successive differentiation* of the given equation to get $y''(0)$, $y'''(0)$, and so on. We illustrate the method using the following example.

Example 9.4. Find a particular solution for an equation with given conditions:

$$y'' - xy' + y - 1 = 0, \quad y(0) = y'(0) = 0.$$

From the given conditions follows that the first two terms in (9.25) are zero. Rearranging terms in the equation, we have

$$y'' = xy' - y + 1, \quad \text{then } y''(0) = -y(0) + 1 = 1.$$

Differentiation of the equation gives

$$y^{(3)} = xy'', \quad y^{(3)}(0) = 0;$$

$$y^{(4)} = xy''' + y'', \quad y^{(4)}(0) = 1;$$

$$y^{(5)} = xy^{(4)} + 2y''', \quad y^{(5)}(0) = 0;$$

$$y^{(6)} = xy^{(5)} + y^{(4)} + 2y^{(4)}, \quad y^{(6)}(0) = 3.$$

---

[1] A real function $f(x)$ is said to be analytic if it can be represented as Taylor series and has derivatives of all orders.

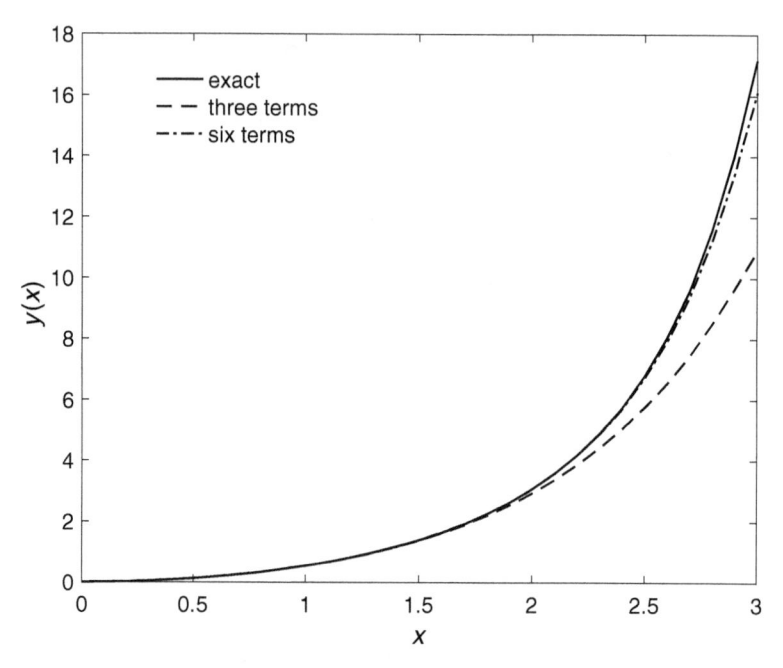

Fig. 9.1. Comparison the exact solution with two series solutions. Solid line is the exact solutions, dash line is the three non-zero term solution, and dash-dot line is the six-non-zero term solution.

Continuing the successive differentiation, we get as many coefficients of the Taylor series as needed. The particular solution to the equation is

$$y(x) = \frac{x^2}{2!} + \frac{x^4}{4!} + \frac{3x^6}{6!} + \cdots + \frac{(2k+1)!!x^{2k+4}}{(2k+4)!} \cdots$$

As one can see from Figure 9.1, the 6-term solution works very well in the considered interval of x. ∎

Note that the method of successive differentiation can work for both linear and nonlinear ODEs.

For finding a *general solution* as a power series solution, we are looking for a solution of equation (9.24) as

$$y(x) = c_0 + c_1 x + c_2 x^2 + \cdots = \sum_{k=0}^{\infty} c_k x^k, \qquad (9.26)$$

where $c_0 = y(0)$, $c_1 = y'(0)$, and $c_k = y^{(k)}(0)/k!$ are coefficients to be found. Differentiating the series above gives

$$y'(x) = 1c_1 + 2c_2 x + 3c_3 x^2 + \cdots = \sum_{k=1}^{\infty} k c_k x^{k-1},$$

$$y''(x) = 2 \cdot 1 c_2 + 3 \cdot 2 c_3 x + 4 \cdot 3 c_4 x^2 + \cdots = \sum_{k=1}^{\infty} k(k-1) c_k x^{k-2}.$$

(9.27)

Next, we represent the coefficients $p(x)$ and $q(x)$ as power series as well

$$p(x) = a_0 + a_1 x + a_2 x^2 + \cdots = \sum_{i=0}^{\infty} a_i x^i,$$

$$q(x) = b_0 + b_1 x + b_2 x^2 + \cdots = \sum_{i=0}^{\infty} b_i x^i.$$

(9.28)

Now, we substitute both the function (9.26) and its derivatives (9.27) together with the coefficients (9.28) written as series into equation (9.24):

$$\sum_{k=1}^{\infty} k(k-1) c_k x^{k-2} + \sum_{i=0}^{\infty} a_i x^i \sum_{k=1}^{\infty} k c_k x^{k-1} + \sum_{j=0}^{\infty} b_j x^j \sum_{k=0}^{\infty} c_k x^k = 0.$$

By cross-multiplying the series and equating to zero the coefficients at the same power of x, we get a *recurrent* system of algebraic equations to determine coefficients c_k:

$$2 \cdot 1 \cdot c_2 + a_0 c_1 + b_0 c_0 = 0;$$

$$3 \cdot 2 \cdot c_3 + 2 a_0 c_2 + (a_1 + b_0) c_1 + b_1 c_0 = 0;$$

$$4 \cdot 3 \cdot c_4 + 3 a_0 c_3 + (2 a_1 + b_0) c_2 + (a_2 + b_1) c_1 + b_2 c_0 = 0;$$

$$\cdots\cdots\cdots\cdots\cdots\cdots\cdots\cdots\cdots$$

$$(k+2)(k+1) c_{k+2} + \sum_{i=1}^{k} [(i+1) a_{k-i} c_{i+1} + b_{k-i} c_i] = 0.$$

The coefficients c_0 and c_1 are remain arbitrary and play a role of given conditions in the Cauchy (initial value) problem. Further, each equation determines the next coefficient starting with c_2 (the first is c_2, the second is c_3, etc.).

For finding two linear independent solutions y_1 and y_2, we use for y_1 the initial conditions

$$y_1(0) = c_0 = 1, \quad y_1'(0) = c_1 = 0,$$

and for the function y_2

$$y_2(0) = c_0 = 0, \quad y_1'(0) = c_1 = 1.$$

Then, functions $y_1(x)$ and $y_2(x)$ are to be linearly independent on any interval including $x = 0$. The general solution of equation (9.24) takes the form

$$y(x) = Ay_1(x) + By_2(x),$$

where A and B are arbitrary constants. If the initial conditions are given as $y(0) = y_0$ and $y'(0) = y_0'$, then $A = y_0$ and $B = y_0'$.

By formal calculations, we have shown how to formally determined the coefficients c_k of the series (9.25). However, it still remains unanswered whether the constructed series will converge and will provide the solution of the equation. From the analysis of series solutions for differential equations follows that if the power series for $p(x)$ and $q(x)$ are convergent for $|x| < R$, the power series for $y(x)$ derived by the above method is also convergent for these values of x and gives a solution of equation. In particular, if $p(x)$ and $q(x)$ are polynomials in x, then the power series obtained is convergent for any x. By using power series for solving a differential equation, we often get so-called special functions, widely used in physics (Bessel functions, hypergeometric functions, etc.)

Note: It is not always possible to derive a general term for the coefficients c_k. In this case, we may just settle for getting the first few terms of the series.

Example 9.5. Solve the equation

$$y'' - xy = 0.$$

This equation is called Airy's equation, and it can be found in quantum mechanics, in aerodynamics, and in studying diffraction of light to name but a few applications. Despite it seemingly simple form, the equation cannot be solved in elementary functions. Substituting series (9.25) and (9.26), the equation we get

$$(2\cdot 1 c_2 + 3\cdot 2 c_3 x + 4\cdot 3 c_4 x^2 + 5\cdot 4 c_5 x^3 \ldots) - x(c_0 + c_1 x + c_2 x^2 + c_3 x^3 \ldots) = 0,$$

and then, equating to zero the coefficients for the same powers of x, we get

$$x^0: \quad 2\cdot 1 c_2 = 0, \quad c_2 = 0;$$
$$x^1: \quad 3\cdot 2 c_3 - c_0 = 0, \quad c_3 = \frac{c_0}{2\cdot 3};$$
$$x^2: \quad 4\cdot 3 c_4 - c_1 = 0, \quad c_4 = \frac{c_1}{3\cdot 4};$$
$$x^3: \quad 5\cdot 4 c_5 - c_3 = 0, \quad c_5 = \frac{c_2}{4\cdot 5};$$
$$\ldots\ldots$$
$$x^k: \quad (k+2)(k+1)c_{k+2} - c_{k-1} = 0 \quad c_k = \frac{c_{k-3}}{(k-1)k}.$$

Then, we have

$$y(x) = c_0 \left[1 + \frac{x^3}{2\cdot 3} + \frac{x^6}{2\cdot 3\cdot 5\cdot 6} + \frac{x^{3k}}{2\cdot 3\cdot 5\cdot 6\ldots(3k-1)3k} + \cdots \right]$$
$$+ c_1 \left[x + \frac{x^4}{3\cdot 4} + \frac{x^7}{3\cdot 4\cdot 6\cdot 7} + \frac{x^{3k+1}}{3\cdot 4\cdot 6\cdot 7\ldots(3k+1)3k} + \cdots \right],$$

where c_0 and c_1 are remain arbitrary constants in the general solution. According to d'Alembert's ratio test (1.13), the series above converges for all $|x| < \infty$. ∎

9.5.2. *Solutions about regular singular points*

So far we considered functions $p(x)$ and $q(x)$ as analytic functions of x. However, many problems in mathematical physics result in differential equations with singularities. For example, Bessel equation

$$x^2 y'' + xy + (x^2 - \nu^2)y = 0, \quad \nu = \text{constant}.$$

This equation can be rewritten in the form $y'' + p(x)y' + q(x)y = 0$, where the coefficients

$$p(x) = \frac{1}{x}, \quad q(x) = \frac{x^2 - \nu^2}{x^2}$$

have *regular singular point* at $x = 0$.

We define a regular singular point when the coefficients have the following structure:

$$p(x) = \frac{p_1(x)}{x}, \quad q(x) = \frac{q_1(x)}{x^2}, \tag{9.29}$$

where $p_1(x)$ and $q_1(x)$ are analytic functions and can be written using power series as

$$p_1(x) = a_0 + a_1 x + a_2 x^2 + \cdots = \sum_{i=0}^{\infty} a_i x^i,$$

$$q_1(x) = b_0 + b_1 x + b_2 x^2 + \cdots = \sum_{i=0}^{\infty} b_i x^i.$$

In other words, a regular singular point[2] for a function $p(x)$ is a pole of *no higher than first order*, and for a function $q(x)$, it is a pole of *no higher than second order*. It is clear that the point $x = 0$ will be singular if at least one of the coefficients a_0, b_0, and b_1 is non-zero. Otherwise, $x = 0$ will be a removable singular point, and the case of analytic coefficients will apply.

Substituting (9.29) into equation (9.24) and multiplying by x^2 we get another form of the original equation as

$$x^2 y'' + x p_1(x) y' + q_1(x) y = 0. \tag{9.30}$$

In this case, the series method above should be extended to find a solution about a regular singular point. One should look for the series solution in the form of a generalized power series, or **Frobenius series**:

$$y(x) = x^\mu \sum_{k=0}^{\infty} c_k x^k, \tag{9.31}$$

---

[2] A singular point that is not regular is said to be an *irregular singular point* of the equation.

where μ is a constant to be determined and it can be real or complex, and where $c_0 \neq 0$. Substituting the Frobenius series together with the series for $p_1(x)$ and $q_1(x)$ into (9.30) gives a set of recurrence relations:

$$x^{\mu+0}: \quad \mu^2 + (a_0 - 1)\mu + b_0 = 0;$$

$$x^{\mu+1}: \quad c_1(\mu^2 + \mu + a_0\mu + a_0) + c_0(a_1\mu + b_1) = 0$$

$$\cdots$$

The first equation is called the *indicial* equation.

The essence of the Frobenius methods can be summarized as follows.

Let

(a) $x = 0$ be a singular regular point of equation (9.30) with coefficients a_0, b_0, and b_1 not equal to zero simultaneously, and the series for $p_1(x)$ and $q_1(x)$ convergent for $|x| < L$,

(b) μ_1 and μ_2 (Re $\mu_1 \geq$ Re μ_2) are roots of the indicial equation

$$\mu^2 + (a_0 - 1)\mu + b_0 = 0. \tag{9.32}$$

Then, we have three possible cases:

Case 1: Distinct root with $\mu_1 - \mu_2$ is not an integer, i.e., $\mu_1 - \mu_2 \neq n$. Then, the system for the coefficients c_k is

$$d_k c_k = F_k(c_0, c_1, c_2, \ldots, c_{k-1}), \quad k = 1, 2, \ldots,$$

where

$$d_k = \begin{cases} k(k + \mu_1 - \mu_2), & \text{if } \mu = \mu_1, \\ k(k - \mu_1 + \mu_2), & \text{if } \mu = \mu_2, \end{cases}$$

and F_k are known functions of the coefficients $c_0, c_1, c_2, \ldots, c_{k-1}$. As we can see, if $\mu_1 - \mu_2$ is not an integer, then $d_k \neq 0$ and the system of equations above lets us create two linear independent solutions of (9.30) type, namely

$$y_1(x) = x^{\mu_1} \sum_{k=0}^{\infty} c_k^{(1)} x^k, \quad \text{and}$$

$$y_2(x) = x^{\mu_2} \sum_{k=0}^{\infty} c_k^{(2)} x^k \quad (c_0^{(1)}, c_0^{(2)} \neq 0) \tag{9.33}$$

with the coefficients $c_k^{(1)}$ and $c_k^{(1)}$ as determined by direct substitution of (9.33) into the original differential equation.

Case 2: Repeated roots $\mu_1 = \mu_2 = \mu$.
The first solution y_1 has the form (9.33)

$$y_1(x) = x^\mu \sum_{k=0}^{\infty} c_k^{(1)} x^k,$$

and the second solution can be represented as

$$y_2(x) = y_1(x) \ln(x) + x^\mu \sum_{k=1}^{\infty} c_k^{(2)} x^k.$$

One can find the coefficients $c_k^{(2)}$ by direct substitution of the above $y_2(x)$ into the original differential equation.

Case 3: Roots are different but by an integer, namely $\mu_1 - \mu_2 = n$. Then, equation (9.30) has only one solution of type (9.31) since the recurrent relation corresponding to the smaller root may or may not lead to a second linear independent solution. Indeed, if $\mu_1 - \mu_2 = n$ where $n \geq 0$ and is an integer, then we can find only one solution corresponding to μ_1. For the $\mu = \mu_2$ at $k = n$, we have $d_n = 0$ and hence we cannot find solutions for $c_n, c_{n+1}, \ldots$. However, in cases $\mu_1 = \mu_2$ and $\mu_1 - \mu_2 = n$, the second solution can be found using Liouville formula (9.6) or the reduction of order method (9.8). Or it can also be found using

$$y_2(x) = c y_1(x) \ln(x) + x^{\mu_2} \sum_{k=0}^{\infty} c_k^{(2)} x^k,$$

where c is a constant that, in some cases, might be zero.

Example 9.6. We consider equation

$$4xy'' + 2y' + y = 0,$$

where $x = 0$ is a regular singular point. The equation can be rewritten as

$$y'' + \frac{1}{2x} y' + \frac{1}{4x} y = 0$$

with $p_1(x) = 1/2$ and $q_1(x) = 1/4$ thus giving $a_0 = 1/2$ and $b_0 = 1/4$. Substituting the Frobenius series (9.31) into the original differential equation and collecting terms at the lowest power of x, namely $x^{\mu-1}$, we get the indicial equation as

$$4\mu(\mu - 1) + 2\mu = 0$$

with two distinct solutions $\mu_1 = 1/2$ and $\mu_2 = 0$. Using $\mu_1 = 1/2$ in the Frobenius series, we can get the recurrence connection for $c_k^{(1)}$ coefficients as

$$c_{k+1}^{(1)} = -\frac{c_k^{(1)}}{(2k+3)(2k+2)}, \quad k = 0, 1, 2, \ldots$$

or

$$c_1^{(1)} = -\frac{c_0^{(1)}}{3!}, \quad c_2^{(1)} = \frac{c_1^{(1)}}{5\cdot 4} = -\frac{c_0^{(1)}}{5!}, \quad c_3^{(1)} = \frac{c_2^{(1)}}{7\cdot 6} = -\frac{c_0^{(1)}}{7!}, \ldots$$

By setting $c_0^{(1)} = 1$, we have the first solution as

$$y_1(x) = x^{\frac{1}{2}} \sum_{k=0}^{\infty} \frac{(-1)^k}{(2k+1)!} x^k.$$

Similarly, the second solution can be derived as

$$y_2(x) = \sum_{k=0}^{\infty} \frac{(-1)^k}{(2k)!} x^k.$$

∎

9.6. Periodic series solutions

Using Fourier series is a useful tool for solving non-homogeneous differential equations. In this section, we consider an application of Fourier series to second-order ODEs with constant coefficients. We consider equation

$$y'' + py' + qy = f(x), \tag{9.34}$$

where $f(x)$ satisfy the Dirichlet conditions: (a) The function must be periodic, (b) it must be single values and continuous, (c) it must have only finished number of maxima and minima within one period, and (d) the integral over one period of $|f(x)|$ must converge. The last three Dirichlet conditions are almost always met in real applications. Then, we can represent $f(x)$ as trigonometric Fourier series (7.13):

$$f(x) = \frac{a_0}{2} + \sum_{k=1}^{\infty}(a_k \cos kx + b_k \sin kx).$$

We assume that $f(x)$ is periodic with a period 2π. Note that it is rather simple to extend the Fourier series expansions to functions of period $2L$. In this case, we replace the variable x on $\pi x/L$ (see Section 7.3.5).

We are looking for a periodic solution of $y(x)$ of equation (9.34) as Fourier series as well

$$y(x) = \frac{A_0}{2} + \sum_{k=1}^{\infty}(A_k \cos kx + B_k \sin kx). \qquad (9.35)$$

Substituting both series into equation (9.34) and equating to zero coefficients in front of $\cos kx$ and $\sin kx$, we generate an infinite system of linear equations for coefficients A_0, A_k, B_k:

$$A_0 q = a_0, \qquad (9.36)$$

$$A_k[(q-k^2)^2 + p^2 k^2] = (q-k^2)a_k - pkb_k, \quad k=1,2,\ldots, \qquad (9.37)$$

$$B_k[(q-k^2)^2 + p^2 k^2] = (q-k^2)b_k - pka_k, \quad k=1,2,\ldots. \qquad (9.38)$$

From equation (9.36) follows that we have one out of the three cases:

(a) If $q \neq 0$, then $A_0 = a_0/q$ with

$$a_0 = \frac{1}{2\pi}\int_0^{2\pi} f(x)dx,$$

(b) if $q = 0$, then a_0 must be zero, and A_0 is an arbitrary constant, and

(c) if $q = 0$ and $a_0 \neq 0$, then there is no periodic solution.

Further, from equations (9.37) and (9.38), we see that we have one of the following situations:

(a) If $p \neq 0$, then we have solutions for the Fourier coefficients as

$$A_k = \frac{(q-k^2)a_k - pkb_k}{(q-k^2)^2 + p^2 k^2} \quad \text{and}$$

$$B_k = \frac{(q-k^2)b_k - pka_k}{[(q-k^2)^2 + p^2 k^2]}. \qquad (9.39)$$

(b) If $p = 0$ and $q \neq k^2$ (for any k), then

$$A_k = \frac{a_k}{q-k^2} = \frac{1}{q-k^2}\frac{1}{\pi}\int_0^{2\pi} f(x)\cos kx\,dx \quad \text{and}$$

$$B_k = \frac{b_k}{q-k^2} = \frac{1}{q-k^2}\frac{1}{\pi}\int_0^{2\pi} f(x)\sin kx\,dx. \qquad (9.40)$$

(c) If for some $k^2 = n^2 = q$ and at the same time $a_n = 0$ and $b_n = 0$, then $A_n \cdot 0 = 0$ and $B_n \cdot 0 =$ and hence A_n and B_n are arbitrary constants, but equation (9.34) does have a periodic solution.

(d) If for some $k^2 = n^2 = q$ but at least one of the coefficients $a_n \neq 0$ or $b_n \neq 0$, then there is no periodic solution.

Example 9.7. Find a periodic solution

$$y'' + 4y = \sum_{k=3}^{\infty} \frac{\cos kx}{k^2}.$$

For this equation, we have $p = 0$, $q = 2^2$, $a_o = 0$, where k starts from 3. Therefore, this is case (b) with solution (9.40):

$$A_k = \frac{1}{(4-k^2)k^2}, \quad B_k = 0, \quad k = 3, 4 \ldots.$$

Then, the periodic solution for the non-homogeneous equations is

$$\sum_{k=3}^{\infty} \frac{\cos kx}{(4-k^2)k^2},$$

and all periodic solutions for the original equations are

$$y(x) = A_2 \cos 2x + B_2 \sin 2x + \sum_{k=3}^{\infty} \frac{\cos kx}{(4-k^2)k^2},$$

where A_2 and B_2 are arbitrary constants. ∎

9.7. Boundary value problem

A second-order linear differential equation

$$y'' + p(x)y' + q(x)y = f(x)$$

has a general solution as indicated in equation (9.10):

$$y(x) = C_1 y_1(x) + C_2 y_2(x) + u(x),$$

where C_1 and C_2 are arbitrary constants. So far, for obtaining a particular solution out of the general one, we called initial conditions, when the function $y(x_0)$ and its derivative $y'(x_0)$ were given as additional condition at the same point x_0. It is natural to use such forms of additional conditions if time is an independent variable, i.e., to study a processes developing in time as a motion of an object. Such forms of additional conditions are called *initial conditions* and the problem itself is called *initial value problem*. However, this is not the only way to specify additional conditions. For example, for vibrations of a string fixed at points $x = a$ and $x = b$, we impose the conditions as values of the function $y(a) = 0$ and $y(b) = 0$ at boundaries. Finding a solution of a differential equation satisfying given boundary conditions is called a *boundary value problem*. Examples of boundary value problems include static equilibrium, oscillations with normal modes, Sturm–Liouville problem, and one-dimensional diffusion problem.

While the boundary value problem is studied in Chapter 10, here we would like to briefly analyze similarities and differences between the initial and boundary value problems.

Let us consider a solution of equation

$$y'' + p(x)y' + q(x)y = f(x) \quad (a \le x \le b)$$

on the interval $[a, b]$ with additional boundary conditions

$$y(a) = y_a, \quad y(b) = y_b.$$

As we know, the solution of the equation has the form

$$y(x) = C_1 y_1(x) + C_2 y_2(x) + u(x),$$

with $y_1(x)$ and $y_2(x)$ being the solutions of the corresponding homogeneous equation and $u(x)$ being some particular solution. Applying the given conditions to the solution gives a system of linear equations for the unknown constants C_1 and C_2:

$$\begin{aligned} C_1 y_1(a) + C_2 y_2(a) &= y_a - u(a) \quad \text{and} \\ C_1 y_1(b) + C_2 y_2(b) &= y_b - u(b). \end{aligned} \quad (9.41)$$

There are two possible situations for the linear system (9.41).

General case: The determinant of the system is not equal to zero. In this case, system (9.41) has a unique solution for any nonhomogeneous term $f(x)$ and values y_a and y_b.

Special case: The determinant is equal to zero. Then, system (9.41) does not have a solution, but for some $f(x)$ and values of y_a and y_b, it may have infinite number of solutions. For example, it is possible to show (addressed later) that for given $f(x)$ and y_a there is only one value of y_b that gives an infinite number of solutions.

Note that for the initial value problem (or Cauchy problem), we always have the general case, that is, there exists a unique solution for the system. Exercise for a reader: Prove it

(*Hint*: Use the condition for linear independence).

Example 9.8. We consider the boundary value problem for the same equation

$$y'' + y = 0$$

but for two sets of boundary conditions:

(1) Consider first

$$0 \leq x \leq \frac{\pi}{2}, \quad y(0) = y_a, \quad y\left(\frac{\pi}{2}\right) = y_b.$$

The general solution of this problem (see Section 9.3.3) is

$$y(x) = C_1 \cos x + C_2 \sin x$$

because the characteristic equation has complex conjugate, pure imaginary roots. Using the boundary conditions, we get

$$C_1 = y_a, \quad C_2 = y_b.$$

Thus, for any values of y_a and y_b, we have the particular solution as

$$y(x) = y_a \cos x + y_b \sin x.$$

(2) Next, we consider the boundary value problem for the same equation but given as

$$0 \leq x \leq \pi, \quad y(0) = y_a, \quad y(\pi) = y_b.$$

Substituting these conditions into the general solution $y(x) = C_1 \cos x + C_2 \sin x$ gives

$$C_1 = y_a, \quad -C_1 = y_b \quad \text{or} \quad C_1 = -y_b.$$

Thus, if $y_a \neq -y_b$, the boundary value problem does not have a solution. If $y_a = -y_b$, then the problem has infinite number of solutions as

$$y(x) = y_a \cos x + C_2 \sin x,$$

where C_2 is any arbitrary constant. ∎

Often, the boundary value problems lead to *eigenvalue problems*, when a solution for a boundary value problem exists only for a set of values of a parameter of an equation. The Sturm–Liouville problem is one most commonly used in physics. The solutions to the Sturm–Liouville problem are associated with eigenfunctions and eigenvalues. The eigenfunctions form an orthogonal set, and the corresponding eigenvalues are often discrete. These properties make Sturm–Liouville problems fundamental in various branches of physics and engineering, including quantum mechanics, heat conduction, and vibration analysis. The Sturm–Liouville problem is analyzed in Section 10.6.

9.8. WKB(J) approximation

Why the (J) in the WKB acronym? It is to pay homage to the legacy of Sir Harold Jeffreys in this and many other areas of applied mathematics and mathematical physics. As an eighteen-year-old student, I (John Adam) bought a copy of the celebrated *Methods of Mathematical Physics*[3] co-written with his wife, Lady Bertha Jeffreys. I vowed that I would try to understand as much of it as possible (and I have tried). It is a magnificent book. But writing WKBJ might be a bridge too far, as they say — the majority of citations in the literature appear to use the WKB form, so I have adapted the acronym accordingly. And it appears that I am not the only one to think Jeffreys deserves more credit than he has been given.

The WKB approximation is a "semiclassical calculation" in quantum mechanics in which the wavefunction is assumed an exponential function with amplitude and phase that slowly varies compared to

---

[3] H. Jeffries and B. S. Jeffries, *Methods of Mathematical Physics*. Cambridge University Press, Cambridge, 1966.

the de Broglie wavelength λ and is then semiclassically expanded. While Wentzel, Kramers, and Brillouin developed this approach in 1926, earlier in 1923, a mathematician, Harold Jeffreys, had already developed a more general method of approximating linear, second-order differential equations; Jeffreys is rarely given his proper credit.

In summarizing the various contributions to approximating a quantum mechanical wavefunction by an oscillatory wave depending on a phase integral, physicist J. Calvert (i) noted the alternative notations for the approximation, such as WKB, BWK, BWKJ, adiabatic, semiclassical, or phase integral, (ii) cited the work by Wentzel, Kramers, Brillouin, Jeffreys, Rayleigh, Liouville, Denham, Langer, and Furry, and (iii) wrote "This would make it the WKBJRLDLF approximation, I suppose".[4]

9.8.1. The WKB(J) approximation: An informal treatment

We consider the following generalized one-dimensional "wave equation" for $y = y(x)$, namely

$$y''(x) + W^2(x)y(x) = 0, \qquad (9.42)$$

where the prime notation represents differentiation with respect to x. In this equation, $y(x)$ could represent, for example, the amplitude of a wavefunction, or electromagnetic field variable, or an acoustic wave, or a water wave; there are many possible contexts that could be considered. The real quantity $W^2(x)$ is assumed to be "slowly varying" in a sense to be defined as follows. The approximate solutions that will be constructed will not be valid in the vicinity of "turning points", that is, values of x for which $W(x) = 0$. Note that if W^2 is a constant (as in the case of a uniform medium), then the general solution of equation (9.42) will be a linear combination of the so-called fundamental solutions $\exp(iWx)$ and $\exp(-iWx)$. This observation suggests a basis for the WKBJ approximation when W^2 is not constant, i.e., when the medium is not uniform.

---

[4]See the first quotation in J. A. Adam, *Rays, Waves, and Scattering*, Chapter 22. Princeton University Press, 2017.

Ignoring any multiplicative constants, let
$$y(x) = \exp[\pm iu(x)], \quad (9.43)$$
which on substitution into (9.42) yields
$$iu'' - (u')^2 + W^2(x) = 0. \quad (9.44)$$
Note that $u'' = 0$ if W is a constant, so if in some sense $W(x)$ is *slowly varying*, the term u'' may well be negligible compared with $(u')^2$, so we can write
$$(u')^2 \approx W^2(x). \quad (9.45)$$
The conditions under which this approximation is valid are derived in the following. From this point on, approximation (9.45) will be treated as exact with solution set
$$u(x) = \pm \int_0^x W(\xi) d\xi \quad (9.46)$$
(ξ is a dummy variable, and in what follows, we use the '$\pm$' to indicate the two linearly independent solutions). Therefore, we have what is referred to as the zeroth-order WKB approximation for $y(x)$:
$$y(x) \approx y_0(x) = \exp\left[\pm i \int_0^x W(\xi) d\xi\right]. \quad (9.47)$$
Thus, the total phase change undergone by the wave is determined by the integral of $W(x)$ over the range of interest in the inhomogeneous medium.

What is the differential equation satisfied by $y_0(x)$? It is easily seen to be
$$y_0''(x) + \left[W^2(x) \mp iW'(x)\right] y_0(x) = 0. \quad (9.48)$$
On comparing this with equation (9.42), we see that the two are approximately equivalent provided the following strict inequality is satisfied:
$$|W'(x)| \ll W^2(x) \quad (9.49)$$
or
$$\left|\frac{W'(x)}{W(x)}\right| \equiv \left|\frac{d\ln W(x)}{dx}\right| \ll W(x). \quad (9.50)$$
This last inequality makes clear the physical meaning of "slowly varying": The relative change in W should be everywhere small compared with W. Obviously, this is unlikely to be satisfied near those

values of x for which $W = 0$. We can take this further; in a medium with refractive index $n(x)$, the equivalent of W is

$$W(x) = k_0 n(x) \equiv \frac{2\pi}{\lambda_0} n(x), \tag{9.51}$$

where k_0 and λ_0 are respectively the wavenumber and wavelength of the "wave" in vacuo. The wavelength in the medium is $\lambda = \lambda_0/n$, so condition 9.50 is equivalent to

$$\frac{\lambda}{2\pi n} \left| \frac{dn}{dx} \right| \ll 1, \tag{9.52}$$

in other words, the relative change of the refractive index over essentially a wavelength should be small.

9.8.2. The next approximation

This is achieved by building on result (9.47) in the following manner. Let

$$y(x) = F(x) y_0(x) = F(x) \exp \left[\pm i \int_0^x W(\xi) d\xi \right]. \tag{9.53}$$

Again, by hypothesis, $F(x)$ is assumed to be slowly varying. On substituting this "ansatz" into equation (9.42), we obtain

$$F''(x) \pm i \left[W'(x) F(x) + 2 W(x) F'(x) \right] = 0. \tag{9.54}$$

Neglecting F'' because of the above assumption it follows that (if for now we assume $W > 0$)

$$\frac{d}{dx} \left[\ln \left(F(x) \sqrt{W(x)} \right) \right] = 0, \tag{9.55}$$

i.e., the product $F(x)\sqrt{W(x)}$ is a constant, or $F(x) \propto W^{-1/2}(x)$. This means that the fundamental set of solutions to equation (9.42) is, in this approximation,

$$y(x) = \frac{1}{\sqrt{W(x)}} \exp \left[\pm i \int_0^x W(\xi) d\xi \right]. \tag{9.56}$$

To derive a condition on $W(x)$ for this to be a good approximation away from $W(x) = 0$, we merely differentiation (9.56) twice to obtain

$$y''(x) + \left\{ \frac{1}{2W} \frac{d^2W}{dx^2} - \frac{3}{4W^2} \left(\frac{dW}{dx} \right)^2 + W^2 \right\} y(x) = 0. \quad (9.57)$$

On comparing this with equation (9.42), it is clear that for the approximation to be valid,

$$\left| \frac{1}{2W} \frac{d^2W}{dx^2} - \frac{3}{4W^2} \left(\frac{dW}{dx} \right)^2 \right| \ll W^2(x). \quad (9.58)$$

Sufficient conditions for this to be satisfied are that both $|W''|$ and $|W'|^2$ are both small compared with W^2.

9.8.3. *General solution*

Clearly, for $W^2(x) > 0$, the general WKB solution for the latest approximation is

$$y(x) = \frac{1}{\sqrt{W(x)}} \left\{ C_1 \exp\left[i \int_0^x W(\xi) d\xi \right] + C_2 \exp\left[-i \int_0^x W(\xi) d\xi \right] \right\}, \quad (9.59)$$

in terms of arbitrary constants C_1 and C_2. Equivalently, this could have been expressed in turns of sine and cosine functions of course. In the case of $W^2(x) < 0$, we may write $V^2(x) = -W^2(x)$ and then express the equivalent of equation (9.59) as

$$y(x) = \frac{1}{\sqrt{V(x)}} \left\{ D_1 \exp\left[\int_0^x V(\xi) d\xi \right] + D_2 \exp\left[-\int_0^x V(\xi) d\xi \right] \right\}. \quad (9.60)$$

So, if $x = a$ is a turning point, i.e., $W(a) = 0$ (and the zero is simple, so $W(x)$ is of one sign for $x \to a^-$ and the other sign for $x \to a^+$), then the solutions (9.59) or (9.59) must be chosen appropriately in each case.

Example 9.9. Airy equation. As an example we use the WKB to construct the leading term of the asymptotic solutions to Airy's equation (discussed in more detail in the following):

$$w'' - xw = 0,$$

where $W^2(x) = -x$. So, the sign of W^2 depends if x is positive or negative. The zeroth-order solution then is

$$y_0(x) = \begin{cases} \exp\left(\pm\frac{2}{3}x^{\frac{3}{2}}\right) & (x > 0), \\ \exp\left(\pm i\frac{2}{3}|x|^{\frac{3}{2}}\right) & (x < 0). \end{cases}$$

Substituting the zeroth approximation into the original Airy equation we have for $x > 0$

$$y_0''(x) = \left[-x \pm \frac{1}{2\sqrt{x}}\right] y_0(x) = 0.$$

Thus, this solution is a good approximation when $|x| \gg 1/(2\sqrt{x})$. Using the next order of the WKB approximation (9.56), we have

$$y(x) = x^{-\frac{1}{4}} \left[c_1 \exp\left(\frac{2}{3}x^{\frac{3}{2}}\right) + c_2 \exp\left(-\frac{2}{3}x^{\frac{3}{2}}\right)\right] \quad (x > 0)$$

and

$$y(x) = x^{-\frac{1}{4}} \left[c_1 \exp\left(i\frac{2}{3}x^{\frac{3}{2}}\right) + c_2 \exp\left(-i\frac{2}{3}x^{\frac{3}{2}}\right)\right] \quad (x < 0),$$

thus exhibiting the expected oscillatory behavior for negative values of x. This is the canonical example for illustrating the effectiveness of the WKB approximation because the results derived using the former can be compared with the exact solutions to the equation. Using (9.58), one can find the regions of x where the WBKB approximation is valid. ∎

Example 9.10. Extended example.
Let $W^2(x)$ in equation (9.42) be defined as

$$W^2(x) = 4(e^{2-x} - 1), \tag{9.61}$$

so to be specific, suppose that we are interested in applying the WKB method to the equation

$$y'' + 4(e^{2-x} - 1)y = 0, \quad 0 \le x < \infty, \tag{9.62}$$

with associated initial conditions $y(0) = 1$, $y'(0) = 0$. The graph of $W^2(x)$ is shown schematically in Figure 9.2 for $x \ge 0$. The region to the left of the first dashed line corresponds to a trigonometric solution; the region to the right of the second dashed line corresponds to exponentially decaying/growing solutions, while in the central region, the WKBJ solution is not valid.

With this choice for W^2, we can compare the WKB(J) solution with the exact solution, expressible in terms of Bessel functions. To see this, we identify the relevant steps, *leaving the details for the reader to complete.*

Exact solution — Step 1: By making the change of variable $\xi = e^{4-2x}$, with $Y(\xi) \equiv y(x)$, and using the chain rule, we can show that equation (9.61) can be written as follows:

$$\xi^2 Y''(\xi) + \xi Y'(\xi) + (\xi^2 - 1)Y(\xi) = 0.$$

This is *Bessel's equation of order one*, with general solution

$$Y(\xi) = A J_1(\xi) + B Y_1(\xi),$$

where A and B are constants to be determined (in Step 2).

Step 2: Applying the initial conditions, we have (we leave it to the reader to show that)

$$A = \frac{\pi e^4}{2} Y_1'(e^4); \quad B = -\frac{\pi e^4}{2} J_1(e^4)$$

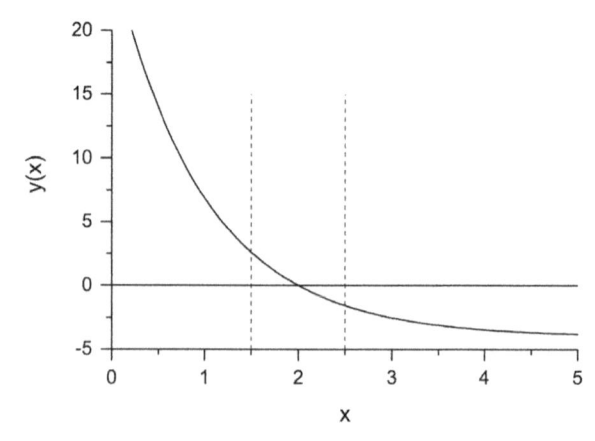

Fig. 9.2. $W^2(x) = 4(e^{2-x} - 1)$.

so that

$$y(x) = \frac{\pi e^4}{2}\left[Y_1'(e^4)J_1(e^{4-2x}) - J_1'(e^4)Y_1(e^{4-2x})\right]. \qquad (9.63)$$

Hint: The Wronskian $W[J_1, Y_1](\xi) = 2/\pi\xi$.

WKB solution: Clearly, the turning point is $x = 2$; for $x < 2$, the solutions are oscillatory, and for $x > 2$, they are exponential in character. Since

$$W^2(x) = 4(e^{2-x} - 1),$$

the integral

$$\mu(x) = \int_x^2 W(\zeta)d\zeta = 2\int_x^2 (e^{2-\zeta} - 1)^{1/2}d\zeta$$

$$= 4\int_0^y \frac{\alpha^2}{1+\alpha^2}d\alpha = 4(y(x) - \arctan y(x)),$$

where

$$y^2(x) = e^{2-x} - 1 = \frac{1}{4}W^2(x).$$

Hence,

$$\mu(x) = 2W(x) - 4\arctan\left(\frac{W(x)}{2}\right). \qquad (9.64)$$

The region $0 < x < 2$: In particular, we can write the WKB solution in $(0, 2)$ as

$$y_{\text{WKB}}(x) = A[W(x)]^{-1/2}\sin\left[\int_x^2 W(\zeta)d\zeta + \beta\right]$$

$$= \left[\frac{W(0)}{W(x)}\right]^{1/2}\frac{\sin[\mu(x) + \beta]}{\sin[\mu(0) + \beta]}, \qquad (9.65)$$

where β is a phase angle to be determined from the second initial condition. The first condition, $y_{\text{WKB}}(0) = 1$, has been used to establish (9.64). After some careful algebra, using the condition $y'_{\text{WKB}}(0) = 0$, the following result is obtained:

$$\beta = \arctan\left\{\frac{4W^3(0)}{16 + 4W^2(0)}\right\} - \mu(0).$$

The connection formulas: We state these without derivation because the analysis would take us outside the scope of this text. The solution in the interval $(0, 2)$ is completely determined, but the connection formulas are required to match the solution for $x > 2$:

$$[W(x)]^{-1/2} \sin\left[\int_x^2 W(\zeta)d\zeta + \beta\right]$$

$$= [W(x)]^{-1/2} \sin\left\{\left(\beta - \frac{\pi}{4}\right) + \left(\int_x^2 W(\zeta)d\zeta + \frac{\pi}{4}\right)\right\}$$

$$= [W(x)]^{-1/2} \left\{\sin\left(\beta - \frac{\pi}{4}\right) \cos\left(\int_x^2 W(\zeta)d\zeta + \frac{\pi}{4}\right)\right.$$

$$\left. + \cos\left(\beta - \frac{\pi}{4}\right) \sin\left(\int_x^2 W(\zeta)d\zeta + \frac{\pi}{4}\right)\right\},$$

$$\Rightarrow\Rightarrow\Rightarrow [V(x)]^{-1/2} \left\{\sin\left(\beta - \frac{\pi}{4}\right) \exp\left(\int_2^x V(\xi)d\xi\right)\right.$$

$$\left. + \frac{1}{2}\cos\left(\beta - \frac{\pi}{4}\right) \exp\left(-\int_2^x V(\xi)d\xi\right)\right\}.$$

(Note the factor of $1/2$ in the final result.) This results in the WKB solution for $x \in (2, \infty)$ as

$$y_{\text{WKB}}(x) = \left[\frac{W(0)}{V(x)}\right]^{1/2} (\sin[\mu(0) + \beta])^{-1}$$

$$\cdot \left\{e^{\rho(x)} \sin\left(\beta - \frac{\pi}{4}\right) + \frac{1}{2}e^{-\rho(x)} \cos\left(\beta - \frac{\pi}{4}\right)\right\},$$

where

$$\rho(x) = \int_2^x V(\xi)d\xi = -2V(x) + 2\ln\frac{2 + V(x)}{2 - V(x)}$$

and

$$V^2(x) = 4(1 - e^{2-x}). \qquad \blacksquare$$

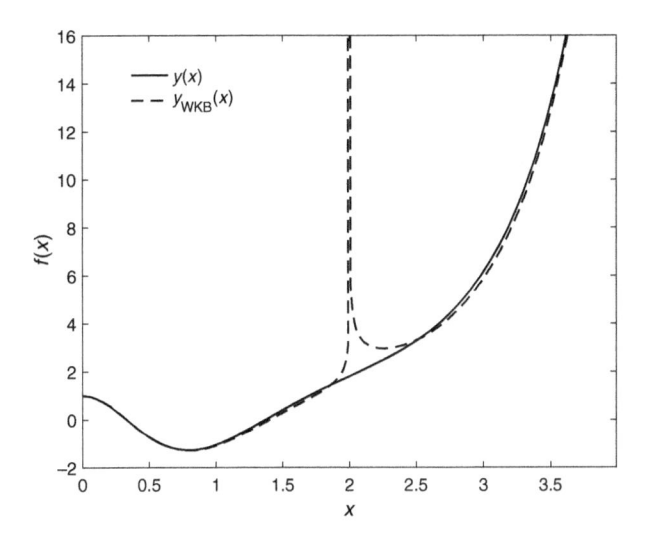

Fig. 9.3. Comparing the exact solution to the WKB solution.

Figure 9.3 shows the WKB solution along with the exact (numerical) solution.

9.9. Some physics examples

9.9.1. *Oscillations*

A large class of motion related to springs, pendulums, and general oscillatory phenomena in physics can be described by equation

$$m\frac{d^2x}{dt^2} = -a\frac{dx}{dt} - kx = 0,$$

where the first term on the right-hand side represents a force of resistance or "drag" force (proportional to velocity, with $a \geq 0$), and the second term is a linear restoring force. Using notations $2b = a/m$ and $\omega_0^2 = k/m$, we can write

$$\frac{d^2x}{dt^2} + 2b\frac{dx}{dt} + \omega_0^2 x = 0,$$

where and b and ω_0 are positive constants. If there is an additional external force, depending on time, acting on the body, then we have a non-homogeneous equation describing forced oscillatory motion

$$\frac{d^2x}{dt^2} + 2b\frac{dx}{dt} + \omega_0^2 x = \frac{1}{m}F(t), \tag{9.66}$$

where $f(t)$ is a continuous function of time t. There are four cases of most common interest describing different motions:

(1) Simple harmonic motion:
$$\frac{d^2x}{dt^2} + \omega_0^2 x = 0.$$

(2) Damped harmonic motion:
$$\frac{d^2x}{dt^2} + 2b\frac{dx}{dt} + \omega_0^2 x = 0.$$

(3) Forced undamped motion:
$$\frac{d^2x}{dt^2} + \omega_0^2 x = f_0 \sin(\omega t + \beta).$$

(4) Forced damped motion:
$$\frac{d^2x}{dt^2} + 2b\frac{dx}{dt} + \omega_0^2 x = f_0 \sin(\omega t + \beta).$$

Note that above we use a notation $f(t)$ as $F(t)/m$.

Example 9.11. Let us consider forced damped motion with external forcing $f_0 \sin(\omega t)$, namely

$$\frac{d^2x}{dt^2} + 2b\frac{dx}{dt} + \omega_0^2 x = f_0 \sin(\omega t).$$

Part 1:

As the first step, we analyze the homogeneous equation

$$\frac{d^2x}{dt^2} + 2b\frac{dx}{dt} + \omega_0^2 x = 0.$$

The solution of this equation gives the free, or proper, vibrations. The characteristic equation is

$$\lambda^2 + 2b\lambda + \omega_0^2 = 0,$$

with the solutions
$$\lambda_{1,2} = \frac{-2b \pm \sqrt{4b^2 - 4\omega_0^2}}{2} = -b \pm \sqrt{b^2 - \omega_0^2}.$$
If there is no resistance from the environment ($b = 0$), then $\lambda_{1,2} = \pm i\omega_0$ and the solution is
$$x(t) = C_1 \cos \omega_0 t + C_2 \sin \omega_0 t = C \cos(\omega_0 t + \varphi).$$
This gives a pure harmonic oscillation of period $\tau = 2\pi/\omega_0$. The constants C_1, C_2, or C and φ are defined by initial conditions $x(t = 0) = x_0$ and $x'(t = 0) = v_0$.

Now, we analyze motion with the resistance.

Case 1 — Aperiodic motion: If $b^2 - \omega_0^2 = \gamma^2$ is positive, then the roots of the characteristic equation are
$$\lambda_{1,2} = -b \pm \gamma,$$
and the solution is
$$x(t) = C_1 e^{(-b+\gamma)t} + C_2 e^{(-b-\gamma)t} = e^{-bt}(C_1 e^{\gamma t} + C_2 e^{-\gamma t}).$$
Since we obviously have here $\gamma < b$, both roots are negative, and x therefore tends to zero as $t \to \infty$.

Case 2 — Special case of aperiodic motion: If $b^2 - \omega_0^2 = \gamma^2 = 0$, then the characteristic equation has equal roots, $\lambda = -b$, and so
$$x(t) = e^{-bt}(C_1 + xC_2).$$
Since the exponential factor tends to zero as $t \to \infty$, this solution also tends to zero.

Case 3 — Damped vibration: Very often the "drag force" is fairly small comparing to the restoring force so that $b^2 - \omega_0^2$ is negative: $b^2 - \omega_0^2 = -\beta^2$. Then, the roots of the characteristic equation are complex conjugate numbers, namely $\lambda_{1,2} = -b \pm i\beta$, and the general solution is
$$x(t) = e^{-bt}(C_1 \cos \beta t + C_2 \sin \beta t).$$
On setting $C_1 = A \cos \varphi$, $C_2 = A \sin \varphi$, the solution can be written in the form
$$x(t) = Ae^{-bt} \cos(\beta t + \varphi),$$
where A is the initial amplitude and φ is the initial phase. This formula represents damped oscillations, with the degree of damping

being characterized by the factor e^{-bt}. The values of the constants C_1 and C_2 or A and φ depend on the initial conditions. Note that the angular frequency of oscillations in this case is $\beta = \sqrt{\omega_0^2 - b^2}$.

Part 2:

Oscillations with the external force $f(t) = f_0 \sin(\omega t)$

$$\frac{d^2 x}{dt^2} + 2b\frac{dx}{dt} + \omega_0^2 x = f_0 \sin(\omega t).$$

Assuming that the "drag" force is less than the restoring force, the solutions of the complementary homogeneous equation are

$$x_1 = e^{-bt} \cos \beta t \quad \text{and} \quad x_2 = e^{-bt} \sin \beta t,$$

where $\beta^2 = \omega_0^2 - b^2$. While the Lagrange method of variation of parameters is a powerful one, here we use the method of undetermined coefficients. We are looking for the particular solution as

$$u(t) = A \cos \omega t + B \sin \omega t,$$

then

$$u'(x) = -A\omega \sin \omega t + B\omega \cos \omega t \quad \text{and}$$
$$u''(x) = -A\omega^2 \cos \omega t - B\omega^2 \sin \omega t.$$

With these u, u', and u'', the original differential equation reads

$$-A\omega^2 \cos \omega t - B\omega^2 \sin \omega t + 2b[-A\omega \sin \omega t + B\omega \cos \omega t]$$
$$+\omega_0^2[A \cos \omega t + B \sin \omega t] = f_0 \sin(\omega t).$$

Collecting terms by $\cos \omega t$ and $\sin \omega t$ gives two equations for the unknown coefficients A and B:

$$-\omega^2 A + 2b\omega B + \omega_0^2 A = 0 \quad \text{and}$$
$$-\omega^2 B - 2b\omega A + \omega_0^2 B = f_0.$$

Solving for the unknown coefficients gives

$$A = -2b\omega \frac{f_0}{(\omega_0^2 - \omega^2)^2 + (2b\omega)^2} \quad \text{and}$$

$$B = (\omega_0^2 - \omega^2) \frac{f_0}{(\omega_0^2 - \omega^2)^2 + (2b\omega)^2}.$$

Therefore, the general solution of the non-homogeneous equation is

$$x(t) = C_1 e^{-bt} \cos \beta t + C_2 e^{-bt} \sin \beta t$$
$$+ \frac{f_0}{(\omega_0^2 - \omega^2)^2 + (2b\omega)^2}[(\omega_0^2 - \omega^2)\sin \omega t - 2b\omega \cos \omega t].$$

It is instructive to analyze a couple of cases:

(1) When $t \to \infty$, then the homogeneous solution $\to 0$ and we have a steady motion driven by the external force.
(2) When $b \to 0$ (no resistance), the solution is

$$x(t) = C_1 \cos \beta t + C_2 \sin \beta t + \frac{f_0}{(\omega_0^2 - \omega^2)^2} \sin \omega t.$$

(3) When $\omega \to \omega_0$ (resonance) but $b \neq 0$,

$$x(t) = C_1 e^{-bt} \cos \beta t + C_2 e^{-bt} \sin \beta t - \frac{f_0}{2b\omega} \cos \omega t$$

and with $t \to \infty$

$$x(t) = -\frac{f_0}{2b\omega} \cos \omega t.$$ ∎

9.9.2. R–L–C electric circuits

For a simple electric circuit with a resistor R, inductor L, capacitor C, and a battery with emf $\mathcal{E}$, Kirchhoff's second law gives the following differential equation for the change in electric charge:

$$L\frac{d^2q}{dt^2} + R\frac{dq}{dt} + \frac{1}{C}q = \mathcal{E}(t).$$

Since the current i in the circuit is defined as the rate of change of the charge q, i.e., $i = dq/dt$, then differentiating the equation above again we can write, for the electric current,

$$L\frac{d^2i}{dt^2} + R\frac{di}{dt} + \frac{1}{C}i = \frac{d}{dt}\mathcal{E}(t).$$

Dividing every term by L and assuming for the emf force $\mathcal{E} = \mathcal{E}_0 F_0 \sin(\omega t + \beta)$ we get the equation that formally looks like the one for forced motion with damping, namely

$$\frac{d^2 i}{dt^2} + \frac{R}{L}\frac{di}{dt} + \frac{1}{CL}i = \frac{1}{L}\mathcal{E}_0 \omega \cos(\omega t + \beta).$$

This equation can be solved the same way that we used in solving the equation for forced oscillations. In the latter equation, the inductor L plays the same role as mass m, the resistor R is equivalent to the drag coefficient, and the capacitor C is like the restoring force with the coefficient $1/k$.

9.10. Mathematical problems

Solve the following homogeneous differential equations:

(1) $y'' - 5y' + 6y = 0$
(2) $y'' + 3y' + 2y = 0$
(3) $y'' - 2y' = 0$
(4) $y'' + 2y' + y = 0$
(5) $4y'' + 4y' + y = 0$
(6) $y'' + y = 0$
(7) $y'' + 2y' + 10y = 0$
(8) $y'' - y' + y = 0.$

Solve the initial value problem for the following homogeneous equations:

(9) $4y'' - 8y' + 5y = 0$, $y(0) = 0$, $y'(0) = 1/2$
(10) $y'' + 4y' + 4y = 0$, $y(0) = 1$, $y'(0) = 3$
(11) $y'' - 5y' + 4y = 0$, $y(0) = y'(0) = 1$
(12) $y'' - 2y' + 3y = 0$, $y(-1) = 1$, $y'(-1) = 3$
(13) $y'' + \omega^2 y = 0$, $y(t_0) = y_0$, $y'(t_0) = v_0$.

Solve the following non-homogeneous equations using both Lagrange's and Euler's methods:

(14) $y'' + y = \cos x$
(15) $y'' - 2y' + 2y = e^x \cos x$
(16) $y'' + 8y' + 25y = 120 \sin 5x$
(17) $y'' + y = 4x \cos x$
(18) $y'' + 2y' = 4e^x(\sin x + \cos x).$

Solve the following non-homogeneous equations:

(19) $y'' + y' = \frac{1}{e^x+1}$
(20) $y'' + y = \frac{1}{\cos^3 x}$
(21) $y'' - y' = e^{2x} \cos e^x$
(22) $y'' - 2y' + y = \frac{x^2+2x+2}{x^3}$.

Solve the initial value problem for the following non-homogeneous equations:

(23) $y'' - 2y' + 3y = 1$, $y(0) = 0$, $y'(0) = 1$
(24) $y'' - 3y' + 2y = e^{3x}$, $y(1) = y_0$, $y'(1) = y_0'$
(25) $y'' - 5y' + 4y = e^x$, $y(0) = y_0$, $y'(0) = y_0'$.

Solve using series solution. Explore convergence if possible (when possible find the solution in closed form):

(26) $y' = \sin xy$
(27) $y'' - xy' - 2y = 0$
(28) $y'' - \sin xy' = 0$, $y(0) = 0$, $y'(0) = 1$
(29) $xy'' + y \sin x = x$, $y(\pi) = 1$, $y'(\pi) = 0$
(30) $y'' - xy' + y - 1 = 0$, $y(0) = y'(0) = 0$
(31) $y'' - y = 1$, $y(0) = 0, y'(0) = 0$
(32) $y'' + y = x$, $y(0) = 1, y'(0) = 0$.

Find periodic solutions, if any:

(33) $y'' + y = \sum_{k=1}^{\infty} (\cos kx)/k^2$
(34) $y'' + y' = \sum_{k=1}^{\infty} (\sin kx)/k^2$
(35) $y'' + 4y = \sin^2 x$.

Few problems with WKB approximation:

(36) Find the form of $W(x)$ such that the left-hand side of equation (9.58) is *identically* zero, i.e.,

$$WW'' = \frac{3}{2}(W')^2.$$

Hint: Divide both sides by WW'.[5]

---

[5]E. Kamke, *Differentialgleichungen: Lösungsmethoden und Lösungen*. Chelsea Pub Co., 1951.

(37) Using recurrence relations for Bessel functions, show that an alternative expression for the exact solution $y(x)$ (9.63) is

$$y(x) = \frac{\pi e^4}{4}[(Y_0(e^4) - Y_2(e^4))J_1(e^{4-2x}) \\ - (J_0(e^4) - J_2(e^4))Y_1(e^{4-2x})].$$

(38) Derive the result (9.64).

9.11. Physics problems

(1) The gravitational force on a particle of mass m inside the earth at a distance r from the center ($r < R$ where R is the radius of earth) is $F = (-mgr)/R$. Write the equation of motion of a particle neglecting the air resistance (if a particle is placed in an evacuated tube through the center of the earth). Find the period of this motion. Note that $g = 9.8$ m/s^2 and $R = 6{,}400$ km.
(2) **Forced undamped motion:** Solve the equation of forced undamped motion with $w \neq w_0$ and the initial conditions for $t = 0$, $x = x_0$, and $v = v_0$. The external force is $f(t) = f_0 \sin(wt+\beta)$. Analyze your solution if $x_0 = 0$, $v_0 = 0$, and $\beta = 0$.
(3) **Resonant forced undamped motion:** Solve the problem above for forced undamped motion but for $w = w_0$.
(4) **Motion driven by a constant force with damping:** Solve the equation of motion for the force $f(t) = f_0$. Assume $b^2 < w_0^2$ and the initial conditions $x = x_0$, $v = v_0 = 0$.
(5) **Forced motion with damping:** Solve the equation of motion for the force $f(t) = f_0 \sin(wt + \beta)$. Assume $b^2 < w_0^2$ and the initial conditions $x = x_0$, $v = v_0 = 0$. After finding the equation of motion, find the steady state motion and also the amplitude, period, and frequency of the steady state motion.
(6) **R–L–C circuit:** If a battery is missing in an R-L-C circuit, (a) find the natural (undamped, i.e., $R = 0$) frequency of vibration, (b) for what values of R the charge and the current will oscillate before subsiding to zero, and (c) find q and i as functions of time if at $t = 0$, $q = q_0$, and $i = 0$. Assume $R^2 < 4L/C$.
(7) **A simple pendulum consists** of a point mass m suspended by a weightless cord of length L. Derive the equation of motion

of the pendulum, that is, the differential equation for θ as a function of t (use the second Newton's law):

$$\frac{d^2\theta}{dt^2} + \frac{g}{l}\sin\theta.$$

Show that for small angles θ this is approximately a simple harmonic motion equation. Then, solve it for the initial conditions at $t = 0$, $\theta = \theta_0$, and $d\theta/dt = 0$. Find the amplitude and period of oscillations.

(8) **Pendulum II**: Consider again the simple pendulum from the problem above but this time do not use $\sin\theta \approx \theta$. Instead try to find an approximate solution using series expansion. Assume the initial conditions at $t = 0$ as $\theta = \theta_0$ and $d\theta/dt = 0$. Compare your solution with the solution from the problem before (expand the solution in 7 into series).

(9) **Applying Archimedes' principle**: A block of wood is floating in water. After an initial depression, the barrel exhibits an up-and-down bobbing motion along a vertical line. Write the differential equation for the vertical motion $y(t)$ of the block if the origin is taken to be on at the surface of the water when the block is at rest. Use Archimedes' principle.

Hint: You will need to find what fraction of the block is under water when the block is at rest.

(10) **Chain I**: A chain of length L hangs from a frictionless and massless pulley of negligible diameter. Initially, the chain is stationary and the two ends of the chain are differing in the initial vertical position by $2h_0$. Write the differential equation describing the motion of the chain. Find how much time will it take for the chain to be out of the pulley. What will be the kinetic energy of the chain at this moment? Check your solution by applying conservation of energy to the chain.

(11) **Vertical motion (again)**: A projectile of mass m kg is thrown vertically upward from the ground with an initial velocity v_0 m/s. The projectile is subject to linear air resistance so that the force of air resistance is given by $f = -bv$. The differential equation describing the motion is

$$m\frac{d^2y}{dt^2} = -b\frac{dy}{dt} - mg.$$

Find its position and velocity as a function of time. How long and how far will it rise? Compare your solution with the solutions from Example 8.7 in Chapter 8.

Answers to mathematical problems

(1) $y(x) = C_1 e^{2x} + C_2 e^{3x}$.
(2) $y(x) = C_1 e^{-x} + C_2 e^{-2x}$.
(3) $y(x) = C_1 + C_2 e^{2x}$.
(4) $y(x) = (C_1 + C_2 x)e^{-x}$.
(5) $y(x) = (C_1 + C_2 x)e^{-x/2}$.
(6) $y = C_1 \cos x + C_2 \sin x$.
(7) $y(x) = C_1 e^{-x} \cos 3x + C_2 e^{-x} \sin 3x$.
(8) $y = C_1 e^{x/2} \cos(\sqrt{3}x/2) + C_2 e^{x/2} \sin(\sqrt{3}x/2)$.
(9) $y(x) = e^x \sin(x/2)$.
(10) $y(x) = (5x + 1)e^{-2x}$.
(11) $y(x) = e^x$.
(12) $y(x) = e^{x+1} \left(\sqrt{2} \sin(\sqrt{2}(x+1)) + \cos(\sqrt{2}(x+1)) \right)$.
(13) $y(x) = y_0 \cos \omega(t - t_0) + (v_0/\omega) \sin \omega(t - t_0)$.
(14) $y(x) = C_1 \cos x + C_2 \sin x + \frac{1}{2} x \sin x$.
(15) $y(x) = e^x (C_1 \sin x + C_2 \cos x) + \frac{1}{2} x e^x \sin x$.
(16) $y(x) = C_1 e^{-4x} \sin 3x + C_2 e^{-4x} \cos 3x - 3 \cos 5x$.
(17) $y(x) = C_1 \cos x + C_2 \sin x + x \cos x + x^2 \sin x$.
(18) $y(x) = C_1 + C_2 e^{-2x} + \frac{6}{5} e^x \sin x - \frac{2}{5} e^x \cos x$.
(19) $y(x) = C_1 + C_2 e^{-x} + x - (e^{-x} + 1) \ln(e^x + 1)$.
(20) $y(x) = C_1 \cos x + C_2 \sin x - \frac{\cos 2x}{2 \cos x}$.
(21) $y(x) = C_1 + C_2 e^x - \cos(e^x)$.
(22) $y(x) = C_1 e^x + C_2 x e^x + \frac{1}{x}$.
(23) $y(x) = \frac{1}{3} \left(2\sqrt{2} e^x \sin(\sqrt{2}x) - e^x \cos(\sqrt{2}x) + 1 \right)$.
(24) $y(x) = \frac{1}{3} \left(e^x (4y_0 - y_0') - e^{4x}(y_0 - y_0') + (e^x + \frac{1}{2} e^{4x}) \right) - \frac{1}{2} e^{2x}$.
(25) $y(x) = e^{x-1} \left(2y_0 - y_0' + \frac{1}{2} e^3 \right) + e^{2(x-1)}(-y_0 + y_0' - e^3) + \frac{1}{2} e^{3x}$.

(26) $y(x) = 1 + 0 + \frac{1}{2!}x^2 + 0 + \cdots$.

(27) $y(x) = C_1 x e^{x^2/2} + C_2\left(1 + x^2 + \frac{1}{3}x^4 + \frac{1}{15}x^6 + \cdots\right)$.

(28) $y(x) = x + \frac{1}{3!}x^3 + \frac{2}{5!}x^5 + \cdots$.

(29) $y(x) = 1 + \frac{1}{2}(x-\pi)^2 + \frac{1}{3\pi}(x-\pi)^3 + \cdots$.

(30) $y(x) = \frac{1}{2!}x^2 + \frac{1}{4!}x^4 + \frac{3}{6!}x^6 + \cdots$.

(31) $y(x) = \frac{1}{2!}x^2 + \frac{1}{4!}x^4 + \frac{1}{6!}x^6 + \cdots$.

(32) $y(x) = 1 - \frac{1}{2}x^2 + \frac{1}{3!}x^3 + \frac{1}{4!}x^4 - \frac{1}{5!}x^5 + \cdots$.

(33) no periodic solution.

(34) $y = C - \sum_{k=1}^{\infty} \frac{\cos kx + \sin kx}{k^3(k^2+1)}$.

(35) no periodic solutions.

Comment: Answers to physics problems

(1) $m\frac{d^2r}{dt^2} = -\frac{mg}{R}r$, for $r(0) = R$, $r'(0) = 0$, $R(t) = R\cos\omega_0 t$, $\omega^2 = g/R$, $T = 2\pi\sqrt{R/g}$.

Chapter 10

Green's Function Method

Green's function method is widely used for solving non-homogeneous ordinary and partial differential equations. Green's functions are named after George Green.[1]

---

[1] George Green was an English mathematician and physicist who lived from 1793 to 1841. He is best known for his pioneering work in the field of mathematical physics and for developing Green's theorem, which is a fundamental result in vector calculus. Green was born in Sneinton, Nottinghamshire, England, and showed an early aptitude for mathematics and physics. Despite not receiving a formal education, he taught himself advanced mathematics and went on to make significant contributions to the field.

Green's theorem relates to the relationship between the line integral of a vector field around a simple closed curve and the double integral of the curl of the vector field over the plane region enclosed by the curve. This theorem is widely used in engineering and physics to analyze electromagnetic fields and fluid flow, among other applications. Green's work laid the foundation for the development of modern vector calculus and is considered a cornerstone of mathematical physics.

In addition to his work on Green's theorem, Green also made significant contributions to the understanding of wave motion and the propagation of waves in various media. He was one of the first to recognize that light, heat, and other forms of electromagnetic radiation could be described as waves, and he made important contributions to the mathematical formulation of wave theory. He also conducted research in the areas of electricity and magnetism and made important contributions to the development of the theory of electromagnetic fields.

Despite his numerous achievements, Green lived a relatively modest life and was not widely recognized during his lifetime. He spent much of his life working as a miller in his family's business and never held a formal academic position. Nevertheless, his work was highly regarded by his contemporaries and he was elected as a fellow of the Royal Society of London in recognition of his

10.1. Dirac's delta function

We begin consideration of Green's functions by exploring the Dirac delta function $\delta(x)$. The function was proposed by Paul Dirac about hundred years after George Green introduced his Green's function method. However, using the delta function makes it easier to operate with Green's functions.

The delta function is a generalized function. After Dirac, the theory of generalized functions was intensively developed by many mathematicians. The rapid development of the theory of generalized functions was stimulated mainly by the needs of mathematical physics, especially the theory of differential equations and quantum physics. Nowadays, the theory of generalized functions is far advanced, has numerous applications in physics and mathematics, and has become well established in mathematics, physics, and engineering.

A generalized function is, as its name implies, a generalization of the classical concept of a function. Such generalization makes it possible to express such idealized concepts in a mathematical form as, for example, the density of a point.

Let us consider a function $\varphi_1(x)$

(1) that has a maximum at $x = 0$,
(2) that is rapidly decreasing on both sides of $x = 0$, and
(3) such that the integral

$$\int_{-\infty}^{\infty} \varphi_1(x) dx = 1.$$

These conditions do not specify the explicit form of $\varphi_1(x)$. There are many functions satisfying the above conditions, for example,

$$\varphi_1(x) = \frac{1}{\pi} \frac{1}{1+x^2}, \tag{10.1}$$

$$\varphi_1(x) = \frac{1}{\sqrt{\pi}} e^{-x^2}. \tag{10.2}$$

---

contributions to mathematics and physics. Today, Green is remembered as a pioneering figure in the field of mathematical physics and his legacy continues to be felt in the ongoing development of this area of study.

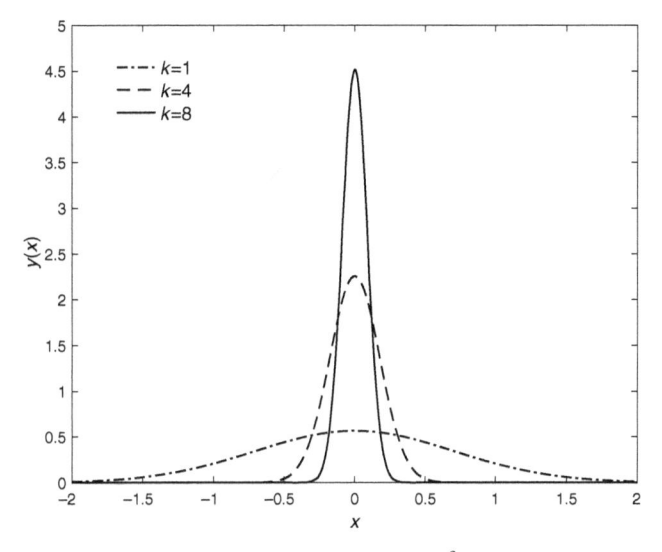

Fig. 10.1. Function $\varphi_k(x) = \frac{k}{\sqrt{\pi}} e^{-(kx)^2}$ for $k = 1, 5, 10$.

Let us increase the heights of the function by a factor of k while simultaneously decreasing its width by the same factor. This transformation can be written as

$$\varphi_k = k\varphi_1(kx).$$

Then, for functions (10.1) and (10.2), it follows

$$\varphi_k(x) = \frac{k}{\pi} \frac{1}{1 + (kx)^2}, \qquad \varphi_k(x) = \frac{k}{\sqrt{\pi}} e^{-(kx)^2}.$$

Figure 10.1 illustrates the behavior of the second function for various values of k. Note that the integral after such transformation does not change:

$$\int_{-\infty}^{\infty} \varphi_k(x) dx = \int_{-\infty}^{\infty} k\varphi_1(kx) dx$$

$$= \int_{-\infty}^{\infty} \varphi_1(kx) d(kx) = \int_{-\infty}^{\infty} \varphi_1(x) dx = 1.$$

Now, we analyze the transformed functions in the limit $k \to \infty$.

Region 1: $x \neq 0$.
For any $x \neq 0$ and $\varphi_1(x) \to 0$ faster than $1/x$, we have $\varphi_k \to 0$. That is the case for the both functions above.

Region 2: $x = 0$.
In this case, $\varphi_1(kx) = \varphi_1(0)$ for any k, and therefore, $\varphi_k(0) = k\varphi_1(0) \to \infty$. Thus, increasing $k \to \infty$, we get a function $\varphi_k(x)$ with following properties:

(1) The function is equal to zero for all $x \neq 0$.
(2) The function $\to \infty$ at $x = 0$.
(3) The integral of this function over the real line is equal to 1.

A function with such properties is called Dirac's delta-function $\delta(x)$. From the properties of $\delta(x)$ follows a very practical relation

$$\int_{-\infty}^{\infty} \delta(x) f(x) dx = f(0). \qquad (10.3)$$

We can show that first by using $\delta(x) = 0$ for all $x \neq 0$, then

$$\int_{-\infty}^{\infty} \delta(x) f(x) dx = \int_{-\epsilon}^{+\epsilon} \delta(x) f(x) dx,$$

where $\epsilon \to 0$. Within the interval 2ϵ function $f(x) \approx f(0)$, then

$$\int_{-\epsilon}^{+\epsilon} \delta(x) f(x) dx = \int_{-\epsilon}^{+\epsilon} \delta(x) f(0) dx = f(0) \int_{-\epsilon}^{+\epsilon} \delta(x) dx = f(0).$$

So, formula (10.3) is based on the three properties of $\delta(x)$ function. Note that if we define $\delta(x)$ function as (10.3), then we can derive all the three properties of $\delta(x)$ from such definition. In a similar way, we can derive another useful property:

$$\int_{-\infty}^{\infty} \delta(x - a) f(x) dx = f(a). \qquad (10.4)$$

The delta function has a number of other interesting properties. In particular,

$$\delta(ax) = \frac{1}{|a|} \delta(x). \qquad (10.5)$$

The property can be proved by using the substitution $ax = x'$ in integration for $|a|\delta(ax)dx$. Here is another practical property of the delta function. If $f(x) = 0$ at only $x = x_0$, then

$$\delta(f(x)) = \frac{1}{f'(x_0)} \delta(x - x_0). \qquad (10.6)$$

This property follows from property (10.5). Indeed, using the Taylor series near $x = x_0$, we write $f(x) = f(x_0) + f'(x_0)(x - x_0)$. Since $f(x_0) = 0$, then $\delta(f(x)) = \delta(f'(x_0)(x - x_0)) = \delta(x - x_0)/f'(x_0)$. If $f(x)$ has more than one root, then in (10.6) we sum over all the roots:

$$\delta(f(x)) = \sum_i \frac{\delta(x - x_i)}{|f'(x_i)|},$$

where x_i are the roots of $f(x)$. It is good to mention another practical feature of the delta function $f(x)\delta(x-a) = f(a)\delta(x-a)$. And finally, a useful property related to the derivative of the delta function is that

$$\int_{-\infty}^{+\infty} f(x)\delta'(x - a)dx = -f'(a).$$

Using the delta function makes it easier to analyze many physics situations. Consider, for example, a thin rod with blocks attached at different points. Let the size of the blocks be very small comparing to the size of the rod, but the masses of the blocks are comparable with the mass of the rod. Then, for finding the total mass of the rod, its center of mass, and its rotational inertia, we need to consider separately the contribution from the blocks and the rod. Let us consider a rod of length L with a linear density $\rho_r(x)$ and a block of mass m_a attached at a point of a. Then, the total mass of the system is

$$m = m_a + \int_0^L \rho_r(x)dx.$$

Using delta function, we can represent the block as having density $\rho_a(x) = m_a\delta(x - a)$. You can easily verify that the integral with this density over the length of the rod gives m_a. Now, we can write the density of the system as

$$\rho(x) = \rho_r(x) + \rho_a(x) = \rho_r(x) + m_a\delta(x - a).$$

Then, the total mass of the system is

$$m = \int_0^L \rho(x)dx.$$

Thus, instead of writing the total mass as a sum of different kind of terms, we write it as an integral. All details about the distribution of masses are stored in a specific form of $\rho(x)$.

10.2. Green's function

10.2.1. A physics example

We start with a physics example. Let a thin flexible rope of length L be stretched along x by a constant force (tension) T. A second force F acts perpendicular to the rope with a distribution $f(x)$ such that

$$F = \int_0^L f(x)dx.$$

We want to find a shape $y(x)$ of the rope shown in Figure 10.2.

We assume that the tension T is much larger than the entire force acting on the rope, i.e., $T \gg F$. In this case, the deviations $y(x)$ of the rope are small and the system is linear, i.e., the total effect from several load forces f_i on the rope is simply a sum of individual deviations. Next, we assume that the applied external force with a magnitude *one* is acting at just one point ξ along x. And the effect of this force is $y(x) = G(x,\xi)$, where $G(x,\xi)$ is called *Green's function* or a function of influence. Now, we consider the external force $f(x)$ acting on the interval from ξ to $\xi + d\xi$. This force has a magnitude $f(\xi)d\xi$ and the deviation of the thread at point x is $G(x,\xi)f(\xi)d\xi$; this follows form the linearity of the system. Adding all such deviations along x from $x = 0$ to $x = L$ we get

$$y(x) = \int_0^L G(x,\xi)f(\xi)d\xi. \qquad (10.7)$$

For the current example, we can derive the explicit form of Green's function $G(x,\xi)$. Using the vertical vector components for the forces

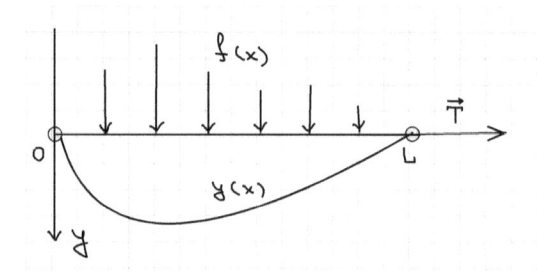

Fig. 10.2. Force $f(x)$ applied to the rope. As the result, the rope takes shape $y(x)$.

in Figure 10.3, we can write for points to the left of ξ

$$-T\sin\alpha = -T\frac{h}{\xi}.$$

Note that here we use the small angle approximation $\sin(\alpha) \approx \tan(\alpha)$. For points on the right, we have the term

$$-T\frac{h}{L-\xi}.$$

Since the thread is in equilibrium, then the net force applied at point ξ is zero, i.e.,

$$1 = T\frac{h}{\xi} + T\frac{h}{L-\xi}.$$

Solving the equation for h gives

$$h = \frac{\xi(L-\xi)}{TL}.$$

Then, the shape of the thread is given as

$$y(x) = h\frac{x}{\xi} = x\frac{(L-\xi)}{TL} \quad \text{for } x < \xi,$$

$$y(x) = h\frac{L-x}{L-\xi} = \xi\frac{(L-x)}{TL} \quad \text{for } x > \xi.$$

Using the explicit form of $y(x)$, we can write Green's function for the problem as

$$G(x,\xi) = \begin{cases} x(L-\xi)/TL & \text{for } x < \xi, \\ \xi(L-x)/TL & \text{for } x > \xi. \end{cases} \tag{10.8}$$

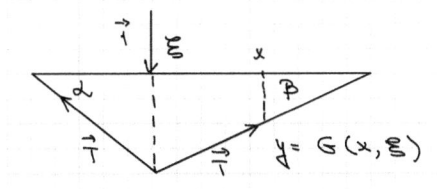

Fig. 10.3. Force of unit magnitude applied at a single point ξ.

Then, for the shape of the thread, we write

$$y(x) = \int_0^L G(\xi, x) f(\xi) d\xi = \frac{L-x}{TL} \int_0^x \xi f(\xi) d\xi$$
$$+ \frac{x}{TL} \int_x^L (L-\xi) f(\xi) d\xi.$$

The same result can be derived by writing the differential equation

$$T \frac{d^2 y}{dx^2} = f(x)$$

for the function $y(x)$ and solving it for the boundary conditions $y(0) = y(L) = 0$. However, it is remarkable that we managed to find a solution to the problem without even writing out the differential equation itself. It was enough for us to know that the law of linearity is valid. It is interesting to note that the same differential equation can be derived using the calculus of variations by minimization of the potential energy of the system (see Chapter 11, equation (11.3)).

10.2.2. *Mathematical approach*

Now, we shall derive equation (10.7) by using the delta function. Let $f(x)$ be an external force acting upon an object within the interval $a \leq x \leq b$. Let $y(x)$ be a result of such influence. One can imagine that $y(x)$ is a result of applying an operator R to the function $f(x)$. For example, an operator of differentiation R transforms x^2 into $2x$, or $R(x^2) = 2x$.

Let R be a linear operator that transforms the function of influence $f(x)$ into the function of response $y(x)$ or $y(x) = Rf(x)$. Since we assume that the operator is a linear one, then $R(f_1 + f_2) = Rf_1 + Rf_2$ and $R(Cf(x)) = CR(f(x))$, where C is a constant.

In the physics example analyzed at the beginning of this section, $G(\xi, x)$ was assumed as the result of a unit force applied to a single point ξ, or the force distributed with density $\delta(\xi - x)$. Then, let $G(\xi, x)$ be a result of such force and we can write

$$G(\xi, x) = R[\delta(x - \xi)].$$

Assume that $f(x)$ can be represented as a sum of forces each acting within a small interval $d\xi$. Then, each term of the sum is

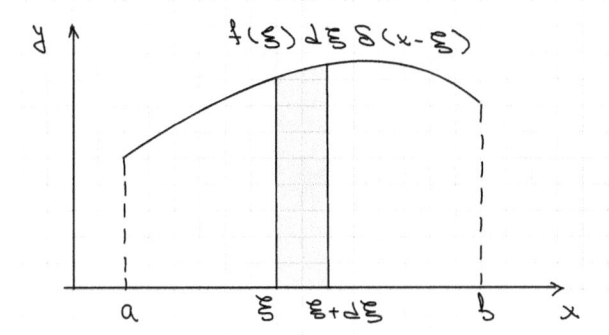

Fig. 10.4. Force acting upon the $[\xi, \xi + d\xi]$ segment.

$f(\xi)d\xi\delta(x - \xi)$. The operator R transforms it into (see Figure 10.4)

$$R[f(\xi)d\xi\delta(x - \xi)] = f(\xi)d\xi R[\delta(x - \xi)] = f(\xi)G(x,\xi)d\xi.$$

Here, we used $R(Cf(x)) = CR(f(x))$ since $f(\xi)$ can be considered as a constant near $x = \xi$. Thus, the total sum of all terms is

$$R\left[\sum f(\xi)d\xi\delta(x - \xi)\right] = \sum R\left[f(\xi)d\xi\delta(x - \xi)\right] = \sum f(\xi)G(x,\xi)d\xi.$$

For $d\xi \to 0$, we replace the sum on integration so that

$$y(x) = R[f(x)] = \int_a^b G(x,\xi)f(\xi)d\xi. \tag{10.9}$$

The function of influence $G(\xi, x)$ can be evaluated analytically or calculated numerically. For complex systems, it can be determined experimentally by measuring a response of the system on a unit force applied at one point. Note that all the considerations above are applied only for linear systems. After Green's function has been found and the linearity of the system has been established, the solution of the problem is written using equation (10.9) for any external influence.

10.3. Green's function solving for first-order ODE

In this section, we apply Green's function method to solving non-homogeneous first-order ordinary linear differential equations.

A first-order linear ordinary differential equation has form (8.5):

$$\frac{dy}{dx} + P(x)y = Q(x). \qquad (10.10)$$

10.3.1. Zero initial condition

We are going to look for a *particular solution* of this equation with a very specific boundary condition, namely that at some x_0 we have $y(x_0) = 0$. Later, we consider how to adapt our solution for this boundary condition to any other boundary value.

We consider a case when the function of influence acts as a delta function at some point for any fixed ξ:

$$\frac{dy}{dx} + P(x)y = \delta(x - \xi). \qquad (10.11)$$

Region I: $x_0 < x < \xi$.

In this region, equation (10.11) is a homogeneous differential equation

$$\frac{dy}{dx} + P(x)y = 0,$$

with a general solution derived before (8.8):

$$y(x) = Cy_1(x) \quad \text{where} \quad y_1(x) = \exp\left(-\int P(x)dx\right). \qquad (10.12)$$

Since we are looking for a particular solution that is $y(x_0) = 0$, we can see that the constant of integration in this case is $C = 0$. Then, for all $x < \xi$, we have

$$y(x) = 0$$

as the particular solution of (10.11).

Region II: $\xi - 0 < x < \xi + 0$.

Next, we integrate equation (10.11) from $x = \xi - 0$ to $\xi + 0$:

$$y(\xi+0) - y(\xi-0) = -\int_{\xi-0}^{\xi+0} P(x)y(x)dx + \int_{\xi-0}^{\xi+0} \delta(x-\xi)dx = 0+1 = 1.$$

Note that since the function $y(x)$ is finite, then the first integral can be disregarded. As we stated before, the solution for $x < \xi$ is equal to 0, then

$$y(\xi + 0) = 1.$$

Region III: $x > \xi$.
For $x > \xi$, equation (10.11) is again a homogeneous one with the solution (10.12)

$$y(x) = Cy_1(x) \quad \text{where} \quad y_1(x) = \exp\left(-\int P(x)dx\right)$$

but with the boundary condition above as $y(\xi + 0) = 1$. Then, we can find the particular solution (namely, the constant of integration) using

$$1 = Cy_1(\xi) \quad \text{or} \quad C = \frac{1}{y_1(\xi)},$$

and the particular solution for $x > \xi$ takes the form

$$y(x) = \frac{1}{y_1(\xi)} y_1(x).$$

Now, we can determine Green's function for the problem as

$$G(x, \xi) = \begin{cases} 0 & x_0 < x < \xi, \\ y_1(x)/y_1(\xi) & x > \xi. \end{cases} \tag{10.13}$$

We recall that a solution for equation (10.10) for any function $Q(x)$ using Green's function can be written as (10.9)

$$y(x) = \int_{x_0}^{\infty} G(x, \xi) Q(\xi) d\xi = \int_{x_0}^{x} G(x, \xi) Q(\xi) d\xi + \int_{x}^{\infty} G(x, \xi) Q(\xi) d\xi.$$

Then, using the explicit form of Green's function (10.13), we finally have a particular solution as

$$y(x) = \int_{x_0}^{x} \frac{y_1(x)}{y_1(\xi)} Q(\xi) d\xi, \tag{10.14}$$

where $y_1(x)$ is a solution of the corresponding homogeneous equation:

$$y_1(x) = \exp\left(-\int P(x)dx\right).$$

Note that in the same way we can show that solution (10.14) is true for $x < x_0$.

10.3.2. Arbitrary initial condition

The solution above is a particular solution of the non-homogeneous differential equation when $y(x_0) = 0$. Now, we are interested to find a solution for the equation for an arbitrary boundary condition.

We know that a general solution of the non-homogeneous equation is a sum of the general solution of the corresponding homogeneous equation and a particular solution of the non-homogeneous equation. Choosing for the particular solution the one we have found for $y(x_0) = 0$, we can write then a general solution of the non-homogeneous equation as

$$y(x) = C y_1(x) + \int_{x_0}^{x} \frac{y_1(x)}{y_1(\xi)} Q(\xi) d\xi. \tag{10.15}$$

If we are looking for a solution satisfying the given initial condition $y = y_0$ at $x = x_0$, then

$$y_0 = C y_1(x_0) + 0, \quad C = \frac{y_0}{y_1(x_0)},$$

and

$$y(x) = y_0 \frac{y_1(x)}{y_1(x_0)} + \int_{x_0}^{x} \frac{y_1(x)}{y_1(\xi)} Q(\xi) d\xi. \tag{10.16}$$

Attention: Both the general solution (10.15) and the particular solution (10.16) were derived before in Chapter 8 using Lagrange's method for solving linear non-homogeneous ODEs (see equations (8.9) and (8.10)). The two solutions are identical.

Example 10.1. A body of mass m is tossed vertically upward with the initial speed of v_0. The body experiences linear air resistance as βv. We want to find the velocity of the body as a function of time. The equation of motion is

$$m \frac{dv}{dt} = -\beta v - mg,$$

where g is the free-fall acceleration. While we can immediately use equation (10.16), we derive the solution using Green's function method. But first, we need the solution of the homogeneous equation $mv' + \beta v = 0$, that is, $v_1(t) = c \exp(-(\beta/m)t)$. For the first region

$t < \tau$ with zero initial condition, we have $v(t) = 0$. For the second region, integrating

$$\frac{dv}{dt} + \frac{\beta}{m}v = \delta(t - \tau)$$

between $\tau - 0$ and $\tau + 0$ gives $v(\tau + 0) = 1$. Using this as the initial condition for the third region, i.e., $t > \tau$, we have the solution as

$$v(t) = e^{-\frac{\beta}{m}(t-\tau)}.$$

Thus, Green's function is

$$G(t, \tau) = \begin{cases} 0 \\ e^{-\frac{\beta}{m}(t-\tau)}. \end{cases}$$

And the general solution is

$$v(t) = v_0 e^{-\frac{\beta}{m}t} + \int_0^t e^{-\frac{\beta}{m}(t-\tau)} g d\tau.$$

The integral can be easily evaluated, and finally, we have

$$v(t) = v_0 e^{-\frac{\beta}{m}t} + \frac{mg}{\beta}\left(e^{-\frac{\beta}{m}t} - 1\right). \qquad \blacksquare$$

10.4. Green's function for second-order ODE: Initial value problem

In this section, we consider non-homogeneous ordinary differential equations with constant coefficients (9.22)

$$y'' + py' + qy = f(x).$$

From Chapter 9, we know that a general solution for can be written as

$$y(x) = C_1 y_1(x) + C_2 y_2(x) + u(x),$$

where $u(x)$ is a particular solution that can be found, for example, using Lagrange's method of variation of parameters or Euler's method of undetermined coefficients (see Chapter 9 for details). The particular solution $u(x)$ can also be found using Green's function.

10.4.1. Simple motion driven by a force

We start with a simple example of motion under and external time dependent force

$$m\frac{d^2y}{dt^2} = f(t). \quad (10.17)$$

The solution of the above equation is completely determined by $f(x)$ together with initial conditions. We set the initial conditions as

$$y(t = t_0) = 0, \quad y'(t = t_0) = 0.$$

Later, as with a first-order ODE, we generalize our solution to arbitrary boundary conditions.

Following the same approach as in Section 10.3, we want to derive Green's function for the equation

$$m\frac{d^2y}{dt^2} = \delta(t - \tau) \quad (10.18)$$

for any choice of τ and given initial conditions.

Region I: $t_0 < t < \tau$.
In this region, the right side of equation above is zero and the solution is $y(t) = at + b$. For given initial conditions, we have $a = 0$ and $b = 0$ with $y(t) = 0$ and $y'(t) = 0$ in this region.

Region II: $\tau - 0 < t < \tau + 0$.
Integrating the equation with delta function from $\tau - 0$ to $\tau + 0$

$$m\int_{\tau-0}^{\tau+0} \frac{d^2y}{dt^2} dt = \int_{\tau-0}^{\tau+0} \delta(t-\tau) dt$$

gives

$$m\left(\frac{dy}{dt}|_{\tau+0} - \frac{dy}{dt}|_{\tau-0}\right) = 1.$$

Since the derivative for $y'(t - \tau) = 0$, we obtain

$$\frac{dy}{dt}|_{\tau+0} = \frac{1}{m}.$$

Thus, the derivative has a discontinuity (not a singularity) at $t = \tau$. However, the function itself is continuous at $t = \tau$ with

$$y(\tau + 0) = y(\tau - 0) = 0$$

because we can disregard change of position in dt time due to the finite velocity $y' = 1/m$.

Region III: $t > \tau + 0$.

For $t > \tau + 0$, equation (10.18) again has the right side equal to zero, but now the initial conditions are different, namely

$$y(\tau + 0) = 0, \quad y'(\tau + 0) = \frac{1}{m}.$$

Integrating the equation gives $y(t) = ct + d$. Applying the initial condition for this region gives $c = 1/m$ and $d = -\tau/m$. Then, the solution for the initial conditions is

$$y = \frac{t}{m} - \frac{\tau}{m} = \frac{1}{m}(t - \tau).$$

Therefore, Green's function for the problem is

$$G(t, \tau) = \begin{cases} 0 & (t_0 < t < \tau), \\ (t - \tau)/m & (\tau < t < \infty). \end{cases} \quad (10.19)$$

Now, we can write the solution of equation (10.17) using Green's function as

$$y(t) = \int_{t_0}^{t} G(t, \tau) f(\tau) d\tau + \int_{t}^{\infty} G(t, \tau) f(\tau) d\tau = \frac{1}{m} \int_{t_0}^{t} (t - \tau) f(\tau) d\tau.$$

$$(10.20)$$

The solution (10.20) can be derived by two more methods.

Method I: We can use rather simple physics consideration, namely the impulse $J = p_f - p_i = mv_f - mv_i$ is equal to a force $f(\tau)$ multiplied by a time interval $d\tau$:

$$m(v_f - v_i) = f(\tau) d\tau.$$

Thus, within a time $d\tau$, a body acquires additional speed $f(\tau)d\tau/m$, and by time t, it would cover a distance $f(\tau)d\tau(1/m)(t - \tau)$. Since the equation of motion is a linear one, then the resulting effect is a sum (integral) of all impacts, i.e.,

$$y(t) = \frac{1}{m} \int_{t_0}^{t} (t - \tau) f(\tau) d\tau.$$

Method II: The other way to derive (10.20) is by double integration of equation (10.17). The first integration gives

$$m[y'(t) - y'(t_0)] = \int_{t_0}^{t} f(\tau)d\tau.$$

Using the given initial conditions $y'(t_0) = 0$ and integrating again, we get

$$my(t) = \int_{t_0}^{t} my'(T)dT = \int_{t_0}^{t} \left(\int_{t_0}^{T} f(\tau)d\tau \right) dT.$$

Note that T changes from t_0 to t and for every T the variable τ changes from t_0 to T. Thus, we integrate within the triangle (see Figure 10.5). When we change the order of integration, then τ changes from t_0 to T, but T changes from τ to t. After changing the order of integration, we get

$$my(t) = \int_{t_0}^{t} d\tau \int_{\tau}^{t} f(\tau)dT = \int_{t_0}^{t} f(\tau)d\tau \int_{\tau}^{t} dT = \int_{t_0}^{t} f(\tau)(t-\tau)d\tau,$$

that is, the solution (10.20) we have obtained earlier.

Note that the solution (10.20) is a particular solution of the equation with zero initial conditions. Now, we consider arbitrary initial or boundary conditions, namely $y(t_0) = y_0$ and $y'(t_0) = v_0$. Then, a general solution can be written as

$$y(t) = y_h(t) + u(t),$$

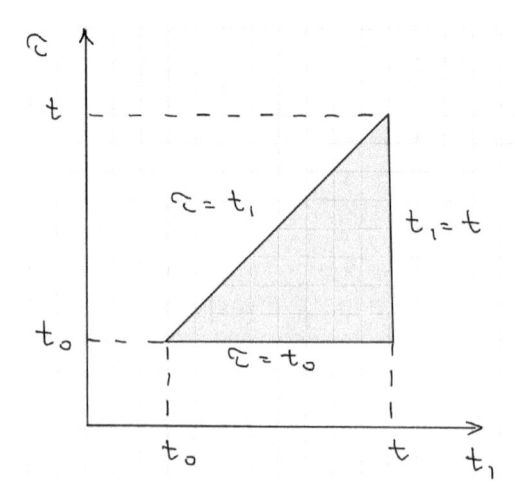

Fig. 10.5. Area of integration for the double integral.

where $u(t)$ is a particular solution of the non-homogeneous equation

$$m\frac{d^2u}{dt^2} = f(t)$$

with zero initial conditions $u(t = t_0) = 0$ and $u'(t = t_0) = 0$, and $y_h(t)$ is a solution of the corresponding homogeneous equation

$$m\frac{d^2y_h}{dt^2} = 0,$$

but now with the given initial conditions, $y_h(t = t_0) = y_0$ and $y'_h(t = t_0) = v_0$. The solution $y_h(t)$ can be easily found by simple integration of the homogeneous equation. The general solution can be written as $y_h(t) = at + b$. For the given initial conditions, $b = y_0$, $a = v_0$, and then,

$$y_h(t) = v_0(t - t_0) + y_0.$$

Therefore, the general solution is

$$y(t) = y_0 + v_0(t - t_0) + \frac{1}{m}\int_{t_0}^{t}(t - \tau)f(\tau)d\tau.$$

We can check the solution by direct substitution into the original equation (10.17).

Note that for a constant force $f(t) = f_0$ we get the well-known solution for motion with constant acceleration:

$$y(t) = y_0 + v_0(t - t_0) + \frac{1}{2}a(t - t_0)^2,$$

with $a = f_0/m$.

10.4.2. Linear non-homogeneous second-order ODE

Now, we explore general linear non-homogeneous second-order differential equations:

$$y'' + p(x)y' + q(x)y = f(x). \qquad (10.21)$$

For finding Green's function for this problem, we need to solve equation

$$y'' + p(x)y' + q(x)y = \delta(x - \xi) \qquad (10.22)$$

with zero initial conditions, namely $y(x = x_0) = 0$ and $y'(x = x_0) = 0$.

Let $y_1(x)$ and $y_2(x)$ be two linear independent solutions of the corresponding homogeneous equation $y'' + p(x)y' + q(x)y = 0$.

Region I: For $x < \xi$, equation (10.22) is a homogeneous equation with the solution

$$y(x) = C_1 y_1(x) + C_2 y_2(x).$$

For the given initial conditions, we have both coefficients $C_1 = 0$ and $C_2 = 0$, so $y(x) = 0$ for $x < \xi$.

Region II: Integrating the equation from $x = \xi - 0$ to $\xi + 0$ gives

$$y'(\xi + 0) - y'(\xi - 0) + p(\xi)[y(\xi + 0) - y(\xi - 0)]$$

$$+ q(\xi) \int_{\xi-0}^{\xi+0} y(x) dx = \int_{\xi-0}^{\xi+0} \delta(x - \xi) dx.$$

Since the third and the fourth terms are equal to zero, and $y'(\xi - 0) = 0$ (do you see why?), we find

$$y'(\xi + 0) = 1.$$

Region III: Now, we need to find solutions of the homogeneous equation for $x > \xi$ with the initial conditions $y(\xi + 0) = 0$ and $y'(\xi + 0) = 1$. The solution of the homogeneous equation is

$$y(x) = C_1 y_1(x) + C_2 y_2(x).$$

For the given initial conditions, we have that

$$y(x = \xi) = C_1 y_1(\xi) + C_2 y_2(\xi) = 0,$$
$$y'(x = \xi) = C_1 y_1'(\xi) + C_2 y_2'(\xi) = 1.$$

Solving for the unknown coefficients C_1 and C_2, we have

$$y(x) = -\frac{y_2(\xi)}{y_1(\xi) y_2'(\xi) - y_1'(\xi) y_2(\xi)} y_1(x)$$

$$+ \frac{y_1(\xi)}{y_1(\xi) y_2'(\xi) - y_1'(\xi) y_2(\xi)} y_2(x).$$

Using the Wronskian for the two functions as

$$W(y_1, y_2) = y_1(\xi) y_2'(\xi) - y_1'(\xi) y_2(\xi),$$

we can write Green's function for equation (10.22) as

$$G(x,\xi) = \begin{cases} 0 & (t_0 < x < \xi), \\ -\dfrac{y_2(\xi)}{W(y_1(\xi)y_2(\xi))}y_1(x) + \dfrac{y_1(\xi)}{W(y_1(\xi)y_2(\xi))}y_2(x) & (\xi < x < \infty). \end{cases}$$

Then, using

$$y(x) = \int_{x_0}^{\infty} G(x,\xi)f(\xi)d\xi = \int_{x_0}^{x} G(x,\xi)f(\xi)d\xi + \int_{x}^{\infty} G(x,\xi)f(\xi)d\xi,$$

the particular solution for the non-homogeneous equation (10.21) takes the form

$$y(x) = -y_1(x)\int_{x_0}^{x} \frac{y_2(\xi)f(\xi)}{W(y_1,y_2)}d\xi + y_2(x)\int_{x_0}^{x} \frac{y_1(\xi)f(\xi)}{W(y_1,y_2)}d\xi. \quad (10.23)$$

The general solution of the original equation (10.21) for arbitrary initial conditions is

$$y(x) = C_1 y_1(x) + C_2 y_2(x) - y_1(x)\int_{x_0}^{x} \frac{y_2(\xi)f(\xi)}{W(y_1,y_2)}d\xi$$

$$+ y_2(x)\int_{x_0}^{x} \frac{y_1(\xi)f(\xi)}{W(y_1,y_2)}d\xi, \quad (10.24)$$

where the coefficients C_1 and C_2 are defined by initial conditions by using $y(x_0) = y_0$ and $y'(x_0) = y'_0$. It is instructive to compare this solution with Lagrange method (9.15), namely

$$y(x) = C_1 y_1(x) + C_2 y_2(x) - y_1(x)\int \frac{y_2(x)f(x)}{W(y_1,y_2)}dx$$

$$+ y_2(x)\int \frac{y_1(x)f(x)}{W(y_1,y_2)}dx.$$

While both solutions look similar, there is a difference in integration. The solution derived by Green's function method involves definite integration from x_0 to x, whereas Lagrange's solution contains indefinite integrals. However, from the second fundamental theorem of calculus follows that

$$\int_{x_0}^{x} f(\xi)d\xi = \Phi(x) - \Phi(x_0),$$

where $\Phi(x)$ is an indefinite integral $\Phi(x) = \int f(x)dx$ and $\Phi(x_0)$ is a number. Thus, using the Lagrange solution leads just to redefining the constants C_1 and C_2 (do you see exactly how?) compared to Green's function method.

Example 10.2. Find the general solution of the following differential equation

$$\frac{d^2y}{dt^2} + \omega_0^2 y = f \sin \omega t$$

for zero initial conditions $y(0) = 0$ and $y'(0) = 0$. The equation represents oscillatory motion in the presence of an external force $f \sin \omega t$. Instead of just applying equation (10.24), we derive the solution following Green's function method. First, we need a solution for the homogeneous equation

$$y'' + \omega_0^2 y = 0,$$

that is,

$$y(t) = c_1 \cos(\omega_0 t) + c_2 \sin(\omega_0 t).$$

Now, we can start deriving Green's function for the equation

$$y'' + \omega_0^2 y = \delta(t - \tau).$$

For the first region with $t < \tau$ with the zero initial condition, we have $y(t) = 0$ (do you see why?). For the second region, we integrated the equation from $\tau - 0$ to $\tau + 0$,

$$\int_{\tau-0}^{\tau+0} y''(t)dt + \omega_0^2 \int_{\tau-0}^{\tau+0} y(t)dt = \int_{\tau-0}^{\tau+0} \delta(t-\tau)dt,$$

that gives us $y'(\tau + 0) - y'(\tau - 0) = 1$. The solution for the third region is simply the solution of the homogeneous equation:

$$y(t) = c_1 \cos(\omega_0 t) + c_2 \sin(\omega_0 t).$$

We can easily find the coefficients c_1 and c_2 by using the initial condition at $\tau + 0$, namely $y'(\tau + 0) = 1$ and $y(\tau + 0) = 0$ (do you see why?). After simple algebra, we have $c_1 = -\sin(\omega_0 \tau)/\omega_0$ and $c_2 = \cos(\omega_0 \tau,)/\omega_0$. Having the coefficients, and using the trigonometric identity $\sin(x-y) = \sin(x)\cos(y) - \sin(y)\cos(x)$, we can write Green's function as

$$G(t,\tau) = \begin{cases} 0, \\ \sin(\omega_0(t-\tau))/\omega_0. \end{cases}$$

Now, we can write the solution as

$$y(t) = \frac{f}{\omega_0} \int_0^t \sin(\omega_0(t-\tau)) \sin\omega\tau \, d\tau.$$

This last integral cam be readily evaluated, and finally, we have for oscillations with zero initial conditions and without friction

$$y(t) = \frac{f}{\omega_0} \left[\frac{\omega_0 \sin(\omega t) - \omega \sin(\omega_0 t)}{\omega_0^2 - \omega^2} \right].$$

For non-zero initial conditions, the general solutions we have are

$$y(t) = C_1 \cos(\omega_0 t) + C_2 \sin(\omega_0 t) + \frac{f}{\omega_0} \left[\frac{\omega_0 \sin(\omega t) - \omega \sin(\omega_0 t)}{\omega_0^2 - \omega^2} \right]. \blacksquare$$

10.5. Green's function for second-order ODE: Boundary value problem

10.5.1. *Introduction to the boundary value problem*

In the previous section (and also in Chapter 9), we concentrated mostly on solving a second-order linear ODE as the initial value problem, when the function of interest and its derivative are given at the same point x_0. However, there are many problems when the two conditions are given at different points of the independent variable, e.g., x_a and x_b. The problem of determining the particular solution of a differential equation that satisfies a given boundary conditions at different points is called a *boundary value* problem. Unlike for the initial value problem, the boundary value problem may not have a solution satisfying given conditions (see Section 9.7). Finding a shape of an elastic rod under a stationary load is an example of a boundary value problem considered in Section 10.2.1.

In the following, we consider boundary value problems on the interval $[0, l]$ for a second-order linear differential equation:

$$y'' + p(x)y' + q(x)y = f(x), \qquad (10.25)$$

where $p(x)$, $q(x)$, and $f(x)$ are continuous functions within $[0, l]$. Let us introduce a function

$$\tilde{p}(x) = \exp\left(\int_{x_0}^{x} p(\xi)d\xi\right)$$

that has the properties $\tilde{p}(x) \neq 0$ and $\tilde{p}(x) > 0$ on $[0, l]$. We can easily check that

$$\frac{d}{dx}\left[\tilde{p}(x)\frac{dy}{dx}\right] = \tilde{p}(x)\frac{d^2y}{dx^2} + \tilde{p}(x)p(x)\frac{dy}{dx}.$$

Then, multiplying both sides of equation (10.25) by $\tilde{p}(x)$ and using the equation above, we can rewrite (10.25) as

$$L[y] \equiv \frac{d}{dx}\left[\tilde{p}(x)\frac{dy}{dx}\right] - \tilde{q}(x)y(x) = \tilde{f}(x) \quad \text{or} \quad L[y] = f(x), \quad (10.26)$$

where $\tilde{q}(x) = -\tilde{p}(x)q(x)$, $\tilde{f}(x) = \tilde{p}(x)f(x)$, and $L[y]$ is the notation for an operator acting on the left side of the equation above. Thus, boundary value problems for a general linear differential equation of second order (10.25) can be reduced to boundary value problems for equation (10.26). If $\tilde{f}(x) = 0$, then we have a homogeneous boundary value problem; otherwise, it is a non-homogeneous boundary value problem.

There are some advantages to work with equation (10.26). From the form of equation (10.26) follows a valuable property called *the Lagrange identity*. Let $y_1(x)$ and $y_2(x)$ be solutions for the corresponding equations:

$$\frac{d}{dx}\left[\tilde{p}(x)\frac{dy_1}{dx}\right] - \tilde{q}(x)y_1(x) = \tilde{f}_1(x),$$

$$\frac{d}{dx}\left[\tilde{p}(x)\frac{dy_2}{dx}\right] - \tilde{q}(x)y_2(x) = \tilde{f}_2(x).$$

Now, we multiply the first equation by $y_2(x)$ and the second equation by $y_1(x)$, and taking the difference of the two, we obtain

$$y_2(x)\frac{d}{dx}\left[\tilde{p}(x)\frac{dy_1}{dx}\right] - y_1(x)\frac{d}{dx}\left[\tilde{p}(x)\frac{dy_2}{dx}\right] = \tilde{f}_1(x)y_2(x) - \tilde{f}_2(x)y_1(x).$$

Since

$$\frac{d}{dx}\left[\tilde{p}(x)\left(y_2\frac{dy_1}{dx} - y_1\frac{dy_2}{dx}\right)\right] = y_2(x)\frac{d}{dx}\left[\tilde{p}(x)\frac{dy_1}{dx}\right]$$
$$- y_1(x)\frac{d}{dx}\left[\tilde{p}(x)\frac{dy_2}{dx}\right]$$
$$+ \tilde{p}(x)\left[\frac{dy_2}{dx}\frac{dy_1}{dx} - \frac{dy_1}{dx}\frac{dy_2}{dx}\right],$$

then we have established Lagrange's identity

$$\frac{d}{dx}\left[\tilde{p}(x)\left(y_2\frac{dy_1}{dx} - y_1\frac{dy_2}{dx}\right)\right] = \tilde{f}_1(x)y_2(x) - \tilde{f}_2(x)y_1(x). \quad (10.27)$$

The integral form of the Lagrange identity is known as Green's formula:

$$\left\{\tilde{p}(x)\left(y_2\frac{dy_1}{dx} - y_1\frac{dy_2}{dx}\right)\right\}_0^l = \int_0^l \left(\tilde{f}_1(x)y_2(x) - \tilde{f}_2(x)y_1(x)\right)dx. \quad (10.28)$$

From Green's formula (10.32) follows that if $y_1(x)$ and $y_2(x)$ are two linear independent solutions for a homogeneous equation, i.e., $\tilde{f}_1(x) \equiv \tilde{f}_2(x) \equiv 0$, then these solutions satisfy equation

$$\tilde{p}(x)\left(y_2\frac{dy_1}{dx} - y_1\frac{dy_2}{dx}\right) = C \quad (10.29)$$

and the Wronskian can simply be written as

$$W(y_1, y_2) = \left(y_2\frac{dy_1}{dx} - y_1\frac{dy_2}{dx}\right) = \frac{C}{\tilde{p}(x)}. \quad (10.30)$$

Since solutions of a homogeneous equations are set within an arbitrary factor, then the constant C can be determined by a normalization of the solutions.

From the definition of the Wronskian, it also follows that if $y_1(x)$ is a solution of a homogeneous equation, then any other linear independent solution can be found from

$$y_1(x)\frac{dy_2(x)}{dx} - \frac{dy_1(x)}{dx}y_2(x) = \frac{C}{\tilde{p}(x)}. \qquad (10.31)$$

If $y_1(x) \neq 0$ on $[0, l]$, then dividing all terms of equation above by $y_1^2(x)$ we get

$$\frac{d}{dx}\left(\frac{y_2(x)}{y_1(x)}\right) = \frac{C}{\tilde{p}(x)y_1^2(x)},$$

and solving the equation gives

$$y_2(x) = C_1 y_1(x) + C y_1(x) \int_0^x \frac{d\xi}{\tilde{p}(\xi)y_1^2(\xi)}.$$

Boundary conditions: Hereafter, we consider boundary value problems defined on the interval $[0, l]$ for linear differential equations of the form (10.25) or (10.26). Most boundary value problems are defined by linear boundary conditions, such as

$$\begin{aligned} a_1 y'(0) + b_1 y(0) &= c_1, \\ a_2 y'(l) + b_2 y(l) &= c_2, \end{aligned} \qquad (10.32)$$

where $a_i^2 + b_i^2 \neq 0$ for $i = 1, 2$.

If $a_1 = a_2 = 0$, then such conditions specify the value of the function at the boundaries. These kinds of conditions are called Dirichlet boundary conditions. The conditions $b_1 = b_2 = 0$ specify the value of the first derivative, and they are called Neumann boundary conditions. The mixed conditions (both the function and the derivative) are called Robin conditions.

If $c_1 = c_2 = 0$, then such boundary conditions are called homogeneous boundary conditions, otherwise they are called non-homogeneous boundary conditions.

It is always possible to reduce non-homogeneous boundary conditions to homogeneous conditions by looking for the unknown function $y(x)$ as a sum of two functions

$$y(x) = h(x) + z(x),$$

where $h(x)$ is any function satisfying given non-homogeneous conditions and $z(x)$ is a new unknown function but *satisfying homogeneous boundary conditions.* For example, for Dirichlet boundary condition $y(0) = a_0$, $y(l) = a_l$, we may choose $h(x)$ as

$$h(x) = a_0 \frac{l-x}{l} + a_l \frac{x}{l}.$$

In a general case for equation

$$p_0 y'' + p_1 y' + p_2 y = f(x), \quad a < x < b$$

with the boundary conditions

$$\alpha_1 y'(a) + \beta_1 y(a) = A_1,$$
$$\alpha_2 y'(b) + \beta_2 y(b) = A_2,$$

the function $h(x)$ can be written as

$$h(x) = \left(\frac{b-x}{b-a}\right)^2 \frac{A_1}{\beta_1} \left(1 - e^{-\frac{\beta_1}{\alpha_1}(x-a)}\right)$$
$$+ \left(\frac{x-a}{b-a}\right)^2 \frac{A_2}{\beta_2} \left(1 - e^{-\frac{\beta_2}{\alpha_2}(b-x)}\right).$$

Then, the equation for the unknown function $z(x)$ takes the form

$$p_0 z'' + p_1 z' + p_2 z = f - p_0 h'' - p_1 h' - p_2 h,$$

with the right-hand side modified compared to the equation for $y(x)$, and the homogeneous boundary conditions are

$$\alpha_1 y'(a) + \beta_1 y(a) = 0,$$
$$\alpha_2 y'(b) + \beta_2 y(b) = 0.$$

Homogeneous boundary value problems $f(x) = 0$: We start with a homogeneous boundary value problem on $0 \le x \le l$:

$$\frac{d}{dx}\left[\tilde{p}(x)\frac{dy}{dx}\right] - \tilde{q}(x) y(x) = 0 \quad y(0) = 0 \; y(l) = 0,$$

with the homogeneous boundary conditions. This problem has always the trivial solution $y(0) = 0$ but we are interested in the existence of non-trivial solutions.

Example 10.3. The problem

$$y''(x) = 0 \quad \text{on} \quad 0 < x < 1 \quad \text{with } y(0) = y(1) = 0$$

has only the trivial solution $y(x) = 0$. But the problem

$$y''(x) + y(x) = 0 \quad \text{on} \quad 0 < x < \pi \quad \text{with } y(0) = y(\pi) = 0$$

has two solutions: $y(x) = 0$ and $y(x) = C \sin x$. ∎

It is helpful to start our analysis for possible other solutions by considering the *initial value problem*

$$\frac{d}{dx}\left[\tilde{p}(x)\frac{dy}{dx}\right] - \tilde{q}(x)y(x) = 0 \quad y(0) = 0, \ y'(0) = y_1.$$

For continuous $\tilde{p}(x)$ and $\tilde{q}(x)$, there is always a unique solution, i.e., $y(x) \neq 0$ for $y_1 \neq 0$. Thus, for all possible y_1, we have a solution as a function of two variables $y(x, y_1)$. If there is such y_1 that $y(l, y_1) = 0$, then there exist a non-trivial solution for the boundary value problem. In numerical solutions for the boundary value problem, such an approach is called a *shooting method*. Note that in this case $Cy(x, y_1)$ is also a solution of the homogeneous boundary value problem.

Non-homogeneous boundary value problems: Now, we consider a non-homogeneous differential equation

$$L[y] \equiv \frac{d}{dx}\left[\tilde{p}(x)\frac{dy}{dx}\right] - \tilde{q}(x)y = \tilde{f}(x), \qquad (10.33)$$

with homogeneous boundary conditions

$$\begin{aligned} \alpha_1 y'(0) + \beta_1 y(0) &= 0, \\ \alpha_2 y'(l) + \beta_2 y(l) &= 0. \end{aligned} \qquad (10.34)$$

In the equation above, $\tilde{p}(x) > 0$ (this follows from the definition of $\tilde{p}(x) = \exp(\int_{x_0}^x p(\xi)d\xi)$ (see (10.25)). We assume that $\tilde{q}(x)$ and $\tilde{f}(x)$ are real and continuous functions.

10.5.2. Case 1: Only trivial solutions for L[y] = 0

We start with the case when the corresponding homogeneous equation $L[y] = 0$ has only trivial solution $y(x) \equiv 0$ for given homogeneous conditions (10.34).

From the linearity of equation (10.33) follows that if a solution of this equation exists, then it is a unique solution. Our goals here are to show that a solution for (10.33) exists and to demonstrate how to find it.

Like for the initial value problem, we consider a special form of the non-homogeneous term, namely $\tilde{f}(x) \equiv \delta(x - \xi)$, thus

$$\frac{d}{dx}\left[\tilde{p}(x)\frac{dy}{dx}\right] - \tilde{q}(x)y = \delta(x - \xi). \tag{10.35}$$

Let the solution of the equation above be $y(x, \xi)$. Integrating equation (10.35) over the interval $[\xi - 0, \xi + 0]$ gives

$$\tilde{p}(\xi + 0)y'(\xi + 0, \xi) - \tilde{p}(\xi - 0)y'0(\xi - 0, \xi) - \int_{\xi-0}^{\xi+0} \tilde{q}(x)y(x, \xi)dx = 1, \tag{10.36}$$

since the integral $\int_{\xi-0}^{\xi+0} \delta(x - \xi)dx = 1$.

Let us use the notation for solution of (10.35) as $y(x, \xi) = \tilde{G}(x, \xi)$. Then, we can rewrite (10.36) as

$$\frac{d}{dx}\tilde{G}(x, \xi)\bigg|_{x=\xi+0} - \frac{d}{dx}\tilde{G}(x, \xi)\bigg|_{x=\xi-0} = \frac{1}{\tilde{p}(\xi)}. \tag{10.37}$$

From this point, we still do not know the form of $\tilde{G}(x, \xi)$ but we have the following properties for this function, namely

(1) $\tilde{G}(x, \xi)$ as a function of x satisfies the homogeneous equation $L[y] = 0$ on $0 < x < \xi$ and $\xi < x < l$,
(2) $\tilde{G}(x, \xi)$ satisfy the boundary conditions (10.34),
(3) $\tilde{G}(x, \xi)$ is continuous on $[0, l]$ but its first derivative has a discontinuity at $x = \xi$, i.e.,

$$\frac{d}{dx}\tilde{G}(x, \xi)\bigg|_{x=\xi-0}^{x=\xi+0} = \frac{1}{\tilde{p}(\xi)}. \tag{10.38}$$

Such a function $\tilde{G}(x,\xi)$ we call Green's function for the boundary value problem. The importance of this function is that a solution for any right-hand side function $\tilde{f}(x)$ can be written using $\tilde{G}(x,\xi)$. Indeed, let us apply Green's formula (10.28) to $y(x)$ and $\tilde{G}(x,\xi)$ on intervals $[0, \xi - \epsilon]$ and $[\xi + \epsilon, l]$:

$$\left[\tilde{p}(x)\left(\tilde{G}(x,\xi)\frac{dy}{dx} - y(x)\frac{d\tilde{G}}{dx}\right)\right]\bigg|_0^{\xi-\epsilon}$$

$$+ \left[\tilde{p}(x)\left(\tilde{G}(x,\xi)\frac{dy}{dx} - y(x)\frac{d\tilde{G}}{dx}\right)\right]\bigg|_{\xi+\epsilon}^{l}$$

$$= \int_0^{\xi-\epsilon} \tilde{G}(x,\xi)\tilde{f}(x)dx + \int_{\xi+\epsilon}^{l} \tilde{G}(x,\xi)\tilde{f}(x)dx.$$

Since both $y(x)$ and $\tilde{G}(x,\xi)$ satisfy homogeneous boundary conditions at $x = 0$ and $x = l$, we find

$$\left[\tilde{p}(x)\left(y(x)\frac{d\tilde{G}}{dx}\right)\right]\bigg|_{\xi-\epsilon}^{\xi+\epsilon} - \left[\tilde{p}(x)\left(\tilde{G}(x,\xi)\frac{dy}{dx}\right)\right]\bigg|_{\xi-\epsilon}^{\xi+\epsilon}$$

$$= \int_0^{\xi-\epsilon} \tilde{G}(x,\xi)\tilde{f}(x)dx + \int_{\xi+\epsilon}^{l} \tilde{G}(x,\xi)\tilde{f}(x)dx.$$

Taking the limit $\epsilon \to 0$ and using the property of Green's function derivative (10.38) and since $y(x)$ and its derivative are continuous, we find that

$$y(\xi) = \int_0^l \tilde{G}(x,\xi)\tilde{f}(x)dx. \tag{10.39}$$

Thus, when we have Green's function $\tilde{G}(x,\xi)$, then we can find a solution for any right-hand side $\tilde{f}(x)$ of the boundary problem just by integrating (10.39).

Now, it is time to construct Green's function for the boundary value problem. We do this by using two solutions for the homogeneous equation, but we apply one solution for the left boundary point and the the other one for the right boundary point.

Let $y_1(x)$ be a solution of a homogeneous equation $L[y_1(x)] = 0$ satisfying only left boundary condition, i.e.,

$$\alpha_1 y_1'(0) + \beta_1 y_1(0) = 0.$$

Thus, we have the problem as an initial value problem. We can choose the initial conditions as

$$y_1(0) = -C\alpha_1, \quad y_1'(0) = C\beta_1.$$

At the beginning, we assumed that the homogeneous equation $L[y(x)] = 0$ for the conditions (10.34) has only trivial solutions. This means that $y_1(x)$ cannot satisfy the right-hand boundary conditions at $x = l$. In the same way, we may introduce $y_2(x)$ that is a solution of $L[y_2(x)] = 0$ satisfying only the right-hand boundary conditions:

$$\alpha_2 y_2'(l) + \beta_2 y_2(l) = 0.$$

The two functions $y_1(x)$ and $y_2(x)$ are linearly independent solutions of $L[y(x)] = 0$ (do you see why?). Thus, we have two solutions of $L[y(x)] = 0$ where each solution satisfies only one of boundary conditions.

Let us look for Green's function $\tilde{G}(x, \xi)$ as

$$\tilde{G}(x, \xi) = \begin{cases} C_1 y_1(x), & 0 \leq x \leq \xi, \\ C_2 y_2(x), & \xi \leq x \leq l. \end{cases}$$

For $x \neq \xi$, Green's function above is a solution of the homogeneous equation satisfying homogeneous boundary conditions. Now, we can find the constants C_1 and C_2 using the properties of $\tilde{G}(x, \xi)$, namely Green's function is continuous, and its first derivative has a discontinuity at $x = \xi$ (10.38). These two properties can be written as a system of linear equations:

$$C_2 y_2(\xi) - C_1 y_1(\xi) = 0,$$
$$C_2 y_2'(\xi) - C_1 y_1'(\xi) = 1/\tilde{p}(\xi).$$

The determinant of the system is equal to Wronskian (10.30)

$$W(y_1, y_2) = C/\tilde{p}(x);$$

this is not equal to zero, and the constant C is determined by normalization. In other words, the linear system of equations has the unique solutions for C_1 and C_2 as

$$C_1 = \frac{y_2(\xi)}{\tilde{p}(\xi)W(y_1(\xi), y_2(\xi))}, \quad C_2 = \frac{y_1(\xi)}{\tilde{p}(\xi)W(y_1(\xi), y_2(\xi))}.$$

Using the solutions of the linear system finally provides us with Green's function as

$$\tilde{G}(x, \xi) = \frac{1}{\tilde{p}(\xi)W(y_1(\xi), y_2(\xi))} \begin{cases} y_2(\xi)y_1(x), & 0 \leq x \leq \xi, \\ y_1(\xi)y_2(x), & \xi \leq x \leq l, \end{cases} \quad (10.40)$$

with the Wronskian

$$W(y_1(\xi), y_2(\xi)) = y_1(\xi)\frac{dy_2(\xi)}{dx} - y_2(\xi)\frac{dy_1(\xi)}{dx}.$$

Note from equation (10.40) that Green's function is symmetric relative to x and ξ, or

$$\tilde{G}(x, \xi) = \tilde{G}(\xi, x). \quad (10.41)$$

Thus, if a homogeneous boundary value problem has only a trivial solution, then a solution to the inhomogeneous boundary value problem exists for any function $f(x)$ on $[0, l]$ and

$$y(x) = \int_0^l \tilde{G}(x, \xi)\tilde{f}(\xi)d\xi$$

$$= \int_0^x \frac{y_2(x)y_1(\xi)}{\tilde{p}(\xi)W(y_1, y_2)}\tilde{f}(\xi)d\xi + \int_x^l \frac{y_1(x)y_2(\xi)}{\tilde{p}(\xi)W(y_1, y_2)}\tilde{f}(\xi)d\xi. \quad (10.42)$$

The physical meaning of Green's function, as noted earlier, is a solution of a boundary problem with a force applied at a single point. It is a good exercise to show that solution (10.42) satisfies the original equation (10.33).

Hint: Differentiate (10.42) twice and use (10.38).

Attention: We have a certain amount of freedom to choose specific forms of $y_1(x)$ and $y_2(x)$; they just must be solutions of the homogeneous equation, where each function satisfies the proper boundary conditions ($y_1(x)$ on the left and $y_2(x)$ the right), and be linear independent. A good choice of $y_1(x)$ and $y_2(x)$ makes it easier to evaluate the integrals in (10.42).

Since $\tilde{f} = \tilde{p}f$, then $\tilde{G}(x,\xi)\tilde{f}(x) = \tilde{G}(x,\xi)\tilde{p}f = G(x,\xi)f(x)$. Hence, Green's functions for the two homogeneous equations $L[y] = 0$ and $y'' + p(x)y' + q(x)y = 0$ are related by $\tilde{G}(x,\xi)\tilde{p}(x) = G(x,\xi)$, and we can write

$$y(x) = \int_0^l G(x,\xi)f(\xi)d\xi$$

$$= \int_0^x \frac{y_2(x)y_1(\xi)}{W(y_1,y_2)} f(\xi)d\xi + \int_x^l \frac{y_1(x)y_2(\xi)}{W(y_1,y_2)} f(\xi)d\xi. \quad (10.43)$$

We can directly derive Green's function $G(x,\xi)$ for the homogeneous equation $y'' + p(x)y' + q(x)y = 0$ in the same way we did for $L[y] = 0$. However, derived by using $L[y] = \tilde{f}(x)$ form we can easily derive both the Lagrange identity (10.32) and Green's formula (10.28).

The solution (10.43) looks somewhat similar to the solution derived by using the Lagrange method (9.15), that is,

$$y(x) = C_1 y_1(x) + C_2 y_2(x) - y_1(x) \int \frac{y_2(x)f(x)}{W(y_1,y_2)} dx$$

$$+ y_2(x) \int \frac{y_1(x)f(x)}{W(y_1,y_2)} dx.$$

The principal difference between the two solutions is the choice of solutions for the homogeneous equations $y_1(x)$ and $y_2(x)$. In Green's function method, the function $y_1(x)$ satisfies the given boundary conditions on the left and $y_2(x)$ on the right. Thus, Green's function (10.40) satisfies both boundary conditions. Instead, functions $y_1(x)$ and $y_2(x)$ in the Lagrange method do not need to satisfy any of the boundary conditions. They only must be solutions of the corresponding homogeneous equation and be linearly independent. The coefficients C_1 and C_2 will adjust the Lagrange solution to match to given boundary conditions.

Example 10.4. We consider a Dirichlet boundary value problem

$$y'' = f(x), \quad 0 < x < l, \quad y(0) = 0, \quad y(l) = 0$$

with *homogeneous boundary conditions*. We can easily see that the solution $y(x) = bx + c$ of the homogeneous equation $y'' = 0$ cannot satisfy the given conditions. Therefore, the homogeneous boundary value problem has only a trivial solution. We set $y_1(x) = b_1 x + c_1 = 0$ at $x = 0$. Then, $c_1 = 0$ and $y_1(x) = b_1 x$. For $y_2(x) = b_2 x + c_2 = 0$ at $x = l$, we get $c_2 = -b_2 l$ and then $y_2(x) = b_2(x - l)$. Then, we are looking for Green's function having the form

$$G(x, \xi) = \begin{cases} b_1 x & 0 \leq x \leq \xi, \\ b_2(x - l) & \xi \leq x \leq l. \end{cases}$$

From the conditions of continuity of the function and discontinuity of the derivative at $x = \xi$, we have the system

$$b_2(\xi - l) - b_1 \xi = 0,$$
$$b_2 - b_1 = 1 \qquad \text{(since } p(x) = 1\text{)}.$$

The solutions of the system are $b_1 = (\xi - l)/l$ and $b_2 = \xi/l$ so that

$$G(x, \xi) = \begin{cases} x(\xi - l)/l & 0 \leq x \leq \xi, \\ (x - l)\xi/l & \xi \leq x \leq l. \end{cases}$$

Then, the solution of the equation is

$$y(x) = \frac{(x - l)}{l} \int_0^x \xi f(\xi) d\xi + \frac{x}{l} \int_x^l (\xi - l) f(\xi) d\xi.$$

As one can see, this is the solution we derived in Section 10.2.1 for a shape of flexible rope when an external force $f(x)$ is applied perpendicular to the rope. ∎

Example 10.5. In this example, we consider a boundary value problem with mixed boundary conditions:

$$y'' + y = x, \quad 0 < x < 2, \quad y(0) = 0, \quad y'(2) = 0.$$

The general solution of the homogeneous problem can be written in two equivalent forms

$$y(x) = c_1 \cos(x) + c_2 \sin(x) = c \cos(x + \varphi),$$

with $y'(x) = -c_1 \sin(x) + c_2 \cos(x) = -c \sin(x + \varphi)$. First, we need to check if the problem has a non-trivial solution by setting

$$y(x = 0) = c_1 \cos(0) + c_2 \sin(0) = 0,$$
$$y'(x = 2) = -c_1 \sin(2) + c_2 \cos(2) = 0.$$

As one can see, the system of equations has only a trivial solution with $c_1 = 0$ and $c_2 = 0$. Thus, the homogeneous equation has only a trivial solution $y(x) = 0$. Now, we need to construct two functions $y_1(x)$ and $y_2(x)$, that are solutions of the homogeneous equation, such that one satisfies the left boundary condition and the other satisfies to the right boundary condition. For $y_1(x) = c_1 \cos(x) + c_2 \sin(x)$, we have

$$y_1(x = 0) = c_1 \cos(x = 0) + c_2 \sin(x = 0) = 0,$$

and we can choose $y_1(x) = \sin(x)$. For the second function, it is more convenient to proceed with the form $y_2(x) = c \cos(x + \varphi)$ rather than $y_2(x) = c_1 \cos(x) + c_2 \sin(x)$. While we eventually get the same answer, these simple forms of $y_1(x)$ and $y_2(x)$ require less work to derive Green's function. Thus, we have

$$y_2'(x = 2) = -c \sin(x + \varphi) = 0.$$

We can achieve the condition by choosing $\varphi = -2$ and then $y_2(x) = \cos(x - 2)$. The Wronskian of the two functions is $W(y_1, y_2) = -\cos(2)$, and from the equation $y'' + y = x$, it follows that $p(x) = 1$. Using (10.40), we can write Green's function of the problem as

$$G(x, \xi) = -\frac{1}{\cos(2)} \begin{cases} \sin(x) * \cos(\xi - 2) & 0 \leq x \leq \xi, \\ \cos(x - 2) * \sin(\xi) & \xi \leq x \leq 2. \end{cases}$$

Then, the solution for $f(x) = x$ is

$$y(x) = -\frac{1}{\cos(2)} \left[\cos(x - 2) \int_0^x \sin(\xi) \xi d\xi + \sin(x) \int_x^2 \cos(\xi - 2) \xi d\xi \right].$$

After evaluating these integrals and using trigonometric identities, we have

$$y(x) = x - \frac{\sin(x)}{\cos(2)}.$$

Now, we shall apply the Lagrange method of variation of parameters (9.15) to the above problem. We choose the linearly independent solutions of the homogeneous equation as $y_1(x) = \cos(x)$ and $y_2(x) = \sin(x)$. As we can see, neither of the functions satisfy the boundary conditions, i.e., $y_1(0) \neq 0$ and $y_2'(2) \neq 0$. With the Wronskian $W(\cos(x), \sin(x)) = 1$, the Lagrange solution is

$$y(x) = C_1 \cos(x) + C_2 \sin(x) - \cos(x) \int \sin(x)\, x\, dx$$
$$+ \sin(x) \int \cos(x)\, x\, dx.$$

Evaluating the above integrals gives

$$y(x) = C_1 \cos(x) + C_2 \sin(x) + x.$$

Applying the boundary conditions $y(0) = 0$ and $y'(2) = 0$, we find that $C_1 = 0$ and $C_2 = -1/\cos(2)$. Consequently, we obtain

$$y(x) = x - \frac{\sin(x)}{\cos(2)}.$$

This is exactly the same solution that we derived above using Green's function method. ∎

It is instructive to consider another example, this time with non-homogeneous boundary conditions.

Example 10.6. Consider the boundary value problem

$$y'' + 4y = 4x, \quad 0 < x < 1, \quad y(0) = 0, \quad y'(1) = 2.$$

As the first step, we need to transform the problem to homogeneous boundary conditions with $y(x) = h(x) + z(x)$, where $h(x)$ is some simple analytical function satisfying the non-homogeneous conditions, and $z(x)$ is still an unknown function that satisfies the homogeneous boundary conditions. For the problem at hand, we may choose $h(x) = 2x$ with $h(0) = 0$ and $h'(x) = 2$. Substituting $y(x) = 2x + z(x)$ into the original equation, we get a new boundary value problem

$$z'' + 4z = -4x, \quad 0 < x < 1, \quad z(0) = 0, \quad z'(1) = 0.$$

Now, we can proceed with choosing $z_1(x)$ and $z_2(x)$ as solutions of the homogeneous equations but satisfying the left boundary condition for $z_1(x)$ and satisfying the right-hand boundary condition for $z_2(x)$.

An analysis similar to one in Example 10.5 shows that we can proceed with

$$z_1(x) = \sin(2x) \quad z_2(x) = \cos(2x - 2).$$

The Wronskian for these two functions is $W(y_1, y_2) = -2\cos(2)$. Then, using (10.40) Green's function can be written as

$$G(x, \xi) = -\frac{1}{2\cos(2)} \cdot \begin{cases} \sin(2x)\cos(2\xi - 2) & 0 \leq x \leq \xi, \\ \cos(2x - 2)\sin(2\xi) & \xi \leq x \leq 1. \end{cases}$$

The solution of the problem is then

$$y(x) = 2x - \frac{1}{2\cos(2)} \left[\cos(2x - 2) \int_0^x \sin(2\xi)(-4\xi)d\xi \right.$$
$$\left. + \sin(2x) \int_x^1 \cos(2\xi - 2)(-4\xi)d\xi \right].$$

Evaluating the integrals, we finally obtain

$$y(x) = x + \frac{\sin(x)\cos(x)}{\cos(2)}. \quad \blacksquare$$

10.5.3. Case 2: Non-trivial solution for L[y] = 0

Now, we consider a case when the homogeneous problem $L[y] = 0$ has a non-trivial solution. We start with a boundary value problem with Cauchy boundary condition:

$$L[y] = \tilde{f}(x), \quad y(0) = 0, \quad y(l) = 0. \quad (10.44)$$

Let $\varphi_0(x)$ be a non-trivial solution of a corresponding homogeneous problem:

$$L[\varphi_0(x)] = 0, \quad \varphi_0(0) = 0, \quad \varphi_0(l) = 0.$$

We can normalize $\varphi_0(x)$ so that

$$\int_0^l \varphi_0^2(x)dx = 1.$$

Note that if the solution $y(x)$ of the boundary value problem exists, then the right side of the equation, i.e., the function $f(x)$, must be orthogonal on $[0, l]$ to the function $\varphi_0(x)$. Indeed, applying Green's formula (10.28) to the functions $y(x)$ and $\varphi_0(x)$, we find that

$$\int_0^l (\varphi_0(x)L[y] - y(x)L[\varphi_0])\,dx = \tilde{p}(x)(\varphi_0(x)y'(x) - y(x)\varphi_0'(x))\big|_0^l$$

$$= \int_0^l \varphi_0(x)\tilde{f}(x)dx = 0;$$

thus, the two functions are orthogonal, i.e.,

$$\int_0^l \varphi_0(x)\tilde{f}(x)dx = 0. \tag{10.45}$$

Note that this is a *necessary condition* for the existence of the solution of the non-homogeneous problem.

Generalized Green's function: Since there is a unique non-trivial solution for the homogeneous equation $L[y] = 0$, then we cannot use only this solution to build Green's function. There is no second linearly independent solution. Therefore, we define a generalized Green's function $G(x, \xi)$ as a solution for the following problem:

(1) Green's function satisfies the non-homogeneous equation

$$L_x[\tilde{G}(x, \xi)] = -\varphi_0(\xi)\varphi_0(x)$$

for $0 < x < \xi$ and $\xi < x < l$.

(2) Green's function satisfies given boundary conditions

$$\tilde{G}(0, \xi) = \tilde{G}(l, \xi) = 0.$$

(3) $\tilde{G}(x, \xi)$ is continuous on $[0, l]$.
(4) The first derivative has discontinuity at $x = \xi$:

$$\frac{d}{dx}\tilde{G}(x, \xi)\bigg|_{\xi-0}^{\xi+0} = \frac{1}{\tilde{p}(\xi)}.$$

(5) $\tilde{G}(x, \xi)$ is orthogonal to the eigenfunction $\varphi_0(x)$ on $[0, l]$:

$$\int_0^l G(x, \xi)\varphi_0(x)dx = 0.$$

Using Green's formula (10.28), one can show that the solution of the original problem can be written as

$$y(\xi) = \int_0^l \tilde{G}(x,\xi)\tilde{f}(x)dx,$$

where $G(x,\xi)$ is the generalized Green's function.

Now, we shall derive the generalized Green's function. Let $w(x)$ be a *particular solution* of the non-homogeneous equation

$$L[w(x)] = -\varphi_0(\xi)\varphi_0(x),$$

on $(0,l)$ as a solution of the *initial value problem* $w(0) = \alpha$ and $w'(0) = \beta$. Then, let $\varphi_1(x)$ be another linearly independent to $\varphi_0(x)$ solution of equation $L[\varphi_1(x)] = 0$; however, note that $\varphi_1(x)$ cannot satisfy the same boundary conditions as $\varphi_0(x)$, i.e., $\varphi_1(0) \neq 0$ and $\varphi_1(l) \neq 0$. We also normalize $\varphi_1(x)$ so that the Wronskian is

$$W(\varphi_1(x), \varphi_0(x)) = 1/\tilde{p}(x).$$

Note that the particular solution $w(x)$ does not satisfy the imposed boundary conditions for the original problem. However, we can chose linear combinations of $w(x)$ and $\varphi_1(x)$ on intervals $[0,\xi]$ and $[\xi,l]$ to satisfy the second condition, namely $\tilde{G}(0,\xi) = \tilde{G}(l,\xi) = 0$. But then we do not have enough freedom to satisfy the conditions 3–5. This is remedied by adding the solution $\varphi_0(x)$ since this solution satisfy given boundary conditions. Then, we seek for Green's function in the following form:

$$\tilde{G}(x,\xi) = w(x) + \begin{cases} C_1\varphi_1(x) + C_3\varphi_0(x) & 0 \leq x \leq \xi, \\ C_2\varphi_1(x) + C_4\varphi_0(x) & \xi \leq x \leq l. \end{cases}$$

Next, we need to find the constants $C_i (i = 1-4)$ to satisfy all the conditions above where

$$\begin{aligned} L[\varphi_0(x)] &= 0 & \varphi_0(0) &= 0,\ \varphi_0(l) = 0, \\ L[\varphi_1(x)] &= 0 & \varphi_1(0) &\neq 0,\ \varphi_1(l) \neq 0, \\ L_x[w(x)] &= -\varphi_0(\xi)\varphi_0(x) & w(0) &= \alpha,\ w'(0) = \beta. \end{aligned}$$

From the condition $G(0, \xi) = G(l, \xi) = 0$, we obtain

$$w(0) + C_1\varphi_1(0) = 0, \quad C_1 = -\varphi_1(0)/w(0),$$
$$w(l) + C_2\varphi_1(l) = 0, \quad C_2 = -\varphi_1(l)/w(l).$$

Since $G(x, \xi)$ is continuous at $x = \xi$, but the first derivative has the discontinuity, we have two more equations:

$$(C_2 - C_1)\varphi_1(\xi) + (C_4 - C_3)\varphi_0(\xi) = 0,$$
$$(C_2 - C_1)\varphi_1'(\xi) + (C_4 - C_3)\varphi_0'(\xi) = 1/p(\xi).$$

Solving this system for $C_2 - C_1$ and $C_4 - C_3$ gives

$$C_2 - C_1 = -\varphi_0(\xi) \quad \text{and} \quad C_4 - C_3 = \varphi_1(\xi).$$

We have already found solutions to C_1 and C_2. We can show that these solutions satisfy to the solution above. Applying Green's formula to $w(x)$ and $\varphi_0(x)$ yields the following:

$$\int_0^l (\varphi_0(x)L[w(x)] - w(x)L[\varphi_0(x)])dx$$
$$= [p(x)(\varphi_0(x)w'(x) - w(x)\varphi_0'(x))]\Big|_0^l = -\int_0^l \varphi_0(x)\varphi_0(\xi)dx.$$

Now, we have

$$-p(l)w(l)\varphi_0'(l) + p(0)w(0)\varphi_0'(0) = -\varphi_0(\xi),$$

and since $W(\varphi_1(x), \varphi_0(x)) = 1/p(x)$, we have

$$p(l)\varphi_0'(l) = 1/\varphi_1(l), \quad p(0)\varphi_0'(0) = 1/\varphi_1(0),$$

then $C_2 - C_1 = -\varphi_0(\xi)$ and both C_1 and C_2 are well defined.

To find C_3 and C_4, we write

$$C_4 = C_3 + \varphi_1(x).$$

Then,

$$G(x, \xi) = w(x) + C_3\varphi_0(x)\begin{cases} C_1\varphi_1(x) & 0 \leq x \leq \xi, \\ C_2\varphi_1(x) + \varphi_1(\xi)\varphi_0(x) & \xi \leq x \leq l, \end{cases}$$

with one unknown constant C_3. From the orthogonality condition (condition 5), we find

$$C_3 \int_0^l \varphi_0^2(x)dx = C_3 = -\int_0^l w(x)\varphi_0(x)dx - C_1 \int_0^\xi \varphi_1(x)\varphi_0(x)dx$$
$$-C_2 \int_\xi^l \varphi_1(x)\varphi_0(x)dx - \varphi_1(\xi)\int_\xi^l \varphi_0^2(x)dx.$$

Thus, we have both coefficients C_3 and C_4.

Note 1: We have considered homogeneous boundary value conditions, however a similar approach can be used to build a generalized Green's function for any boundary conditions.

Note 2: We assumed that the homogeneous equation $L[y(x)]$ has only one linear independent solution $\varphi_0(x)$, but it is possible to generalize our consideration for two linear independent solutions (but no more than two since we have a second order differential equation). In this case, we have

$$\int_0^l \varphi_i(x)\tilde{f}(x)dx = 0, \quad i = 1, 2.$$

Example 10.7. In this example, we illustrate how to build a generalized Green's function for solving the flowing boundary value problem:

$$y'' + y = f(x), \quad 0 < x < \pi, \quad y(0) = 0, \ y(\pi) = 0.$$

The homogeneous problem $y'' + y = 0$ has a solution

$$\varphi_0(x) = \sqrt{\frac{2}{\pi}} \sin x,$$

where the coefficient has been chosen to satisfy the normalization condition. The solution of the non-homogeneous equation only exists if

$$\int_0^\pi f(x) \sin x\, dx = 0.$$

For the generalized Green's function, we need two more functions, namely $\omega(x)$ and $\varphi_1(x)$. We can find the first function from the equation

$$\omega'' + \omega = -\frac{2}{\pi}\sin\xi \sin x.$$

Using methods for solving a second-order ODE, we get a *particular solution* as

$$\omega(x) = \frac{1}{\pi}(\sin\xi)x\sin x.$$

Next, we need a particular solution of the homogeneous equation $y'' + y = 0$ that does not satisfy the given boundary conditions. Namely, for $\varphi_1(x)$, we can chose $\varphi_1(x) = \cos x$. This function is linearly independent with $\sin x$ and does not satisfy the given boundary conditions. Then, we have

$$G(x,\xi) = \frac{1}{\pi}(\sin\xi)x\cos x + \begin{cases} C_1\cos x + C_3\sin x, & 0 \le x \le \xi, \\ C_2\cos x + C_4\sin x, & \xi \le x \le \pi. \end{cases}$$

Now, we need to find all the coefficients C_i. From the given boundary conditions $G(0,\xi) = 0$ and $G(\pi,\xi)$, we have that $C_1 = 0$ and $-\sin\xi - C_2 = 0$, or $C_1 = 0$ and $C_2 = -\sin\xi$. From the conditions for Green's function and its derivative, it follows that

$$C_3\sin\xi + \sin\xi\cos\xi - C_4\sin\xi = 0,$$

$$\sin^2\xi + C_4\cos\xi - C_3\cos\xi = 1,$$

from which $C_4 = C_3 + \cos\xi$ and

$$G(x,\xi) = \frac{1}{\pi}(\sin\xi)x\cos x$$

$$+ C_3\sin x \begin{cases} 0, & 0 \le x \le \xi, \\ -\sin\xi\cos x + \cos\xi\sin x, & \xi \le x \le \pi. \end{cases}$$

We can find the constant C_3 from the orthogonality condition of $G(x,\xi)$ to $\varphi_0(x)$. After evaluating the corresponding integrals, we find that

$$C_3 = -\frac{\pi - \xi}{\pi}\cos\xi - \frac{1}{2\pi}\sin\xi,$$

so finally

$$G(x,\xi) = \frac{1}{\pi}(x\sin\xi\cos x + \xi\cos\xi\sin x)$$

$$-\frac{1}{2\pi}\sin\xi\sin x - \begin{cases}\cos\xi\sin x, & 0 \le x \le \xi,\\ \sin\xi\cos x, & \xi \le x \le \pi.\end{cases}$$

With this Green's function, we can easily find a solution for $f(x)$ satisfying the orthogonality condition with $\varphi_0(x)$:

$$\int_0^\pi f(x)\varphi_0(x)dx = 0.$$

For example, for $f(x) = \cos x$, we see that

$$y(x) = \frac{x}{2}\sin x - \frac{\pi}{4}\sin x;$$

this is the solution for $y'' + y = \cos(x)$ with $y(0) = 0$ and $y(\pi) = 0$. We can easily check the solution by substituting it into the original differential equation.

It is instructive to analyze the physics behind the equation. The generalized Green's function describes harmonic vibrations of a uniform elastic rod fixed at the ends under a special type of external force. Namely, the frequency of the external force is resonant, i.e., it is the same as the frequency of natural vibrations of the rod, when the amplitude of the natural vibrations is given by $\varphi_0(x)$. However, the space distribution of the external force is such that the amplitude of forced oscillations remains bound. This is achieved by fulfilling the conditions of orthogonality: Both the right side of the equation, as well as the generalized Green's function, and the solution of the non-homogeneous boundary value problem itself are orthogonal to the function $\varphi_0(x)$. It is important to note that in this example the generalized Green's function $G(x,\xi)$ describes steady-state oscillations of the elastic rod under a periodic force of the amplitude $\frac{2}{\pi}\sin\xi$ applied at point ξ.

Now, we try to use the Lagrange method of variation of parameters to the above problem, which previously seemed challenging for Green's function method. Choosing two linearly independent solutions (for the corresponding homogeneous equation) as $y_1(x) = \cos(x)$ and $y_2(x) = \sin(x)$, we can readily find that

$$y(x) = C_1 \cos(x) + C_2 \sin(x)$$
$$- \cos(x) \int \sin(x) \cos(x) dx + \sin(x) \int \cos(x) \cos(x) dx.$$

After integrating, we have

$$y(x) = C_1 \cos(x) + C_2 \sin(x) + \cos(x) + \frac{1}{2} x \sin(x)$$
$$\equiv C_1 \cos(x) + C_2 \sin(x) + \frac{1}{2} x \sin(x).$$

Applying the boundary conditions, we find

$$y(x) = C_2 \sin(x) + \frac{1}{2} x \sin(x),$$

where C_2 can have any value ∎

Practical notes: As we noted earlier, a boundary value problem may have a unique solution, or infinitely many solutions, or no solution at all. The *Fredholm Alternative* states that a non-homogeneous boundary value problem $L[y] = f(x)$ with given boundary conditions has a unique solution **EITHER** if the corresponding homogeneous equation $L[y] = 0$ has only trivial solutions for the boundary conditions **OR** if there are non-trivial solutions $y_0(x)$ for $L[y] = 0$, then there are infinitely many solutions when

$$\int_a^b y_0(x) f(x) dx = 0,$$

and there is no solution when

$$\int_a^b y_0(x) f(x) dx \neq 0.$$

Green's function method works very well for both initial and boundary value problems. The major advantage of the method comes when a non-homogeneous part of a differential equation may vary from problem to problem. Otherwise, the Lagrange method variation of parameters provides a good alternative for solving boundary value problems.

10.6. Sturm–Liouville problem and Green's function

10.6.1. Sturm–Liouville problem

A special case of a boundary value problem is the eigenvalue problem for homogeneous differential equations of the form

$$L[y] + \lambda \mu(x) y(x) = 0, \qquad (10.46)$$

with homogeneous boundary conditions

$$\alpha_1 y'(0) + \beta_1 y(0) = 0,$$
$$\alpha_2 y'(l) + \beta_2 y(l) = 0. \qquad (10.47)$$

Consequently, we need to find values of the parameter λ, such that the homogeneous problem has non-trivial solutions on $[0, l]$. Here, $\mu(x) > 0$ is a weight or density function that is continuous on $[0, l]$. The $L[y]$ operator was defined earlier (10.26) as

$$L[y] = \frac{d}{dx}\left[\tilde{p}(x)\frac{dy}{dx}\right] - \tilde{q}(x) y.$$

Equation (10.46) together with the boundary conditions (10.47) is known as a Sturm–Liouville equation or Sturm–Liouville problem. Values of the parameter λ when equation (10.46) has non-trivial solutions are called eigenvalues, and non-trivial solutions corresponding to them are called eigenfunctions. Many problems in physics and engineering are formulated in terms of eigenvalue problems, e.g., vibrations of various systems.

Eigenfunctions of the Sturm–Liouville problems are widely used for solving many mathematical problems, including boundary value problems for ordinary and partial differential equations. Studying the full Sturm–Liouville problem is beyond the scope of this book. Therefore, we mention here only the most important properties of eigenvalues and eigenfunctions:

(1) The Sturm–Liouville problem has infinite number of eigenvalues which can be numerated as

$$|\lambda_1| \leq |\lambda_2| \leq \cdots |\lambda_n| \leq \cdots.$$

(2) Every eigenvalue corresponds to a unique eigenfunction, defined to within an arbitrary multiplicative constant.

(3) In the case of boundary conditions, $y(0) = 0$ and $y(l) = 0$, and $\tilde{q} \geq 0$, all the eigenvalues are positive, i.e., $\lambda_n > 0$.

(4) Eigenfunctions corresponding to different eigenvalues are orthogonal on $[0, l]$ with the weight function $\mu(x)$, namely $\int_0^l y_m(x) y_n(x) \mu(x) dx = 0$ for $m \neq n$. In practical applications, eigenfunctions are often normalized $\int_0^l y_n^2(x) \mu(x) dx = 1$. These two conditions, orthogonality and normalization, can be written as

$$\int_0^l y_m(x) y_n(x) \mu(x) dx = \delta_{m,n}, \tag{10.48}$$

where $\delta_{m,n}$ is the Kronecker delta defined as

$$\delta_{m,n} \equiv \begin{cases} 0 & \text{for } m \neq n, \\ 1 & \text{for } m = n. \end{cases}$$

(5) The eigenfunctions form a *complete set*, i.e., a continuous function $g(x)$ with $g(0) = 0$ and $g(l) = 0$, can be expanded in *absolutely and uniformly convergent series* as

$$g(x) = \sum_{n=1}^{\infty} c_n y_n(x), \tag{10.49}$$

where

$$c_n = \int_0^l g(x) y_n(x) \mu(x) dx. \tag{10.50}$$

Properties (2), (3), and (4) can be easily derived by using the Sturm–Liouville equation (for property (4), we also need Green's formula (10.28)). We leave proofs to a curious reader.

10.6.2. Green's function expansion

Now, we consider a boundary value problem

$$L[y] = f(x), \tag{10.51}$$

with homogeneous boundary conditions $y(0) = 0$ and $y(l) = 0$. Let φ_n be eigenfunctions of the Sturm–Liouville problem

$$L[\varphi_n] + \lambda_n \varphi_n = 0, \tag{10.52}$$

where φ_n is subject to *the same boundary conditions* as $y(x)$. As it was previously stated (the third property), for these boundary conditions, all eigenvalues are positive, i.e., $\lambda_n > 0$. For simplicity, we assume that $\varphi_n(x)$ functions are orthonormal (orthogonal and normalized), i.e.,

$$\int_0^l \varphi_m(x)\varphi_n(x)dx = \delta_{m,n}. \tag{10.53}$$

Since the eigenfunctions $\varphi_n(x)$ form a complete set, the solution $y(x)$ may be written as

$$y(x) = \sum_{n=1}^{\infty} a_n \varphi_n(x), \tag{10.54}$$

where a_n constants are to be determined. Substituting this into the original equation (10.49) gives

$$L\left[\sum_{n=1}^{\infty} a_n \varphi_n(x)\right] = \sum_{n=1}^{\infty} a_n L[\varphi_n(x)] = \sum_{n=1}^{\infty} a_n(-\lambda_n \varphi_n(x)) = f(x).$$

Multiplying both sides by φ_m and integrating from 0 to l, we get

$$-\sum_{n=1}^{\infty} \int_0^l a_n \lambda_n \varphi_m(\xi)\varphi_n(\xi)d\xi = \int_0^l \varphi_m(\xi)f(\xi)d\xi.$$

Using the orthogonality of the eigenfunctions, we can express the coefficients c_n as

$$a_n = -\frac{\int_0^l \varphi_n(\xi)f(\xi)d\xi}{\lambda_n \int_0^l \varphi_n(\xi)\varphi_n(\xi)d\xi} = -\frac{1}{\lambda_n}\int_0^l \varphi_n(\xi)f(\xi)d\xi. \tag{10.55}$$

Then, the solution $y(x)$ of the boundary problem can be written as

$$y(x) = \sum_{n=1}^{\infty} -\frac{1}{\lambda_n}\left(\int_0^l \varphi_n(\xi)f(\xi)d\xi\right)\varphi_n(x)$$

$$= \int_0^l \left(\sum_{n=1}^{\infty} -\frac{1}{\lambda_n}\varphi_n(\xi)\varphi_n(x)\right)f(\xi)d\xi. \tag{10.56}$$

Here, we have interchanged sums and integrals since according to the property (5) of the eigenfunctions, the sum is uniformly convergent.

Thus,
$$y(x) = \int_0^l G(\xi, x) f(\xi) d\xi, \qquad (10.57)$$
where
$$G(\xi, x) = -\sum_{n=1}^{\infty} \frac{1}{\lambda_n} \varphi_n(\xi) \varphi_n(x). \qquad (10.58)$$

Hence, we have derived a representation of Green's function as an expansion over eigenfunctions of the Sturm–Liouville problem for *given boundary conditions*. The series expansion above is equivalent to Green's function derived earlier (10.42) by solving $L[y] = \delta(x - \xi)$.

In many applications, it is common to see a non-homogeneous equation written in somewhat different from from $L[y] = f(x)$, namely
$$L[y] + \mu y = f(x), \qquad (10.59)$$
where μ is not an eigenvalue of the corresponding homogeneous equation. There are two ways to construct Green's function: (1) by solving corresponding eigenvalue problem $L[y] + (\mu + \lambda)y = 0$ or (2) by solving $L[\varphi] + \lambda\varphi = 0$ and then expanding $y(x)$ from (10.59) over the complete set of $\varphi_n(x)$. In both cases, we arrive at the same solution for Green's function as
$$G(\xi, x) = \sum_{n=1}^{\infty} \frac{\varphi_n(\xi)\varphi_n(x)}{\mu - \lambda_n}. \qquad (10.60)$$

Example 10.8. In this example, we examine the same boundary problem that we solved previously in Example 10.4, namely
$$y''(x) = f(x), \quad 0 < x < l, \quad y(0) = 0, \quad y(l) = 0.$$
The Sturm–Liouville equation
$$\varphi''(x) + \lambda\varphi(x) = 0$$
satisfying given boundary conditions produces eigenfunctions (normalized on $[0, l]$) as
$$\varphi_n(x) = \sqrt{\frac{2}{l}} \sin\left(\frac{n\pi}{l} x\right),$$
with eigenvalues $\lambda_n = n^2 \pi^2 / l^2 > 0$. There is no solution for $\lambda = 0$ satisfying the boundary conditions (do you see why?). Thus, Green's

function in this case is

$$G(x,\xi) = -\frac{2}{l}\sum_{n=1}^{\infty}\frac{\sin(n\pi x/l)\sin(n\pi\xi/l)}{n^2\pi^2/l^2}$$

$$= -\frac{2l}{\pi^2}\sum_{n=1}^{\infty}\frac{\sin(n\pi x/l)\sin(n\pi\xi/l)}{n^2}.$$

This Green's function must be identical to the one derived before in Example 10.4, namely

$$G(x,\xi) = \begin{cases} x(\xi-l)/l & 0 \le x \le \xi, \\ (x-l)\xi/l & \xi \le x \le l. \end{cases}$$

We can verify it by expanding the latter in a Fourier sine series. The alternative way is to compare the two Green's functions by calculating their values at some points for x and ξ, for example, at the middle point, i.e., $x = \xi = l/2$. For Green's function from Example 10.4, we have $G\left(\frac{1}{2},\frac{1}{2}\right) = -\frac{l}{4}$. Now, we work with Green's function as series. For $x = \xi = l/2$, we have for the sine functions $\sin(n\pi/2)$, that is, zero for even values of n and $+1$ or -1 for odd values, specifically, $+1$ for $n = 1 + 4k$ and -1 for $n = 3 + 4k$, where $k = 0, 1, 2, 3, \ldots$. Hence,

$$G\left(\frac{1}{2},\frac{1}{2}\right) = -\frac{2l}{\pi^2}\sum_{n=1}^{\infty}\frac{1}{n^2} \quad n = 1, 3, 5, \ldots.$$

The sum over odd values of n is $\pi^2/8$ (see the reference[2]), then $G\left(\frac{1}{2},\frac{1}{2}\right) = -\frac{l}{4}$ which is in agreement with Green's function from Example 10.4. ∎

10.7. Exercises

Evaluate

(1) $\int_{-\infty}^{\infty} x^2 \delta(x-3)dx$

---

[2]I. S. Gradshteyn and I. M. Ryzhik, *Table of Integrals, Series, and Products*, 7th edn. Elsevier, 2007.

Simplify by writing delta functions in terms of $\delta(x - x_i)$:

(2) $\delta(2x - 8)$.
(3) $\delta(x^2 + x - 2)$.

Equations

(4) Verify the solution (10.20) using Lagrange's method of variation of parameters.
(5) Compare the particular solution derived using Green's function (10.23) and the one from Lagrange's method (9.14).

Find the solution of each of the following differential equations subject to the indicated *initial condition*:

(6) $y' + y = e^x$, $\quad y(1) = 1/e$.
(7) $y' + 2y = 4x$, $\quad y(0) = -2$.
(8) $y' + y = \cos x$, $\quad y(0) = 3/2$.
(9) $x'' = 1$, $\quad x(0) = -2$, $\quad x'(0) = 0$.
(10) $x'' = \sin t$, $\quad x(0) = 0$, $\quad x'(0) = 1$.
(11) $x'' = e^t$, $\quad x(-\infty) = 0$, $\quad x'(-\infty) = 0$.
(12) $y'' - y = 1$, $\quad y(0) = 0$, $\quad y'(0) = 0$.
(13) $y'' + y = t$, $\quad y(0) = 1$, $\quad y'(0) = 0$.

Solve the following *boundary value* problems using both Green's function method and Lagrange's method of variation of parameters:

(14) $y'' = e^{-x}$ $\qquad 0 < x < 1 \quad y(1) = 0, \ y'(1) = 0$.
(15) $y'' + y = \cos x$ $\qquad 0 < x < \pi/2 \quad y'(0) = 0, \ y'(\pi/2) = 0$.
(16) $y'' - y' = \sin x$ $\qquad 0 < x < \pi \quad y'(0) = 0, \ y(\pi) = 0$.
(17) $xy'' + y' = 1/(x \ln x) \quad 1 < x < e \quad y(1) = 0, \ y'(e) = 0$.

Find eigenvalues and their corresponding eigenfunctions:

(18) $y'' + \lambda y = 0$, $\quad y'(0) = 0, \ y'(l) = 0$.
(19) $y'' + \lambda y = 0$, $\quad y'(0) = 0, \ y(l) = 0$.
(20) $y'' + \lambda y = 0$, $\quad y(0) = 0, \ y'(l) + \beta y(l) = 0$.
(21) Derive formula (10.60) by solving the eigenvalue problem $L[y] + (\mu + \lambda) y = 0$ and by solving $L[\varphi] + \lambda \varphi = 0$ and then expanding $y(x)$ over the complete set of $\varphi_n(x)$ functions.

10.8. Problems

Problem 1: Derive Green's function for the equation of motion

$$m\frac{dv}{dt} = -\beta v + f(t).$$

Solve the equation for the following forces: (a) $f(t) = -mg$ (you can compare your result with Example 8.7) and (b) $f(t) = \alpha e^{-\beta t}$. Assume the initial condition as $v(t = 0) = v_0$.

Problem 2: Derive Green's function for the differential equation

$$\frac{d^2y}{dt^2} + 2b\frac{dy}{dt} + \omega_0^2 = f(t)$$

and solve it for given initial conditions $y(0) = A$, $y'(0) = 0$ for the following $f(t)$:

(1) $f(t) = f_0$ (constant)
(2) $f(t) = f_0 e^{at}$
(3) $f(t) = f_0 \sin \omega t$.

Problem 3: Atwood machine is a standard example in general physics courses when the mass of the cable is assumed to be negligible. Consider an Atwood machine system where M is the cable mass, L is the cable length that connects two masses such as $m_2 > m_1$, as shown in Figure 10.6. Initially, the system is stationary, and two ends of the cable are differing in the initial vertical position by $2h_0$. Assume that the pulley is massless and frictionless.

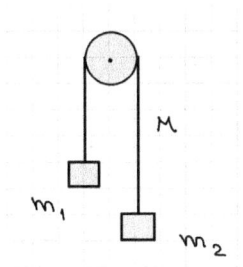

Fig. 10.6. Atwood machine with a cable of mass M.

(1) Write a differential equation describing motion of the system.
(2) Derive Green's function and find a general solution of the equations of motion.
(3) Find the acceleration and it compare with a well-known analytic solution when $M \to 0$.
(4) Find how much time will it take for block 1 to reach the pulley.
(5) What will be the kinetic energy of the chain at this moment? Check your solution by applying conservation of energy to the cable + the blocks.

Answers and solutions

(1) 9.
(2) $\frac{1}{2}\delta(x-4)$.
(3) $\frac{1}{3}\delta(x-1) + \frac{1}{3}\delta(x+2)$.
(6) $y = e^x/2 - (e^2/2 - 1)e^{-x}$.
(7) $y = 2x - 1 - e^{-2x}$.
(8) $y = e^{-x} + (1/2)(\sin x + \cos x)$.
(9) $x = t^2/2 - 2$.
(10) $x = 2t - \sin t$.
(11) $x = \int_{-\infty}^{t}(t-\tau)e^\tau d\tau = e^t$.
(12) $y = (1/2)(e^t + e^{-t}) - 1$.
(13) $y = t + \cos t - \sin t$.
(14) $y = e^{-x} + x - \frac{x}{e} - 1$.
(15) $y = \frac{1}{2}(x\sin x + \cos x)$.
(16) $y = \frac{1}{2}(e^x - \sin(x) + \cos(x) - e^\pi + 1)$.
(17) $y = \ln x(\ln \ln x - 1)$.
(18) $y_n = c\cos\frac{n\pi}{l}x$, $\lambda_n = \left(\frac{n\pi}{l}\right)^2$, $n = 0, 1, 2, \ldots$.
(19) $y_n = c\cos\frac{\pi}{l}\left(n + \frac{1}{2}\right)x$, $\lambda_n = \left(\frac{(n-1/2)\pi}{l}\right)^2$, $n = 1, 2, 3, \ldots$.
(20) $y_n = c\sin\sqrt{\lambda_n}x$, $\sqrt{\lambda_n} = \beta\tan\sqrt{\lambda_n}l$.

Chapter 11

Calculus of Variations

11.1. Introduction

The calculus of variations, founded by Euler and Lagrange,[1] is of utmost importance in both theoretical and applied physics. For example, Lagrangian and Hamiltonian formulations of classical mechanics are based on calculus of variations. Variational methods are widely used in quantum mechanics for calculating energy levels of various systems, and very often these methods are the most effective ones.

---

[1] Joseph-Louis Lagrange (1736–1813) was a distinguished Italian-French mathematician and physicist who left an indelible mark on a wide range of scientific disciplines. Born in Turin, Italy, Lagrange's early mathematical talents quickly propelled him into the academic sphere, where he became known for his groundbreaking work in calculus, number theory, and celestial mechanics. He made significant contributions to the fields of algebra and analysis, introducing concepts such as Lagrange multipliers and the theory of functions. His formulation of the principle of least action, known as the Lagrangian mechanics, revolutionized classical mechanics and laid the foundation for modern theoretical physics.

Lagrange's influence extended beyond mathematics and physics, as he played a pivotal role in the development of the Enlightenment era. His rational and systematic approach to scientific problems greatly impacted the way scholars approached complex issues across various disciplines. Additionally, his comprehensive treatise on the theory of functions paved the way for significant advancements in the understanding of functions and calculus, solidifying his status as a key figure in the evolution of modern mathematics. Lagrange's remarkable versatility and innovative thinking continue to inspire scientists and mathematicians to this day, leaving an enduring legacy that resonates across multiple domains of knowledge.

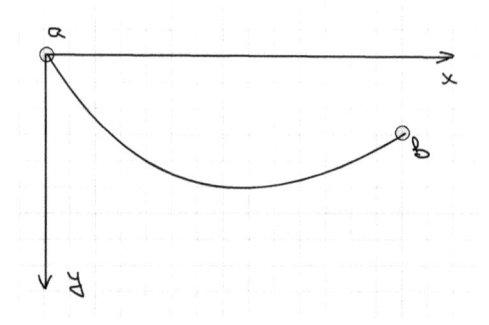

Fig. 11.1. The brachistochrone problem.

The essence of the calculus of variation in application to physics can be well illustrated by considering the brachistochrone problem. The problem was posed by Johann Bernoulli in 1696. Let a particle of mass m slide down under the influence of gravity without friction along a path L from point a to point b. If points a and b are fixed and the particle starts from rest, how should we choose the shape of L for the time of descent to be minimum?

Let us assume that the path is described by a single-valued function $y(x)$, as shown in Figure 11.1 (do you see why the function is single valued?). The function satisfies the conditions $y(a) = y_a$ and $y(b) = y_b$. Using conservation of energy, the speed of the particle at any point on $y(x)$ can easily be determined as

$$\frac{mv^2}{2} = mg(y_a - y), \quad v = \sqrt{2g(y_a - y)}.$$

The horizontal component of the velocity is

$$\frac{dx}{dt} = \frac{dx}{ds}\frac{ds}{dt} = v\frac{dx}{ds} = v\frac{dx}{\sqrt{dx^2 + dy^2}} = \sqrt{2g(y_a - y)}\frac{1}{\sqrt{1 + y'^2}}$$

or

$$dt = \frac{\sqrt{1 + y'^2}}{\sqrt{2g(y_a - y)}} dx.$$

Then, the total time of travel is

$$T = \int_a^b \frac{\sqrt{1 + y(x)'^2}}{\sqrt{2g(y_a - y(x))}} dx.$$

In Chapter 6, using the first kind of line integrals, we calculated times for various paths. But now we need to find that function $y(x)$ for which the integral above (called a *functional*, $T[y]$) takes the smallest value. This is a common problem using calculus of variations.

11.2. From few degrees of freedom to many

From calculus, we know that to find values for which a function $f(x)$ of a single variable x attains a maximum or minimum, we need to solve the equation $df(x)/dx = 0$. Thus, the problem has one degree of freedom. When a function $f(x, y)$ depends on two independent variables (see Chapter 5), then there are two degrees of freedom. From $\partial f(x,y)/\partial x = 0$ and $\partial f(x,y)/\partial y = 0$, we can find stationary points of the function. In the calculus of variations, we deal with an infinite number of degrees of freedom since we are looking for a *function*, not for a point in space.

We start our analysis with a system similar to the one we analyzed before in Section 10.2.1. Suppose that a set of n particles are connected to each other by identical flexible ropes. The first particle and the last one are fixed on x-axis and they cannot move. In the absence of external forces, the rope with particles is stretched along x by a constant force (tension) T. We assume that *small* external forces F_i are applied to the particles perpendicular to the x-axis. By small we mean that the tension T is much larger than the entire force acting on the rope $T \gg \sum_i F_i$.

For simplicity, we also assume that the particles can only move in the vertical direction along y-axis, and the tension T remains *constant*. Our goal is to find the new equilibrium state of the chain described by deviations y_i from the x-axis (Figure 11.2). In Chapter 10, we analyzed the system using Green's function method. Here, we explore a different approach.

We shall examine potential energy of the chain of particles counting from the initial configuration (stretched along x). There are two forces acting on every particle, namely the tension and the external forces. Since change in potential energy can be written as negative work done by a force, we have for the elastic energy $U_{el} = T\Delta l$, where Δl is the expansion of the rope and the tension T is assumed constant. The expansion of the distance between particles i and $i+1$

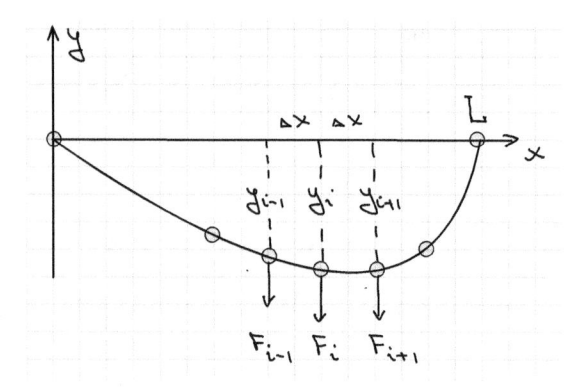

Fig. 11.2. A set of particles connected by springs.

can be written as

$$\Delta l_i = \sqrt{(\Delta x)^2 + (y_{i+1} - y_i)^2} - \Delta x = \Delta x \left(\sqrt{1 + \frac{(y_{i+1} - y_i)^2}{(\Delta x)^2}} - 1 \right).$$

For small deviations $(y_{i+1}-y_i)^2/\Delta x^2$, we can use $\sqrt{(1+\beta)} \approx 1+\beta/2$, then

$$\sqrt{1 + \frac{(y_{i+1} - y_i)^2}{(\Delta x)^2}} \approx 1 + \frac{(y_{i+1} - y_i)^2}{2\Delta x^2} \quad \text{and finally} \quad \Delta l_i = \frac{(y_{i+1} - y_i)^2}{2\Delta x}.$$

The change in the elastic potential energy of all particles can be written as

$$U_{\text{el}} = T \sum_i \frac{(y_{i+1} - y_i)^2}{2\Delta x} = \frac{T}{2\Delta x} \sum_i (y_{i+1} - y_i)^2.$$

The potential energy related to the external force is simply the negative work done by the force:

$$U_{\text{ext}} = -\sum_i F_i y_i.$$

Thus, the change in total potential energy is

$$U = \frac{T}{2\Delta x} \sum_i (y_{i+1} - y_i)^2 - \sum_i F_i y_i. \tag{11.1}$$

Equilibrium corresponds to minimum in potential energy or the net force on every particle is zero:

$$\frac{\partial U}{\partial y_i} = 0 \quad (i = 1, 2, \ldots, n).$$

In each row of the system of equations (11.1), we need only three terms for the partial derivative y_i (do you see why?), namely

$$\frac{\partial U}{\partial y_i} = \frac{\partial}{\partial y_i} \left(\frac{T}{2\Delta x}[(y_i - y_{i-1})^2 + (y_{i+1} - y_i)^2] - F_i y_i \right)$$

$$= -\frac{T}{\Delta x}(y_{i+1} - 2y_i + y_{i-1}) + F_i = 0.$$

Thus, the equilibrium condition for the set of particles is

$$\frac{T}{\Delta x}(y_{i+1} - 2y_i + y_{i-1}) - F_i = 0 \quad (i = 2, 3, \ldots, n-1). \tag{11.2}$$

Remember that the first and the last particles are fixed, i.e., $y_1 = 0$, $y_n = 0$. Thus, we have a system of $n-2$ linear equations[2] or the system has $n-2$ degrees of freedom.

Now, we progress to infinite degrees of freedom by letting $n \to \infty$. We return to the equilibrium condition (11.2), and we use $F_i = f_i \Delta x$, where is f_i the force density distribution along x. Then, we have

$$T \frac{(y_{i+1} - 2y_i + y_{i-1})}{\Delta x^2} - f_i = 0.$$

Note that $(y_{i+1} - 2y_i + y_{i-1})/\Delta x^2$ is the central difference approximation to second-order derivatives.[3] Then, in the limit $\Delta x \to 0$, we have the following ordinary differential equation:

$$T \frac{d^2 y}{dx^2} = f(x), \tag{11.3}$$

---

[2] It is interesting to note that the system (11.2) can be solved analytically if all external forces $F_i = F$ are equal. After some work, we can derive the following solution as

$$y_i = -(i-1)(n-i)\frac{\Delta x F}{2T} \quad (i = 1, 2, \ldots, n).$$

[3] J. D. Hoffman, *Numerical Methods for Engineers and Scientists*. McGraw-Hill, 1992.

with the boundary conditions $y(0) = 0$ and $y(L) = 0$. In a special case when $f(x) \equiv f_0$, the solution of the above equation is

$$y(x) = \frac{f_0}{T}\frac{x^2}{2} + C_1 x + C_2.$$

Applying given boundary conditions, we have $C_1 = -f_0 L/2T$ and $C_2 = 0$. And finally, the shape of the string is

$$y(x) = \frac{f_0}{2T} x(x - L).$$

Now, we return to the potential energy U. We can rewrite equation (11.1) as

$$U = \sum_i \left[\frac{T}{2}\left(\frac{y_{i+1} - y_i}{\Delta x}\right)^2 + f_i y_i \right] \Delta x.$$

As $n \to \infty$, $\Delta x \to 0$ and we can represent $(y_{i+1} - y_i)/\Delta x$ as the derivative at point i, i.e., y'_i. Then, we may write for the potential energy

$$U = \sum_i \left[\frac{T}{2} y'^2_i - f_i y_i \right] \Delta x.$$

And in the limit $n \to \infty$, the sum becomes integral!

$$U = \int_0^L \left[\frac{T}{2} y'^2 + f(x) y \right] dx, \qquad (11.4)$$

where $f(x)$ is the density distribution of the force acting on the system. Thus, to find the shape of the rope, we need to find a function $y(x)$ that minimizes the value of the integral (11.4) with given conditions $y(0) = 0$ and $y(L) = 0$. Hence, the problem of finding the shape $y(x)$ of the elastic rope, under conditions $y(0) = 0$ and $y(L) = 0$, can be solved either by solving the differential equation (11.3) or by finding such a function $y(x)$ that achieves an extremum (not necessarily a minimum) value of the integral (11.4) under the same conditions.

11.3. A necessary condition for an extremum

In the following, we use the term *functional* when a scalar value corresponds to a given function $y(x)$. In this section, we concentrate on functionals of the form

$$J[y] \equiv J = \int_a^b F(x, y, y')dx. \tag{11.5}$$

While F is a function of three variables x, y, and y', however, $y(x)$ and $y'(x)$ are functions of x, then the integral is actually a function of just one variable x, thus, the functional J has a certain numerical value for a given function $y(x)$. For example, the value of the functional

$$J = \int_0^1 (y'^2 + 2y\cos(x))dx$$

has the value $J \approx 1.7635$ for $y(x) = x$ and $J \approx 5.9506$ for $y(x) = e^x$.

We assume that $F(x, y, y')$ in (11.5) is a continuous function of its three variables and have continuous derivatives up to the second order, and the function $y(x)$ has a continuous derivative. We also set the values of the function at the ends of the integration intervals as

$$y(a) = y_a, \quad y(b) = y_b. \tag{11.6}$$

Definition. We say that the functional J reaches *relative* maximum (minimum) for $y(x)$ on $[a, b]$ if the value of the functional is larger (smaller) than that of any other function $\tilde{y}(x)$ in the ϵ vicinity of $y(x)$, i.e.,

$$|\tilde{y}(x) - y(x)| \leq \epsilon.$$

Such a definition of a relative extremum is completely identical to the definition of relative extremum of a function $f(x)$ in calculus.

Let $y(x)$ be a function that gives the extremum value to the functional J ($y(x)$ is the function that we want to find). And suppose $\mu(x)$ is any function that is continuous within $[a, b]$ and equal to zero at the end points. Then, the function $y(x) + \alpha\mu(x)$ satisfies the same boundary conditions as $y(x)$. The functional

$$J(\alpha) = \int_a^b F(x, y(x) + \alpha\mu(x), y'(x) + \alpha\mu'(x))dx \tag{11.7}$$

is a function of the parameter α. Therefore, since $y(x)$ gives an extremum to the functional, then the function $J(\alpha)$ must have an extremum at $\alpha = 0$, and therefore its derivative must be equal to zero at $\alpha = 0$. Then, we get the **necessary condition for a local extremum**

$$\left.\frac{dJ(\alpha)}{d\alpha}\right|_{\alpha=0} = \int_a^b \frac{d}{d\alpha} F(x, y(x) + \alpha\mu(x), y'(x) + \alpha\mu'(x))dx = 0. \tag{11.8}$$

The condition (11.8) is only a necessary one for an extremum. Sufficient conditions are rather complicated, but in most practical applications, they are not required since we can deduce them from the statement of a problem in hand.

For our further consideration, we need the **fundamental lemma of the calculus of variations**. If the integral

$$\int_a^b f(x)\mu(x)dx = 0, \tag{11.9}$$

where $f(x)$ is a continuous function on $[a, b]$ and $\mu(x)$ is *any* continuous function with continuous derivatives satisfying conditions $\mu(a) = 0$ and $\mu(b) = 0$, then $f(x) \equiv 0$ everywhere in $[a, b]$. The lemma can be easily proved assuming the opposite. So, we assume that $f(x) > 0$ within some interval $[x_1, x_2]$. Now, we define the function $\mu(x)$ as

$$\mu(x) = \begin{cases} 0 & \text{for } a \leq x \leq x_1, \\ (x - x_1)^2(x - x_2)^2 & \text{for } x_1 \leq x \leq x_2, \\ 0 & \text{for } x_2 \leq x \leq b. \end{cases}$$

Such function $\mu(x)$ satisfies all conditions imposed in the lemma. Then, the integral

$$\int_{x_1}^{x_2} f(x)(x - x_1)^2(x - x_2)^2 dx > 0$$

since we integrate a positive function in $[x_1, x_2]$. But by the condition of the lemma, such an integral must be equal to zero. Therefore, it is only possible if $f(x) \equiv 0$.

11.4. Euler–Lagrange equation

Now, we return to equation (11.8). Differentiating under the integral by using the rule of partial differentiation (5.5), we have

$$\frac{d}{d\alpha} F(x, y(x) + \alpha\mu(x), y'(x) + \alpha\mu'(x)) = \frac{\partial F}{\partial y}\mu(x) + \frac{\partial F}{\partial y'}\mu'(x).$$

Using the following notations for partial derivatives as

$$\frac{\partial F}{\partial y} \equiv F_y, \quad \frac{\partial F}{\partial y'} \equiv F_{y'},$$

we can write

$$\left.\frac{dJ(\alpha)}{d\alpha}\right|_{\alpha=0} = \int_a^b [F_y(x, y, y')\mu(x) + F_{y'}(x, y, y')\mu'(x)]dx = 0.$$

Integrating the second term by parts gives

$$\int_a^b F_{y'}(x, y, y')\mu'(x)dx = F_{y'}(x, y, y')\mu(x)|_a^b$$
$$- \int_a^b \frac{d}{dx}[F_{y'}(x, y, y')]\mu(x)dx,$$

and we have

$$\left.\frac{dJ}{d\alpha}\right|_{\alpha=0} = F_{y'}(x, y, y')\mu(x)|_a^b$$
$$+ \int_a^b \left[F_y(x, y, y') - \frac{d}{dx}F_{y'}(x, y, y')\right]\mu(x)dx. \quad (11.10)$$

Since $\mu(a) = 0$ and $\mu(b) = 0$, then the first term is equal to zero and we can rewrite the derivative of the functional in compact form as

$$\left.\frac{dJ(\alpha)}{d\alpha}\right|_{\alpha=0} = \int_a^b \left[F_y - \frac{d}{dx}F_{y'}\right]\mu(x)dx.$$

Applying the fundamental theorem of calculus of variations to the above integrals shows that the function $y(x)$ giving the extremum to

the functional
$$J = \int_a^b F(x, y, y')dx$$
must satisfy the differential equation
$$F_y - \frac{d}{dx}F_{y'} = 0, \tag{11.11}$$
where F_y and $F_{y'}$ are partial derivatives of the function F. This is Euler–Lagrange equation for a function F of the form $F(x, y, y')$.

It is practical to evaluate explicitly the derivative in the second term of Euler–Lagrange equation. From Chapter 5, we know how to differentiate a function $f(x, y(x), z(x))$. Using (5.6), we can write
$$\frac{d}{dx}f(x, y(x), z(x)) = \frac{\partial f}{\partial x} + \frac{\partial f}{\partial y}\frac{dy}{dx} + \frac{\partial f}{\partial z}\frac{dz}{dx}.$$
Applying the rule of the differentiation above to the second term in (11.11), we have
$$\frac{d}{dx}F_{y'} = F_{xy'} + F_{yy'}\frac{dy}{dx} + F_{y'y'}\frac{dy'}{dx},$$
where $F_{xy'}$, $F_{yy'}$, and $F_{y'y'}$ are second-order partial derivatives. Then, we can write Euler–Lagrange equation in the explicit form as
$$F_{y'y'}y'' + F_{yy'}y' + F_{xy'} - F_y = 0. \tag{11.12}$$
Thus, we reduce the original problem finding an extremum for functional (11.5) to the second-order differential equation, and its general solution has two arbitrary constants, which are determined by the two boundary conditions $y(a) = y_a$ and $y(b) = y_b$. As we know (see Chapter 9, Section 9.7), a boundary value problem for a second-order equation can have a single solution, many solutions, or none depending on given boundary conditions.

It is important to note that we have not previously imposed the condition that the second derivative $y''(x)$ does exist. Let us now prove that if $F_{y'y'} \neq 0$, then the second derivative exists and it is continuous. The total derivative $\frac{d}{dx}F_{y'}$ is the limit of
$$\frac{F_{y'}(x + \Delta x, y + \Delta y, y' + \Delta y') - F_{y'}(x, y, y')}{\Delta x}$$
$$= [F_{xy'}]_{av} + [F_{y,y'}]_{av}\frac{\Delta y}{\Delta x} + [F_{y'y'}]_{av}\frac{\Delta y'}{\Delta x},$$

where $[F_{xy'}]_{av}$, $[F_{y,y'}]_{av}$, and $[F_{y'y'}]_{av}$ are average values between (x, y, y') and $(x + \Delta x, y + \Delta y, y' + \Delta y')$. In the limit $\Delta x \to 0$,

$$\frac{\Delta y'}{\Delta x} = y'' = \frac{\frac{d}{dx}F_{y'} - F_{xy'} - y'F_{yy'}}{F_{y'y'}}.$$

Thus, the function $y(x)$ has a continuous second derivative when $F_{y'y'} \neq 0$, and we do can use the explicit form (11.12).

Comment: Note that a product of $J'(0)\alpha$ is a differential of the function $J(\alpha)$ at $\alpha = 0$. It is commonly called as *a first variation* of the functional with a notation δJ. Using another notation $\delta y = \alpha \mu(x)$, we can rewrite (11.10) as

$$\delta J = J'(0)\alpha = [F_{y'}\delta y]_a^b + \int_a^b \left(F_y - \frac{d}{dx}F_{y'}\right) \delta y \, dx. \qquad (11.13)$$

Example 11.1. Here we apply the Euler–Lagrange equation to the potential energy integral (11.4)

$$U = \int_0^L \left[\frac{T}{2}y'^2 + f(x)y\right] dx,$$

with $F(x, y, y') = Ty'^2/2 + f(x)y$. Then, from (11.12), we have differential equation

$$Ty'' = f(x)$$

that is exactly the differential equation (11.3) derived from physics consideration for a set of masses connected by an elastic rope. ∎

11.4.1. *Special cases of Euler–Lagrange equation*

There is a number of special cases of the Euler–Lagrange equation, when one, or two, of variables in $F(x, y, y')$ is missing:

(1) $F = F(x, y)$, i.e., F does not depend on y'.
   In this case, the Euler–Lagrange equation takes the form

$$F_y = 0. \qquad (11.14)$$

This equation in not differential but algebraic. It determines one or a finite number of functions (lines) that may or may not satisfy the boundary conditions. For example, for $F(x,y) = (x-y)y$, $F_y(x,y) = x - 2y = 0$, and then $y = x/2$.

(2) $F = F(y')$ or F depends only on y'.
Then, we have

$$y'' F_{y'y'} = 0. \tag{11.15}$$

From the condition $F_{y'y'} \neq 0$ follows that $y'' = 0$ and the solution is a straight line

$$y(x) = C_1 x + C_2,$$

where C_1 and C_2 are defined by given boundary conditions $y(a) = y_a$ and $y(b) = y_b$.

(3) $F = F(x, y')$ or F does not depend on y.
Euler–Lagrange equation then reduces to

$$\frac{d}{dx} F_{y'}(x, y') = 0, \tag{11.16}$$

thus we have

$$F_{y'}(x, y') = C,$$

where C is an arbitrary constant. This is a differential equation for y'.

(4) $F = F(y, y')$ or F does not depends explicitly on x.
In this case, from (11.12) follows

$$F_{y'y'} y'' + F_{yy'} y' - F_y = 0.$$

We can show that this equation leads to

$$F - y' F_{y'} = C. \tag{11.17}$$

This is the first-order differential equation for y', where C is an arbitrary constant. Indeed,

$$\frac{d}{dx}(F - y' F_{y'}) = F_y y' + F_{y'} y'' - F_{y'} y'' - F_{y'y} y'^2 - F_{y'y'} y' y''$$

$$= -y'(F_{y'y'} y'' + F_{y'y} y' - F_y).$$

Thus,
$$\frac{d}{dx}(F - y'F_{y'}) = 0.$$

Example 11.2. Find a function making the following integral stationary:
$$\int_0^{2\pi} (y'^2 - y^2)dx, \quad y(0) = 1, \ y(2\pi) = 1.$$

The function $F = y'^2 - y^2$ does not have explicit dependence on the independent variable x. Then, we can use the version of Euler–Lagrange equation as $F - y'F_{y'} = C$ which results in $-y'^2 - y^2 = C$. This is a first-order nonlinear ordinary differential equation. Differentiating both sides gives a second-order linear differential equation, namely $y'' + y = 0$ with the general solution $y(x) = c_1 \cos(x) + c_2 \sin(x)$. Applying the given boundary conditions gives the solution to the problem as $y(x) = \cos(x) + c_2 \sin(x)$, where c_2 is an arbitrary constant. ∎

Example 11.3. Find a function making the following integral stationary:
$$\int_{-1}^{1} (y'^2 - 2xy)dx, \quad y(-1) = -1, \ y(1) = 1.$$

The function $F = (y'^2 - 2xy)$ is a function of all three variables with $F_{y'y'} \neq 0$, therefore we use the explicit form of Euler–Lagrange equation (11.12). After carrying out all partial differentiations, now we have a second-order differential equation $2y'' + 2x = 0$. The general solution of the equation is $y(x) = -x^3/6 + c_1 x + c_2$. Applying given boundary conditions gives
$$y(x) = -\frac{1}{6}x^3 + \frac{7}{6}x.$$
∎

Example 11.4. Now, we can apply our analysis to the brachistochrone problem described in the introduction of this chapter. Mathematically, we formulate the problem as finding the path $y(x)$ which minimizes the integral
$$T = \int_0^b \frac{\sqrt{1 + y(x)'^2}}{\sqrt{y(x)}} dx,$$

subject to the boundary conditions $y(0) = 0$ and $y(b) = y_b$ (with the y-axis directed down). The integrand F does not depend explicitly on x, therefore we can apply one of special cases for Euler–Lagrange equations, namely the one for $F(y, y')$, when
$$F - y' F_{y'} = 0.$$
After evaluating the derivative $F_{y'}$, we have
$$\frac{\sqrt{1 + y'^2}}{\sqrt{y}} - \frac{y'^2}{\sqrt{y(1 + y'^2)}} = C.$$
We can easily rewrite the above equation as
$$y(1 + y'^2) = C_1.$$
This is a nonlinear ordinary differential equation, and it is not obvious how to solve it. First, we rewrite it as
$$dx = \sqrt{\frac{y}{C_1 - y}} dy.$$
Then, we introduce a parametric representation for y as $y = C_1(1 - \cos t)/2 = C_1 \sin^2(t/2)$ with $y' = C_1 t' \sin(t)/2$. Thus, we can write
$$dx = C_1 \sin^2(t/2) dt.$$
Integration gives
$$x = \frac{C_1}{2}(t - \sin t) + C_2,$$
and then
$$y = \frac{C_1}{2}(1 - \cos t).$$
The solution represents a cycloid, that is, the locus of a point on the rim of a circle of radius $C_1/2$ rolling along a straight line. The constants C_1 and C_2 are defined by the positions of the end points a and b. By setting the initial point at the origin of the coordinate, we have $C_2 = 0$ with $t = 0$. The radius $C_1/2$ and t_b can be found from the given boundary for point b, namely
$$x_b = \frac{C_1}{2}(t_b - \sin t_b), \quad y_b = \frac{C_1}{2}(1 - \cos t_b).$$
It is interesting to note that if $y_b < 2x_b/\pi$, then the trajectory descends below y_b to pick up more speed to minimize total time. Figure 11.3 shows an example of such brachistochrone path. ∎

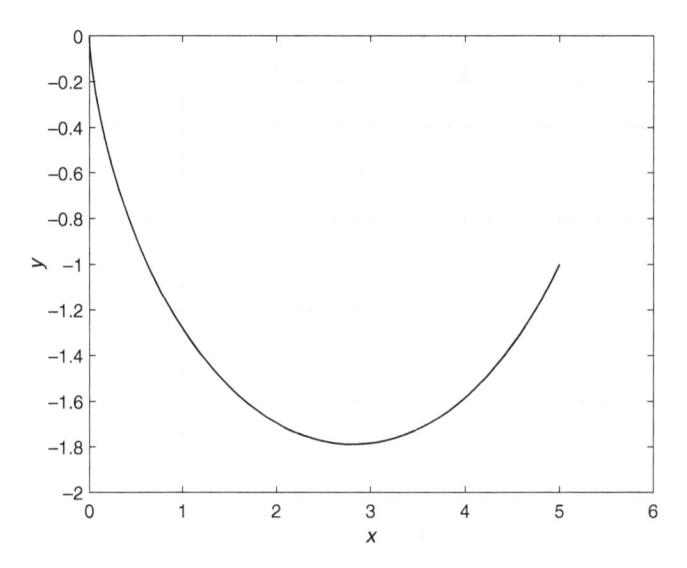

Fig. 11.3. The brachistochrone path for $x_b = 5$ and $y_b = -1$.

11.5. Generalizations of the Euler–Lagrange equation

There are very many situations when we need to go beyond the classical Euler–Lagrange equation. In this section, we consider some variations of the main problem, namely when (1) a functional depends on derivatives of higher than first order, (2) a functional has more than one unknown function, and (3) an unknown function depends on more than one independent variable (multiple integration).

11.5.1. *The Euler–Poisson equation*

Suppose we have a functional depending on higher-order derivatives:

$$J = \int_a^b F(x, y(x), y'(x), y''(x), \ldots, y(x)^{(n)}), \qquad (11.18)$$

with boundary conditions

$$y(a) = y_a, \; y'(a) = y'_a, \ldots, \; y^{(n-1)}(a) = y_a^{(n-1)},$$
$$y(b) = y_b, \; y'(b) = y'_b, \ldots, \; y^{(n-1)}(b) = y_b^{(n-1)},$$

and the function F has continuous derivatives up to the $(n+1)$st order. Using the same approach that we used to derive the Euler–Lagrange equation, namely $y(x) = \alpha\mu(x)$, we find for the derivative

$$J'(0) = \int_a^b [F_y \mu(x) + F_{y'} \mu'(x) + \cdots F_{y^{(n)}} \mu^{(n)}(x)]dx.$$

Using integration by parts, we have

$$J'(0) = \int_a^b \mu(x) \left[F_y - \frac{d}{dx} F_{y'} + \frac{d^2}{dx^2} F_{y''} \right.$$
$$\left. - \cdots + (-1)^n \frac{d^n}{dx^n} F_{y^{(n)}} \right] dx = 0.$$

Then, according to fundamental lemma of calculus of variations (11.9), the unknown function $y(x)$ in (11.18) is a solution of Euler–Poisson equation:

$$F_y - \frac{d}{dx} F_{y'} + \frac{d^2}{dx^2} F_{y''} - \cdots + (-1)^n \frac{d^n}{dx^n} F_{y^{(n)}} = 0. \qquad (11.19)$$

The general solution depends on $2n$ arbitrary constants that are determined by given boundary conditions. We should remember that deriving Euler–Poisson equation we assumed that both the function $y(x)$ and its derivatives up to $y^{(2n)}$ are continuous.

Example 11.5. Solve the following variational problem

$$J = \int_0^1 y''^2(x) dx,$$

when $y(0) = 0$, $y'(0) = 1$, $y(1) = 1$, $y'(1) = 1$. The Euler–Poisson equation in this case takes the form

$$\frac{d^2}{dx^2}(2y'') = 0, \quad \text{or} \quad y'''' = 0.$$

The general solution of the above equation is $y = a + bx + cx^2 + dx^3$. From the boundary conditions follows $a = 0$, $b = 1$, $c = 0$, and $d = 0$, then the solution of the problem is just $y(x) = x$. ∎

11.5.2. The case of two functions

In this case, the functional depends on two functions, such as

$$J = \int_a^b F(x, y(x), y'(x), z(x), z'(x))dx, \qquad (11.20)$$

with given boundary conditions

$$y(a) = y_a, \quad y(b) = y_b, \quad z(a) = z_a, \quad z(b) = z_b,$$

and the function F has second partial derivatives. In this case, we can follow the same approach that we used for deriving Euler–Lagrange equation. We consider $y(x)+\alpha\mu_y(x)$ and $z(x)+\beta\mu_z(x)$ with $\mu_y(x)$ and $\mu_z(x)$ equal to zero at the end points. After differentiating (11.20), $\partial J/\partial \alpha$ and $\partial J/\partial \beta$, setting the partial derivatives to zero at $\alpha = 0$ and $\beta = 0$, and using the fundamental lemma of calculus of variations, we have the generalized Euler–Lagrange equation in the case of two functions as a system of the *two* partial equations:

$$\begin{aligned} F_y - \frac{d}{dx}F_{y'} &= 0, \\ F_z - \frac{d}{dx}F_{z'} &= 0. \end{aligned} \qquad (11.21)$$

For n functions, the system above is easily written as a system of n equations.

Example 11.6. Find functions $y(x)$ and $z(x)$ making stationary the following functional:

$$J = \int_0^1 (y'^2 + z'^2)dx,$$

when $y(0) = 0$, $y(1) = 1$, $z(0) = 1$, $z(1) = 0$. Then, the system of Euler–Lagrange equations takes the form

$$2y'' = 0,$$
$$2z'' = 0.$$

The general solution of the system can be easily found as

$$y = a + bx, \quad z = c + dx.$$

Applying the boundary conditions results in

$$y = x,$$
$$z = 1 - x.$$
∎

11.5.3. The case of double integrals

Let the functional depend on a function $f(x,y)$, which is a function of two independent variables x and y:

$$J = \iint_D F(x, y, f, f_x, f_y) dx dy, \qquad (11.22)$$

where f_x and f_y are partial derivatives, D is the area of integration in a plane, the function F has second partial derivatives, and the boundary conditions are given as $f(x,y)|_L = \varphi(x,y)$. Similar to functions of a single variable, we consider $f(x,y) + \alpha\mu(x,y)$, where $\mu(x,y) = 0$ on the boundary. Then, we can show that the function $f(x,y)$ can be found from the equation

$$F_f - \frac{\partial}{\partial x} F_{f_x} - \frac{\partial}{\partial y} F_{f_y}, \qquad (11.23)$$

or, expanding the partial derivative

$$F_{f_x f_x} f_{xx} + 2 F_{f_x f_y} f_{xy} + F_{f_y f_y} f_{yy}$$
$$+ F_{f_x f} f_x + F_{f_y f} f_y + F_{x f_x} + F_{y f_y} - F_f = 0. \qquad (11.24)$$

This is a second-order partial differential equation, and its solution should satisfy boundary conditions given on a contour enclosing the area of integration.

11.6. Variational problem with constraints

We have already considered a problem of finding stationary points of a function $f(x,y)$ subject to an additional condition $g(x,y) = C$ (see Partial Differentiation, Section 5.7). In the calculus of variations, problems with constraints are solved in the same way as for functions of few variables.

Let us consider the problem of finding an extremum of functional

$$J = \int_a^b F(x, y, y') dx \qquad (11.25)$$

for given boundary conditions

$$\begin{aligned} y(a) &= y_a, \\ y(b) &= y_b, \end{aligned} \qquad (11.26)$$

with the additional condition that the integral

$$J_C = \int_a^b G(x, y, y') dx = C \qquad (11.27)$$

has the given value C. Such problems are called isoperimetric problems.

The problem is reduced to the usual problem of calculus of variations without additional conditions, by Euler's theorem. This states if $y(x)$ provides a solution for (11.25) with the condition (11.27) and given boundary conditions, and $y(x)$ does not give an extremum for integral J_C, then there is such constant λ that $y(x)$ is a solution for the usual variational problem:

$$\int_a^b H(x, y, y') dx, \quad \text{where } H = F + \lambda G. \qquad (11.28)$$

As one can see, the equation above reminds us of Lagrange's method of undetermined multipliers. Truly, if we divide the integration interval $[a, b]$ into a large number N of intervals and replace the integrals (11.25) and (11.27) by the sums depending on the values y_i at points i, then we have the problem of finding conditional stationary points for a function of N variables using the method of undetermined multipliers.

The solution $y(x)$ for (11.28) can be found from the Euler–Lagrange equation for H, namely

$$H_y - \frac{d}{dx} H_{y'} = 0. \qquad (11.29)$$

The differential equation above has three arbitrary constants; these can be determined using the boundary conditions (11.26) and the condition (11.27).

Example 11.7. Find a solution for the following isoperimetric problem:

$$\int_0^1 y'^2 dx \quad \text{with } y(0) = 1, \quad y(1) = 6,$$

with the condition

$$\int_0^1 y(x) dx = 3.$$

In this case, we have $H(y, y') = y'^2 + \lambda y$, and the Euler–Lagrange equation for the case of "missing" x is $H - y' H_{y'} = A$, where A is an arbitrary constant. Then, we have

$$-y'^2 + \lambda y = A.$$

Differentiating both sides and solving second-order differential equation give

$$-2y'y'' + \lambda y' = 0 \quad \text{or} \quad y'(-2y'' + \lambda) = 0.$$

Since $y' \neq 0$, we have $y'' = -\lambda/2$ with the solution

$$y(x) = -\frac{\lambda}{2} x^2 + C_2 x + C_1.$$

Using the boundary conditions together with the condition $\int_0^1 y(x) dx = 3$, we get $\lambda = -6$, $C_2 = 2$, and $C_1 = 1$, thus

$$y(x) = 3x^2 + 2x + 1. \qquad \blacksquare$$

Example 11.8. Dido's problem, also known as the isoperimetric problem, is a classical problem in mathematics named after Dido, the legendary founder and queen of Carthage. The problem can be stated as follows: For a given length of a rope L, find a shape of the rope that would enclose the biggest surface area of land between two points on a shore assuming that the points on the shore are a straight line (see Figure 11.4). In this case, the surface area is simply an integral:

$$S = \int_a^b y(x) dx.$$

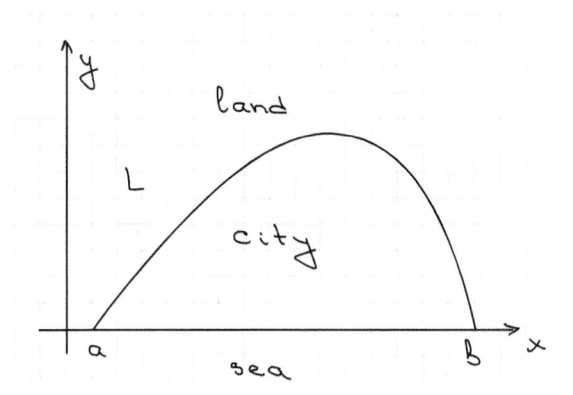

Fig. 11.4. The Princess Dido's problem.

The boundary conditions are $y(a) = y(b) = 0$ and the length of the rope is defined as

$$L = \int_a^b \sqrt{1 + y'^2(x)}\,dx.$$

Then, according to Euler's theorem, the shape $y(x)$ is a solution of the functional

$$\int_a^b \left(y + \lambda\sqrt{1 + y'^2(x)}\right)dx.$$

Since the functional does not depends explicitly on x, the Euler–Lagrange equation takes the form

$$F - y'F_{y'} = C_1,$$

where C_1 is a constant. Then, we have

$$y + \lambda\sqrt{1 + y'^2(x)} - \frac{\lambda y'^2}{\sqrt{1 + y'^2(x)}} = C_1.$$

After some straightforward algebra

$$y - C_1 = -\frac{\lambda}{\sqrt{1 + y'^2(x)}},$$

and then

$$y' = \frac{\sqrt{\lambda^2 - (y - C_1)^2}}{y - C_1}.$$

or

$$\frac{y - C_1}{\sqrt{\lambda^2 - (y - C_1)^2}} dy = dx.$$

Integrating both sides, we have

$$(y - C_1)^2 + (x - C_2)^2 = \lambda^2,$$

that is, the equation of a circle with the center located at (C_2, C_1) and a radius λ. All the constants are determined by the boundary conditions and the length of the rope that should be larger than the distance between the boundary points, or $L > (b - a)$. ∎

11.7. A variational problem with variable end points

So far we have considered variational problems with fixed boundary conditions when $y(a)$ and $y(b)$ were merely numbers. There is a class of problems when the boundary conditions are given as functions, e.g. $g(x)$ and $h(x)$, and the points a and b along x are not prescribed, but require to be found (see Figure 11.5). An example of such a problem is to find a shortest distance between two curves given by the functions $g(x)$ and $h(x)$. We consider the problem with variable boundary conditions for a functional

$$J = \int_{x_a}^{x_b} F(x, y, y') dx, \qquad (11.30)$$

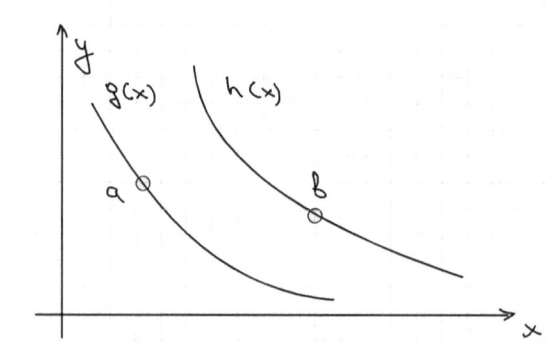

Fig. 11.5. Variable boundaries.

where x_a and x_b are unknown; however, we are given that the left point x_a is located on $g(x)$ and the right point x_b is located on $h(x)$. As we know, the solution to Euler–Lagrange equation has two arbitrary constants that are defined by given boundary conditions. Now, the constants together with x_a and x_b are determined using two equations from the conditions called *transversality condition*

$$[F + (g' - y')F_{y'}]_{x=x_a} = 0,$$
$$[F + (h' - y')F_{y'}]_{x=x_b} = 0 \tag{11.31}$$

and two equations for the boundary points

$$y(x_a) = g(x_a),$$
$$y(x_b) = h(x_b). \tag{11.32}$$

Thus, equations (11.30)–(11.32) form the necessary conditions for solving a problem with variable end points.

Note that if one of the boundary points (let's say the right one) can only move vertically, i.e., $x_b = b$ and $h'|_{x=x_b} = \infty$, then the condition of transversality is

$$F_{y'}|_{x=x_b} = 0.$$

And if one of end points (let's say the left one) can only move horizontally, i.e., $g' = 0$, then the condition transversality is

$$[F - y'F_{y'}]_{x=x_a} = 0.$$

Example 11.9. Find a shortest distance between parabola $y = x^2$ and a straight line $y = x - 5$.

The distance is given by the integral

$$J = \int_{x_a}^{x_b} \sqrt{1 + y'^2}\, dx.$$

The left boundary is located on $g(x) = x^2$ and the right one is on $h(x) = x - 5$. The Euler–Lagrange equation for $F(y')$ has a general solution $y = C_1 x + C_2$, where C_1 and C_2 are two unknown constants.

The transversality conditions take the form

$$\left[\sqrt{1+y'^2} + (2x - y')\frac{y'}{\sqrt{1+y'^2}}\right]_{x=x_a} = 0 \quad \text{and}$$

$$\left[\sqrt{1+y'^2} + (1 - y')\frac{y'}{\sqrt{1+y'^2}}\right]_{x=x_b} = 0,$$

with $y' = C_1$. For the end points, we have

$$C_1 x_a + C_2 = x_a^2 \quad \text{and}$$
$$C_1 x_b + C_2 = x_b - 5.$$

The four equations above define all four constants as $C_1 = -1$, $C_2 = 3/4$, $x_a = 1/2$, and $x_b = 23/8$. Therefore, the shortest line connecting the two functions is a straight line $y = -x + 3/4$ and the shortest distance is

$$\int_{1/2}^{23/8} \sqrt{1 + (-1)^2} dx = \frac{19\sqrt{2}}{8}. \qquad \blacksquare$$

11.8. Direct methods for solving variational problems

Very often it is difficult, if not impossible, to solve the Euler–Lagrange equation analytically. However, it is possible to find an approximate solution in a relatively straightforward way. Suppose we are looking for a solution as a function depending on several parameters $y = f(x, C_1, C_2, \ldots, C_n)$ in such way that function f satisfies given boundary conditions. Then, after integrating analytically, the functional depends explicitly on the n parameters. Thus, the problem is reduced to finding a stationary point of a function of n variables. Certainly, the solution is an approximation to an exact solution, but with a good choice of the parameters, it can nevertheless be a very good approximation.

Example 11.10. Assume that we want to find the solution for

$$J = \int_0^1 (y'^2 + y^2) dx,$$

with $y(0) = 0$ and $y(1) = 1$. While the problem can be solved analytically, we look for a trial solution in the form

$$y(x) = x + Cx(1-x) \quad \text{with } y'(x) = 1 + C - 2Cx.$$

The trial function satisfies given boundary conditions. Substituting the function into the integral gives

$$J = \int_0^1 [(1 + C - 2Cx)^2 + (x + Cx - Cx^2)^2] dx = \frac{4}{3} + \frac{1}{6}C + \frac{11}{30}C^2.$$

This is a function of the parameter C. Setting the derivative to zero gives

$$\frac{1}{6} + \frac{11}{15}C = 0.$$

Then, $C \approx 0.227$ and the approximate solution can be written as $y = x - 0.227x(1-x) = 0.773x + 0.227x^2$. Comparing this solution with the exact solution

$$y = \frac{e^x - e^{-x}}{e - e^{-1}}$$

shows that the two solutions agree within a few percent, as illustrated in Figure 11.6. And the values of the functional calculated with the two solutions differ by less than 0.1%. We can increase the accuracy of the approximate solution by increasing number of terms for the trial function. ∎

11.9. The principle of least action

The principle of least action is widely used in practically all areas of physics. While many ideas about "least actions" have been expressed since ancient Greek times, probably one of the most known examples in physics is Fermat's principle.

We start with a rather simple example — one-dimensional motion of an object subject to a conservative force $F = -dU(x)/dx$, where $U(x)$ is potential energy. Then, using Newton's second law, we have

$$m\frac{d^2x}{dt^2} = -\frac{dU(x)}{dx}.$$

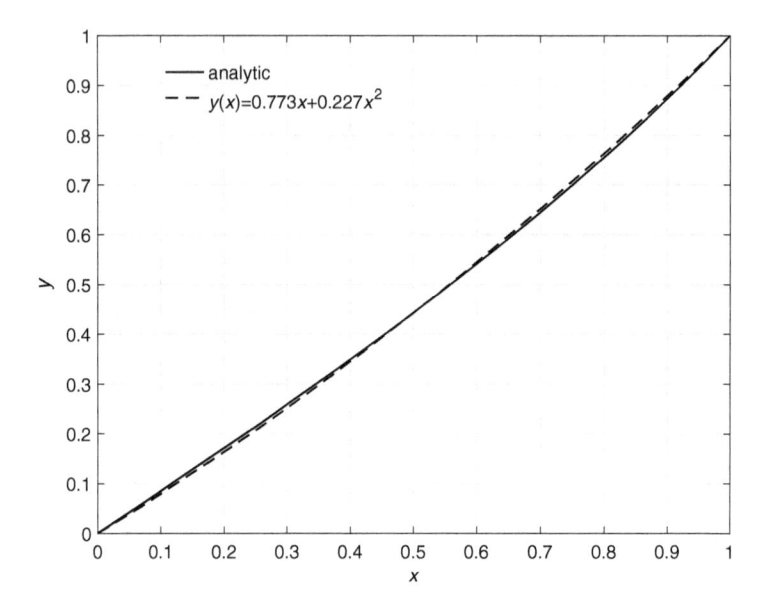

Fig. 11.6. Comparing the analytic and approximate solutions.

Using the notation $x' = dx/dt$, we can rewrite the above equation as

$$\frac{dU}{dx} + \frac{d}{dt}(mx') \quad \text{or} \quad \frac{d}{dx}(-U(x)) - \frac{d}{dt}\left[\frac{d}{dx'}\left(\frac{mx'^2}{2}\right)\right] = 0.$$

This equation reminds us of the Euler–Lagrange equations, but we do not have the same function in the two terms. However, since $-U(x)$ does not depend on t, we can add it to the term on the right, and since $mx'^2/2$ does not depend on x, we add it to the left term (note that the corresponding derivatives will be zero), thus we obtain

$$\frac{\partial}{\partial x}\left[\frac{mx'^2}{2} - U(x)\right] - \frac{d}{dt}\left(\frac{\partial}{\partial x'}\left[\frac{mx'^2}{2} - U(x)\right]\right) = 0.$$

And this is exactly the Euler–Lagrange equation

$$F_x - \frac{d}{dt} F_{x'} = 0.$$

for the function

$$F(t, x, x') = \frac{mx'^2}{2} - U(x).$$

Then, the functional, corresponding to the Euler–Lagrange equation, is

$$J = \int_{t_i}^{t_f} \left[\frac{mx'^2}{2} - U(x) \right] dt = 0.$$

Note that the first term is the kinetic energy K of the object. The function

$$L = K - U \qquad (11.33)$$

is called the Lagrangian. Then, the variational problem is to find stationary values for the integral

$$S = \int_{t_i}^{t_f} L \, dt; \qquad (11.34)$$

this is called the *action* for motion from the initial point to the final one. If the time interval is sufficiently small, the integral above provides not only stationary but also a minimal value for real motion. The ability to find this motion, based on the variational problem, is called *the principle of least action*. The principle of least action is valid for any isolated system (without energy loss or dissipation).

11.10. Fermat's principle of least time

What is the path taken by a 'ray' of light that travels from a point A to a point B in a nonuniform medium? According to Fermat's principle, it travels along a path in the medium that minimizes the time to travel between A and B. The time to travel along a path Γ joining points A and B is

$$T_\Gamma = \int_\Gamma \frac{ds}{c} = \int_\Gamma \frac{n(s)}{c_0} ds, \qquad (11.35)$$

where s is the arc length along Γ, c_0 is the speed of light in vacuum, c is the speed of light in the medium, and $n = c_0/c$ is the refractive index of the medium. In a two-dimensional medium (x, y) suppose that the refractive index varies only in the y-direction, i.e., $n = n(y)$. The ray path Γ can be parametrized by the x-coordinate (in particular) so that the end points of the path are $A(a, y(a))$ and $B(b, y(b))$. Then, $T_\Gamma \to T(y)$, where

$$T(y) = \frac{1}{c_0} \int_a^b n(y)\sqrt{1 + (dy/dx)^2}\, dx. \qquad (11.36)$$

Mathematical note: $T(y)$ is a *functional* of the mapping $y : x \longrightarrow y(x)$. Simply put, it is a function of a function — changing y changes $T(y)$.

We can write the integrand of (11.36) as $F(y, y'; x)$, even though in this case there is no explicit functional dependence on x. As shown earlier in this chapter, the Euler–Lagrange equation for F is

$$\frac{d}{dx}\left(\frac{\partial F}{\partial y'}\right) - \frac{\partial F}{\partial y} = 0, \qquad (11.37)$$

$$\begin{aligned}
\frac{dF}{dx} &= \frac{\partial F}{\partial x} + y'\frac{\partial F}{\partial y} + y''\frac{\partial F}{\partial y'} \\
&= \frac{\partial F}{\partial x} + y'\frac{d}{dx}\left(\frac{\partial F}{\partial y'}\right) + y''\frac{\partial F}{\partial y'} \\
&= \frac{\partial F}{\partial x} + \frac{d}{dx}\left[y'\left(\frac{\partial F}{\partial y'}\right)\right]
\end{aligned}$$

so that

$$\frac{d}{dx}\left[F - y'\left(\frac{\partial F}{\partial y'}\right)\right] = \frac{\partial F}{\partial x}. \qquad (11.38)$$

In this case, $F = n(y)\sqrt{1 + (y')^2}$ is independent of x, so equation (11.38) reduces to

$$\frac{d}{dx}(n(1 + (y')^2)^{-1/2}) = 0 \quad \text{or}$$

$$\frac{n(y)}{\sqrt{1 + (y')^2}} = K(x_0, y_0), \qquad (11.39)$$

K being the constant determined from the initial conditions, e.g., the point (x_0, y_0) from which the ray proceeds in the medium. A useful physical interpretation of K is that it is equal to the value of the refractive index when $y' = 0$, i.e., when the tangent to the ray becomes horizontal (this is assumed to occur once only here, but sometimes, the ray path is sinusoid-like, as with the Novaya Zemlya phenomenon). This is in fact Snell's law of refraction for a medium continuously varying in the y-direction. To see this, consider the angle $\theta(y)$ between the ray tangent and the vertical (y) direction; then $\sin\theta = (1+(y')^2)^{-1/2}$ and equation (11.39) can be written as

$$n(y)\sin\theta(y) = K(x_0, y_0). \tag{11.40}$$

Equation (11.39) leads to a nice proof of the so-called *Mirage Theorem*[4] (or concavity theorem). If we solve for y'^2 and differentiate the result with respect to x, then for $y' \neq 0$,

$$y'' = K^{-2}n(y)n'(y) \propto n'(y). \tag{11.41}$$

From this, it follows that if the refractive index increases/decreases upward ($n' > 0, n' < 0$), then the ray path is concave upward/downward ($y'' > 0, y'' < 0$). Clearly, since the signs of n' and y'' are always the same, the sign of one of (n', y'') determines the sign of the other.

11.11. Exercises and problems

Exercises

(1) Find functions making following integrals stationary:
   (a) $J = \int_0^{2\pi}(y'^2 - y^2)dx$, $y(0) = 1$, $y(2\pi) = 1$
   (b) $J = \int_1^2(y'^2 + 2yy' + y^2)dx$, $y(1) = 1$, $y(2) = 0$
   (c) $J = \int_0^\pi(y'^2 + 4y\cos x - y^2)dx$, $y(0) = 0$, $y(\pi) = 0$
   (d) $J = \int_0^1(y'^2 - y^2 - y)e^{2x}dx$, $y(0) = 0$, $y(1) = e^{-1}$
   (e) $J = \int_{-1}^1(y'^2 - 2xy)dx$, $y(-1) = -1$, $y(1) = 1$.
(2) Find functions making following integrals stationary:
   (a) $J = \int_0^1(y''^2 + 2y'^2 + y^2)dx$, $y(0) = 0$, $y(1) = 0$, $y'(0) = 1$, $y'(1) = -\sinh(1)$

---

[4]J. A. Adam, *Rays, Waves, and Scattering*. Princeton University Press, Princeton, 2017.

(b) $J = \int_0^1 (y'^2 + y''^2)dx$, $y(0) = 0$, $y(1) = \sinh(1)$, $y'(0) = 1$, $y'(1) = -\cosh(1)$.

(3) Find functions $y(x)$ and $z(x)$ making following integral stationary:

(a) $J = \int_0^{\pi/4}(2z - 4y^2 + y'^2 - z'^2)dx$, $y(0) = 0$, $y(\pi/4) = 1$, $z(0) = 0$, $z(\pi/4) = 1$

(b) $J = \int_0^{\pi/2}(y'^2 - 2yz + z'^2)dx$, $y(0) = 0$, $y(\pi/2) = 1$, $z(0) = 0$, $z(\pi/2) = 1$.

(4) Solve the isoperimetric problems:

(a) $J = \int_0^1(x^2 + y'^2)dx$, $y(0) = 0$, $y(1) = 0$, under condition $\int_0^1 y^2 dx = 2$

(b) $J = \int_0^1 y'^2 dx$, $y(0) = 0$, $y(1) = 1/4$, under condition $\int_0^1 (y - y'^2)dx = 1/12$.

(5) Find the shortest distance between a circle $x^2 + y^2 = 1$ and a straight line $x + y = 4$.

(6) Find approximate solution for the functional
$J = \int_0^1 (y'^2 + y^2)dx$, $y(0) = 0$, $y(1) = 1$ and compare with the exact solution:

(a) for a trial function $y = x + C\sin\pi x$
(b) for a trial function $y = x + C_1 x(1-x) + C_2 x^2(1-x)$.

(7) Find approximate solution for the functional:
$J = \int_0^1(y'^2 + xy^2)dx$, $y(0) = 0$, $y(1) = 1$ for a trial function $y = x + Cx(1-x)$.

Problems

(1) Find the shortest path between points (x_a, y_a) and (x_b, y_b) in a plane.
(2) The catenary problem, also known as the hanging chain problem[5]: A heavy chain of length L is suspended between points

---

[5]The problem has a long history, and its mathematical analysis dates back to the 17th century. The catenary was first mathematically described by Jacob Bernoulli in 1691 and later solved by his brother Johann Bernoulli and Leibniz.

(a, y_a) and (b, y_b). Find its shape in equilibrium. Note that you can make your work easier by assuming that $x_a = 0, y_a = 0$, and $x_b = l, y_b = 0$. Clearly, $l < L$.

Hint: The equilibrium corresponds to the lowest potential energy.

Answers and solutions

(1) (a) $y(x) = \cos x + C \sin x$
    (b) $y(x) = \frac{\sinh(2-x)}{\sinh(1)}$
    (c) $y(x) = (C + x) \sin x$
    (d) $y(x) = \frac{1}{2}[e^{-x} - 1 + xe^{-x}(e + 1)]$
    (e) $y(x) = \frac{7}{6}x - \frac{1}{6}x^3$.

(2) (a) $y(x) = (1 - x)\sinh(x)$
    (b) $y(x) = \sinh(x)$.

(3) (a) $y(x) = \sin(2x)$, $z(x) = -\frac{x^2}{2} + \frac{\pi^2 + 32}{8\pi}$
    (b) $y(x) = \sin(x)$, $z(x) = \sin(x)$.

(4) (a) $(x) = \pm 2\sin(n\pi x)$
    (b) $y(x) = \frac{1}{4}(2x - x^2)$.

(5) $2\sqrt{2} - 1$.

(6) (a) $y(x) = x - 0.0588\sin(\pi x)$, $I = 1.315$ (error 0.2%)
    (b) $y(x) = x - 0.146x(1 - x) - 0.163x^2(1 - x)$.

(7) $y(x) = x - \frac{1}{7}x(1 - x)$.

Chapter 12

Functions of Complex Variables

12.1. Introduction

A function is a rule of correspondence between an independent variable (or variables) and a dependent variable, e.g., $f(x)$ in case of real numbers. A function of a complex variable, or a complex function, is a function that takes complex numbers as inputs and produces complex numbers as outputs. In other words, we define a function of a complex variable as a function that assigns a complex number w to a point z in a complex plane. Since each complex number $z = x + iy$ is characterized by a pair of real numbers, then a function $w = f(z)$ of a complex variable can be expressed in terms of a pair of real functions of the real variables x and y as

$$f(z) = u(x,y) + iv(x,y), \qquad (12.1)$$

where $u(x,y)$ is a real part and $v(x,y)$ is an imaginary part of the function $f(z)$.

Functions of complex variables, or complex analysis, have a broad range of applications in science and engineering. Here are some examples. Quantum states in quantum mechanics are often described as complex-valued wavefunctions. In fluid dynamics, potential flow theory uses complex variables to model fluid flow around objects. Signal processing in engineering is based on the complex Fourier transform.

In Section 2.4, we defined an exponential function of a complex argument; in particular, we wrote (2.24)

$$e^z = e^{x+iy} = e^x \cdot e^{iy} = e^x(\cos y + i \sin y). \qquad (12.2)$$

As one can see, the exponential function has the real part $u(x,y) = e^x \cos y$ and the imaginary part $v(x,y) = e^x \sin y$. By defining the exponential function (12.2), we can define trigonometric functions for a complex argument: thus

$$\sin z = \frac{e^{iz} - e^{-iz}}{2i},$$

$$\cos z = \frac{e^{iz} + e^{-iz}}{2}.$$

As we see later, all of the above elementary functions of a complex variable have a derivative as a function of a complex variable. In other words, each of these functions has a specific limit

$$\frac{f(z + \Delta z) - f(z)}{\Delta z}$$

as the complex number Δz tends to zero. Such functions are said to be *analytic*. This chapter is devoted to developing the theory of what is called analytic functions of a complex variable.

A function $f(z)$ can be defined either on the entire complex plane or only in a specific region of the plane of the complex variable z, such as a circle, rectangle, and annulus. For any given region, we distinguish between its *internal* points and *contour* points or *boundary*. For instance, in the case of a circle with a center at the origin and a radius of one, the internal points are characterized by the condition

$$|z| < 1, \quad \text{or} \quad x^2 + y^2 < 1,$$

and the contour is a circle described as

$$|z| = 1, \quad \text{or} \quad x^2 + y^2 = 1.$$

A defining feature of an internal point is that not only the point itself but also its surroundings entirely belong to the region. Specifically, a point P is considered an internal point of the region if a sufficiently small circle centered at P is entirely contained within the region.

Contour points, conversely, are not considered internal points of the region. However, within any arbitrarily small neighborhood of a contour point, there exist internal points belonging to the region. A set of all boundary points forms the boundary of the region.

Additionally, we assume that our region maintains continuity (or integrity), meaning it does not break up into separate pieces, i.e., any two points within the region can be connected by a line that is entirely located inside the region. This property is called *convexity*. Henceforth, when discussing the region, we typically refer to the collection of internal points. If a boundary is present, we classify the region as *closed*. Generally, functions $f(z)$ can be single-valued (one to one correspondence that we can consider as a mapping from z-plane of complex numbers to w-plane) or multi-valued. In this chapter, we concentrate on single-valued functions.

Continuity: Let us consider a function $f(z)$ defined within some region D, i.e., at any point z inside D, $f(z)$ has a certain complex value. Let then z_0 be a point inside the region D. A function $f(z)$ is called *continuous* at point z_0 if

$$\lim_{z \to z_0} f(z) = f(z_0), \tag{12.3}$$

or for any given positive ε, there is a positive number δ such that $|f(z) - f(z_0)| < \epsilon$ if only $|z - z_0| < \delta$. A function is considered continuous within region D if and only if it is continuous at every point within D. This definition of continuity extends to both interior and boundary points of the set. When defining continuity at any point z_0 on the boundary, it is essential to consider that the point z can approach z_0 in any manner, but it should remain within the closed region D.

Let us write again $f(z)$ and z by using the real and imaginary parts:

$$z = x + iy,$$
$$f(z) = u(x, y) + iv(x, y).$$

Assuming that $z_0 = x_0 + iy_0$, the condition $z \to z_0$ is equivalent to $x \to x_0$ and $y \to y_0$. The definition of continuity at the point z_0 implies that as z approaches z_0, i.e., as x approaches x_0 and y

approaches y_0, the function $f(z)$ must converge to $f(z_0)$:

$$\lim_{\substack{x \to x_0 \\ y \to y_0}} [u(x,y) + iv(x,y)] \to u(x_0, y_0) + iv(x_0, y_0)$$

or

$$\lim_{\substack{x \to x_0 \\ y \to y_0}} u(x,y) \to u(x_0, y_0) \quad \text{and} \quad \lim_{\substack{x \to x_0 \\ y \to y_0}} v(x,y) \to v(x_0, y_0).$$

Consequently, the continuity of the function of a complex variable $f(z) = u(x,y) + iv(x,y)$ implies that both its real part $u(x,y)$ and imaginary part $v(x,y)$ are continuous. Conversely, if $u(x,y)$ and $v(x,y)$ are continuous at a point (x_0, y_0), then the complex function $f(z) = u(x,y) + iv(x,y)$ is also continuous at that point (x_0, y_0).

Example 12.1. Let $f(z) = z^2 + z$, then

$$f(z) = (x+iy)^2 + z = x^2 - y^2 + x + i2xy + iy = u(x,y) + iv(x,y),$$

where $u(x,y) = x^2 - y^2 + x$ and $v(x,y) = 2xy + y$. The function is defined on the entire complex plane z. The function is single-valued, since for every value of z there is only one value of $f(z)$. The real $u(x,y)$ and imaginary $v(x,y)$ parts of the function are continuous functions of x and y, therefore $f(z) = z^2 + z$ is a continuous function on the entire complex plane for any finite values of (x,y). ∎

12.2. Derivatives and the Cauchy–Riemann conditions

The derivative of a function of a complex variable with respect to its argument can be formally defined in a manner analogous to derivatives for functions of real variables, namely

$$f'(z) = \lim_{\Delta z \to 0} \frac{f(z + \Delta z) - f(z)}{\Delta z}. \tag{12.4}$$

Given that $\Delta z = \Delta x + i \Delta y$, the limit $\Delta z \to 0$ involves a double limit, and there are infinitely many ways by which $\Delta z = (\Delta x + i \Delta y) \to 0$. Nonetheless, our focus lies solely on derivatives that remain independent of the specific approach taken as $\Delta z \to 0$.

Let us mark inside a region D a point A with coordinate $z + \Delta z = (x + \Delta x) + i(y + \Delta y)$ and a point B with a coordinate

$z = x + iy$. We consider two specific approaches for B converging to A as $\Delta z \to 0$. In the first way, we establish $\Delta z = \Delta x, \Delta y = 0$, i.e., we approach A along the real x-axis. In the second way, we set $\Delta z = i\Delta y, \Delta x = 0$, or Δz approaches zero along the imaginary y-axis.[1] Now, we apply the definition of the derivative (12.4) to the above two approaches:

$$f'(z) = \lim_{\Delta z \to 0} \frac{f(z + \Delta z) - f(z)}{\Delta z}$$

$$= \lim_{\substack{\Delta x \to 0 \\ \Delta y \to 0}} \frac{[u(x + \Delta x, y + \Delta y) - u(x, y)] + i[v(x + \Delta x, y + \Delta y) - v(x, y)]}{\Delta x + i\Delta y}. \tag{12.5}$$

Then, for $\Delta z = \Delta x$, $\Delta y = 0$ follows

$$f'(z) = \lim_{\Delta x \to 0} \left[\frac{u(x + \Delta x, y) - u(x, y)}{\Delta x} + i\frac{v(x + \Delta x, y) - v(x, y)}{\Delta x} \right].$$

If the functions $u(x, y)$ and $v(x, y)$ possess partial derivatives with respect to x, then we can express

$$f'(z) = \frac{\partial u(x, y)}{\partial x} + i\frac{\partial v(x, y)}{\partial x}. \tag{12.6}$$

When Δz approaches zero along the imaginary axis $\Delta x = 0$ with $\Delta z = i\Delta y$, $\Delta x = 0$, then

$$f'(z) = \lim_{\Delta y \to 0} \frac{1}{i} \left[\frac{u(x, y + \Delta y) - u(x, y)}{\Delta y} + i\frac{v(x, y + \Delta y) - v(x, y)}{\Delta y} \right],$$

and finally,

$$f'(z) = -i\frac{\partial u(x, y)}{\partial y} + \frac{\partial v(x, y)}{\partial y}. \tag{12.7}$$

Comparing the two derivatives (12.6) and (12.7) for $f'(z)$ we obtain the conditions for the derivative $f'(z)$ to be path independent when

---

[1] It would be interesting to also approach the point from an arbitrary direction, e.g., at an angle with the x-axis.

$\Delta z \to 0$. These conditions are for the partial derivatives, specifically,

$$\frac{\partial u(x,y)}{\partial x} = \frac{\partial v(x,y)}{\partial y},$$
$$-\frac{\partial u(x,y)}{\partial y} = \frac{\partial v(x,y)}{\partial x}. \qquad (12.8)$$

These are *necessary* conditions for a complex function $f(z)$ to possess a derivative at point z. Importantly, they are not only necessary but also sufficient conditions for $f(z)$ to have a derivative. The conditions (12.8) are commonly referred to as the *Cauchy–Riemann conditions*.[2] If a function $f(z)$ has derivatives at all points within a certain domain D, and these derivatives are continuous in D, then such a function is referred to as an *analytic function* within the domain D. The Cauchy–Riemann conditions provide a bridge between functions of real and complex variables.

It's noteworthy that relations (12.8) enable us to derive various expressions for the derivative of a analytic function of a complex variable such as

$$\begin{aligned} f'(z) &= u_x(x,y) + iv_x(x,y) = v_y(x,y) - iu_y(x,y) \\ &= u_x(x,y) - iu_y(x,y) = v_y(x,y) + iv_x(x,y). \end{aligned} \qquad (12.9)$$

---

[2] Augustin-Louis Cauchy (1789–1857) was a French mathematician who made profound contributions to a diverse array of mathematical fields and left an indelible mark on the development of analysis and other areas of mathematics. Born in Paris, Cauchy's mathematical prowess was evident from a young age, and he quickly rose to prominence in the academic world. His work encompassed a wide range of topics, including complex analysis, number theory, elasticity, and mathematical physics. Cauchy is particularly renowned for his rigorous approach to mathematical analysis, advocating for the concept of limits and laying the foundation for modern calculus.

Cauchy's legacy extends beyond his work in analysis. He introduced the concept of a Cauchy sequence and the Cauchy–Riemann equations, which became cornerstones of complex analysis. His rigorous treatment of mathematical concepts and dedication to establishing solid foundations greatly influenced the way mathematics is practiced and understood today. Cauchy's impact extended beyond mathematics into other disciplines, such as physics, where he made significant contributions to the understanding of wave propagation and elasticity. His emphasis on mathematical rigor and precision set a new standard for mathematical inquiry, shaping the course of 19th-century mathematics and leaving an enduring legacy that continues to shape the field to this day.

The conditions (12.8) are not the only possible form of the Cauchy–Riemann relations. For example, the real and imaginary parts of the analytic function $f(z) = u(\rho, \varphi) + iv(\rho, \varphi)$ of a complex variable $z = \rho e^{i\varphi}$ are connected by

$$\frac{\partial u}{\partial \rho} = \frac{1}{\rho}\frac{\partial v}{\partial \varphi}, \quad \frac{1}{\rho}\frac{\partial u}{\partial \varphi} = -\frac{\partial v}{\partial \rho}, \qquad (12.10)$$

where ρ and φ are polar coordinates of a point (x, y).

The concept of an analytic function is a cornerstone concept in the theory of functions of a complex variable. This is attributed to the distinctive role played by the class of analytic functions, not only in resolving numerous mathematical problems but also in various applications.

Example 12.2. Show that the exponential function

$$e^z = e^x(\cos y + i \sin y)$$

is analytic in the x, y-plane. Indeed, in this case,

$$u(x, y) = e^x \cos y, \quad v(x, y) = e^x \sin y.$$

Then,

$$\frac{\partial u}{\partial x} = e^x \cos y, \quad \frac{\partial u}{\partial y} = -e^x \sin y,$$

$$\frac{\partial v}{\partial x} = e^x \sin y, \quad \frac{\partial v}{\partial y} = e^x \cos y.$$

These partial derivatives are continuous, and they satisfy the Cauchy–Riemann conditions. Evaluating the derivative of the function $(e^z)'$ by using (12.6), we obtain

$$(e^z)' = e^x \cos y + ie^x \sin y = e^x(\cos y + i \sin y) = e^z,$$

i.e., $(e^z)' = e^z$. By the same way, we can show that $(e^{ax})' = ae^{ax}$. We have derived the same rule for differentiating the exponential function of a complex variable as in the case of a real variable. Now, it is easy to show that derivatives of $\sin z$ and $\cos z$ are also evaluated by the same rules as for a real variable. Indeed, by applying the rule of differentiation for the exponential function, we obtain

$$(\sin z)' = \left(\frac{e^{iz} - e^{-iz}}{2i}\right)' = \frac{e^{iz} + e^{-iz}}{2} = \cos z,$$

$$(\cos z)' = \left(\frac{e^{iz} + e^{-iz}}{2}\right)' = i\frac{e^{iz} - e^{-iz}}{2} = -\sin z. \quad \blacksquare$$

Properties of analytic functions: The definition of the derivative (12.4) enables us to write a number of properties of analytic functions of a complex variable, such as the following:

(1) If the function $f(z)$ is analytic in the region D, then it is continuous in that region.
(2) If $f_1(z)$ and $f_2(z)$ are analytic functions in region D, then their sum and product are also analytic functions in region D, and the function $g(z) = \frac{f_1(z)}{f_2(z)}$ is an analytic function wherever $f_2(z) \neq 0$.
(3) Differentiating the first equation in (12.8) term by term with respect to x, the second with respect to y, and then subtracting the second from the first, we obtain

$$\frac{\partial^2 u(x,y)}{\partial x^2} + \frac{\partial^2 u(x,y)}{\partial y^2} = 0. \qquad (12.11)$$

In the same way, from equations (12.8), we derive

$$\frac{\partial^2 v(x,y)}{\partial x^2} + \frac{\partial^2 v(x,y)}{\partial y^2} = 0. \qquad (12.12)$$

From this, it is evident that the real and imaginary parts of the analytic function $f(z)$ must satisfy the Laplace equation.[3]

(4) Let the function $u(x,y)$ be given as the real part of the analytic function $f(z)$. Then, the imaginary part of this function is determined up to an additive constant. Indeed, the total differential of the unknown function $v(x,y)$ is defined as (5.4)

$$dv = v_x(x,y)dx + v_y(x,y)dy.$$

---

[3] The Laplace equation is a partial differential equation. In two dimensions, the Laplace equation is given by $\frac{\partial^2 u(x,y)}{\partial x^2} + \frac{\partial^2 u(x,y)}{\partial y^2} = 0$. Solutions to the Laplace equation are called harmonic functions.

By using the Cauchy–Riemann conditions, we obtain
$$dv = -u_y(x,y)dx + u_x(x,y)dy.$$
Then, the function $v(x,y)$ can be found as a line integral
$$v(x,y) = \int_L (-u_y dx + u_x dy) + C,$$
where L is a path from the origin $(0,0)$ to (x,y) and C is a constant. This is a second-kind line integral. Methods for evaluating such integrals can be found in Chapter 6, Section 6.3.2 (see equations (6.20) and (6.21)). Since $u(x,y)$ satisfies Laplace's equation, then the integral above is path independent (see Section 6.3.3). Then, we can choose a path from point $(0,0)$ to (x,y) consisting of two parts: (I) along the x axis where $y = 0$ and $dy = 0$ and then (II) along the y axis with $dx = 0$, as illustrated in Figure 12.1. Then, the path integral above can be rewritten as
$$v(x,y) = \int_I (-u_y dx + u_x dy) + \int_{II} (-u_y dx + u_x dy) + C$$
$$= -\int_I u_y(x,0) dx + \int_{II} u_x dy + C.$$
Another method to determine the imaginary part $v(x,y)$ involves two steps: First, by using $v_x = -u_y$, we find
$$v(x,y) = -\int u_y dx + Y(y),$$
where $Y(y)$ is some function of y. Then, we differentiate the above result as $v_y(x,y)$, and by applying the Cauchy–Riemann condition, we use $v_y(x,y) = u_x$ to find $Y(y)$.

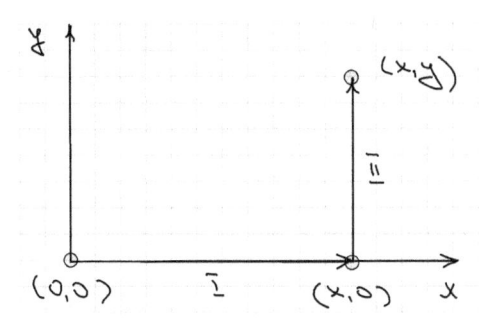

Fig. 12.1. Evaluating the path integral from $(0,0)$ to (x,y).

Example 12.3. Consider the function $u(x, y) = x^2 - y^2 + x$. Show that this function is a harmonic function (satisfies Laplace's equation), and find the imaginary part $v(x, y)$ of the corresponding analytic function.

Indeed, by evaluating second-order partial derivatives, we obtain

$$\frac{\partial^2 u(x, y)}{\partial x^2} + \frac{\partial^2 u(x, y)}{\partial y^2} = 2 - 2 = 0,$$

that is, $u(x, y)$ satisfies Laplace's equation. Let us find the function $v(x, y)$, such that $u(x, y) + iv(x, y)$ is an analytic function. Partial derivatives for $u(x, y)$ are $u_x(x, y) = 2x + 1$ and $u_y(x, y) = -2y$. By using the first method, we obtain

$$v(x, y) = \int_I 0 dx + \int_{II} (2x + 1) dy + C = 2xy + y + C.$$

Now, we employ the second method outlined earlier. From the Cauchy–Riemann conditions, we have $u_x = 2x + 1 = v_y$ and $u_y = -2y = -v_x$. Then,

$$v(x, y) = \int 2y dx + Y(y) = 2xy + Y(y).$$

Evaluating the partial derivative for the above result, and using $v_y = u_x = 2x + 1$, we obtain

$$v_y = 2x + Y_y(y) = 2x + 1.$$

Then, $Y_y(y) = 1$ and $Y(y) = y + C$. Finally, we obtain

$$v(x, y) = 2xy + y + C,$$

that is, the same result we derived by using the first method. We do the final check to be sure that $v(x, y)$ is a harmonic function:

$$\frac{\partial^2 v(x, y)}{\partial x^2} + \frac{\partial^2 v(x, y)}{\partial y^2} = 0 + 0 = 0.$$ ∎

12.3. Integrals

Consider a line L in Figure 12.2 given on the complex plane. Utilizing the parametric representation of the curve L, we define the

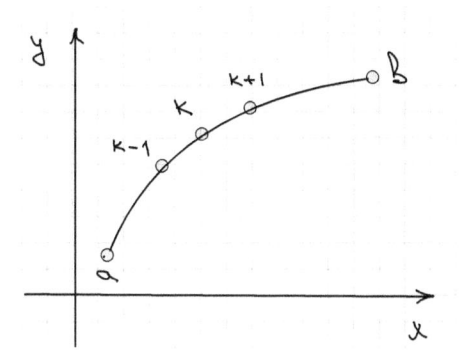

Fig. 12.2. Evaluating the path integral from $(0,0)$ to (x,y).

coordinates (x, y) of each of its points using the parametric equations as $x(t)$ and $y(t)$ with $t_a \leq t \leq t_b$. By having $x(t)$ and $y(t)$, we can specify a complex function $z(t) = x(t) + iy(t)$ of a real variable t. Let for every point z on the curve C us define a function $f(z)$.

We partition the line L into partial n segments with points $z_0, z_1, z_3, \ldots, z_n$ corresponding to increasing values of the parameter t, such that $t_{k+1} > t_k$. We can define a sum of the products as

$$\sum_{k=1}^{n} f(z_k^*)(z_k - z_{k-1}), \qquad (12.13)$$

where $z_k^* = x_k^* + iy_k^*$, such that $x_k^* = x(\tau_k)$ and $y_k^* = y(\tau_k)$ with $t_{k-1} \leq \tau_k \leq t_k$. The limit of this sum as n approaches infinity, and with an infinitesimal decrease in the length of each partial segment, is called the contour integral of the function $f(z)$ over the line L:

$$\int_L f(z) dz = \lim_{n \to \infty} \sum_{k=1}^{n} f(z_k^*)(z_k - z_{k-1}). \qquad (12.14)$$

Let us write the function as a sum of the real and imaginary parts, namely

$$f(z_k^*) = u(x_k^*, y_k^*) + iv(x_k^*, y_k^*).$$

Then, the sum (12.13) can be written as

$$\sum_{k=1}^{n} f(z_k^*)(z_k - z_{k-1}) = \sum_{k=1}^{n} [u(x_k^*, y_k^*) + iv(x_k^*, y_k^*)]$$
$$\times [(x_k - x_{k-1}) + i(y_k - y_{k-1})]$$

or

$$\sum_{k=1}^{n} f(z_k^*)(z_k - z_{k-1})$$

$$= \sum_{k=1}^{n} [u(x_k^*, y_k^*)(x_k - x_{k-1}) - v(x_k^*, y_k^*)(y_k - y_{k-1})]$$

$$+ i \sum_{k=1}^{n} [v(x_k^*, y_k^*)(x_k - x_{k-1}) + u(x_k^*, y_k^*)(y_k - y_{k-1})].$$

As n approaches infinity, the real and imaginary parts of the sums above correspond to the integral sums of line integrals of the second kind, respectively. Consequently, the integral (12.14) can be expressed as the sum of two line integrals of the second kind

$$\int_L f(z)dz = \int_L u(x,y)dx - v(x,y)dy + i \int_L v(x,y)dx + u(x,y)dy. \tag{12.15}$$

Above, we assumed that the line L has end points, but this definition is also applicable when integrating over closed contours. Such line integrals we call *contour integrals*.

Earlier in Section 6.3.3, we demonstrated that the second-kind line integrals

$$\int_{ab} P(x,y)dx + Q(x,y)dy \tag{12.16}$$

are path independent, i.e., how do we connect points a and b when

$$\frac{\partial P(x,y)}{\partial y} = \frac{\partial Q(x,y)}{\partial x}. \tag{12.17}$$

If $u(x,y)$ and $v(x,y)$ are analytic functions, then according to the Cauchy–Riemann conditions, it follows that the integrals given in (12.15) are path-independent. Using this property of analytic functions can considerably simplify the evaluation of integrals involving complex variables.

The integral (12.15) has exactly the same properties as the ordinary second-kind line integral considered in Chapter 6. Let us highlight the main properties of these integrals:

(1) The constant factor can be factored out from the integral sign:
$$\int_L a f(z)dz = a \int_L f(z)dz, \qquad (12.18)$$
where a is a complex constant.
(2) The integral of a sum of two functions is equal to the sum of the integrals:
$$\int_L [f_1(z) + f_2(z)]dz = \int_L f_1(z)dz + \int_L f_2(z)dz. \qquad (12.19)$$
(3) When the direction of the integration along the line changes, the value of the integral changes sign:
$$\int_{AB} f(z)dz = - \int_{BA} f(z)dz. \qquad (12.20)$$
(4) If you divide the integration path into several parts, the value of the integral over the entire path is equal to the sum of the integrals over its individual paths:
$$\int_{AB+CD} f(z)dz = \int_{BA} f(z)dz + \int_{CD} f(z)dz. \qquad (12.21)$$

Example 12.4. Evaluate
$$\int_L z^2 dz$$
along the upper semicircle with the center at $z = 0$ ranging from -1 to $+1$.

Since $z^2 = x^2 - y^2 + i2xy$, then we can write (12.15) as
$$\int_L (x^2 - y^2)dx - 2xy\,dy + i \int_L 2xy\,dx + (x^2 - y^2)dy.$$

The path of integration can be parameterized as $y = \sqrt{1-x^2}$ with $y'(x) = -x/\sqrt{1-x^2}$. Then, by using (6.21), we obtain the integral
$$\int_{-1}^{1} (4x^2 - 1)dx + i \int_{-1}^{1} \frac{(3x - 4x^3)}{\sqrt{1-x^2}}dx.$$

Evaluating the first integral gives $2/3$, and the second integral is equal to zero since the integrand is an odd function integrated between

−1 and 1. Thus, we obtain that for the given path

$$\int_L z^2 dz = \frac{2}{3}.$$

If we note that z^2 is an analytic function, then by choosing the path of integration as a straight line connecting $x = -1$ and $x = 1$ with $y = 0$ on such line, we immediately obtain

$$\int_{-1}^{1} x^2 dx = \frac{2}{3},$$

which is a much easier way to evaluate the integral. ∎

Example 12.5. Evaluate integral

$$\int_C \frac{dz}{z - z_0},$$

where the path C is a circle of radius ρ centered at point z_0 and the direction of integration is counterclockwise as shown on Figure 12.3. It is very convenient to represent z in polar form as $z = z_0 + \rho e^{i\varphi}$ with $0 \leq \varphi \leq 2\pi$. Then, for the integral, we obtain

$$\int_C \frac{dz}{z - z_0} = \int_0^{2\pi} \frac{i\rho e^{i\varphi} d\varphi}{e^{i\varphi}} = i \int_0^{2\pi} d\varphi = 2\pi i.$$

It is interesting that the value of the integral is independent of ρ and the location of point z_0. However, the value of the integral depends on the direction of integration. For integrating in the clockwise direction, i.e., from 2π to 0, we obtain that the integral is equal to $-2\pi i$. ∎

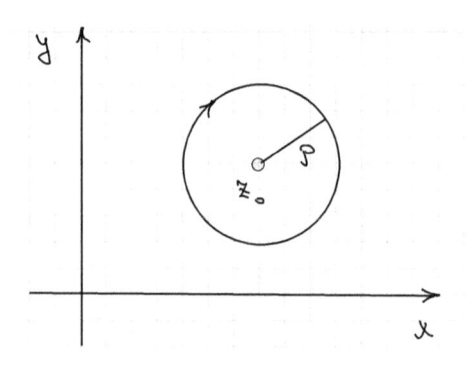

Fig. 12.3. The closed path of integration.

12.3.1. *Cauchy's integral theorem*

Since the value of a contour integral depends on the direction of integration, we define the positive direction of integration when the internal region bounded by the given closed contour C remains to the left of the direction of movement. Integration in the positive direction is denoted as $\int_{C^+}$ and integration in the negative direction as $\int_{C^-}$.

Let C be a closed contour enclosing a simply connected region D, where the function $f(z)$ is analytic.[4] We assume that C is a simple closed contour, meaning that C does not intersect itself. In Chapter 6, Section 6.5.3, we introduced Green's theorem connecting line integrals of the second kind to double integrals as

$$\int_L P(x,y)dx + Q(x,y)dy = \iint_S \left(\frac{\partial Q}{\partial x} - \frac{\partial P}{\partial y}\right) dxdy. \qquad (12.22)$$

Cauchy's integral theorem states that if an analytic function $f(z)$ is in a simply connected region, then integral of $f(z)$ around contour C is zero:

$$\int_C f(z)dz = 0. \qquad (12.23)$$

The proof is very straightforward. According to (12.15), we can write

$$\int_C f(z)dz = \int_C u(x,y)dx - v(x,y)dy + i\int_C v(x,y)dx + u(x,y)dy. \qquad (12.24)$$

Since the function $f(z)$ is analytic everywhere inside the contour C, then the functions $u(x,y)$ and $u(x,y)$ in the region bounded by this contour have continuous partial derivatives of the first order. Therefore, we can apply Green's theorem to the above integral. Besides, the partial derivatives of $u(x,y)$ and $v(x,y)$ are connected by the Cauchy–Riemann conditions. Then, we obtain

$$\int_C u(x,y)dx - v(x,y)dy = \iint_S \left[-\frac{\partial v}{\partial x} - \frac{\partial u}{\partial y}\right] dxdy = 0$$

---

[4] In complex analysis, a region in the complex plane is said to be simply connected if it is connected (meaning that any two points in the region can be joined by a curve entirely contained within the region) and its interior does not have any holes or isolated points.

and

$$\int_C v(x,y)dx + u(x,y)dy = \iint_S \left[-\frac{\partial u}{\partial x} + \frac{\partial v}{\partial y}\right] dxdy = 0.$$

Cauchy's integral theorem was formulated above for a simply connected region. However, it can be generalized to the case of multiply connected regions. In that case, the complete boundary of the region consists of several closed contours: external C and internal $C_1, C_2, \ldots, C_n$ contours. Let us assume for our consideration that there is only one internal contour, as in Figure 12.4. Let make a cut connecting the outer contour C with the inner contour C_1. Then the area enclosed by C and C_1 is a simply connected region. The integrals over the cut l cancel each other out since we just integrate twice but in the opposite directions. Therefore, by applying Cauchy's theorem to the multiply connected region, we obtain

$$\int_{C+} f(z)dz + \int_{C-} f(z)dx = 0. \qquad (12.25)$$

Note that we integrated over C in the positive direction but over C_1 in the negative direction to keep the region D on the left.

It is important to note that for Cauchy's theorem, the function $f(z)$ must be analytic everywhere in a region enclosed by a contour C. For example, let us consider a region as an annulus with $2 < |z| < 4$. Function $f(z) = 1/z$ is analytic in the annulus. However, integral

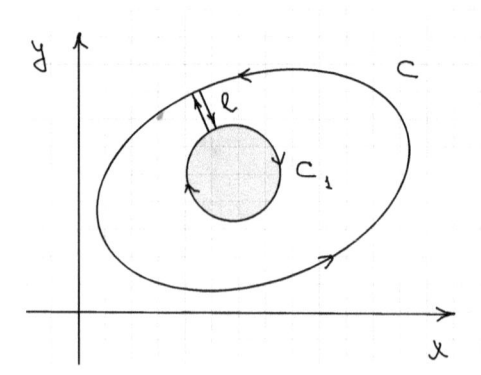

Fig. 12.4. Cauchy's theorem for multiply connected regions.

over a contour of radius $r = 3$ is not equal to zero but

$$\int_{|z|=3} \frac{1}{z} dz = 2\pi i.$$

This happens because the contour includes a point where the function is not analytic, i.e., $f(z) = 1/z$ is not analytic at $z = 0$.

12.3.2. Indefinite integral

For an analytic function $f(z)$ on a simply connected region, we can define a function

$$F(z) = \int_{z_0}^{z} f(\xi) d\xi, \quad (12.26)$$

where ξ denotes the variable of integration, taken along a path inside the region between points z_0 and z. Let us consider two paths of integration C_1 and C_1' as shown in Figure 12.5. From Cauchy's integral theorem follows that integration along closed contours $C_1 + C_2$ or $C_1' + C_2$ must be zero:

$$\int_{C_1} f(z)dz + \int_{C_2} f(z)dz = 0 \quad \text{and} \quad \int_{C_1'} f(z)dz + \int_{C_2} f(z)dz = 0.$$

Then, the value of the integral (12.26) does not depend on the path on integration between points z_0 and z.

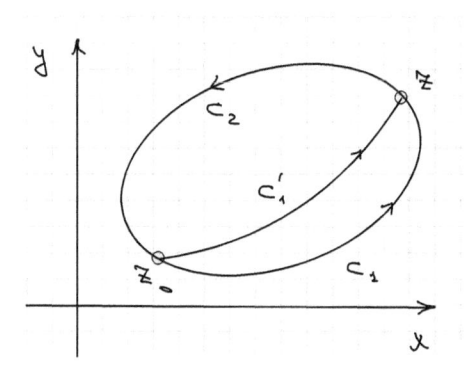

Fig. 12.5. Various paths of integration.

Next, we consider a difference

$$F(z+\Delta z) - F(z) = \int_{z_0}^{z+\Delta z} f(\xi)d\xi - \int_{z_0}^{z} f(\xi)d\xi = \int_{z}^{z+\Delta z} f(\xi)d\xi,$$

where integration can be performed, for example, along a straight line connecting points z and $z+\Delta z$. This integration path is convenient because in this case we have

$$\int_{z}^{z+\Delta z} d\xi = \Delta z.$$

Adding and subtracting $f(z)$ to the integrand $f(\xi)$ we obtain

$$F(z+\Delta z) - F(z) = \int_{z}^{z+\Delta z} [f(\xi) + f(z) - f(z)d\xi]$$

$$= f(z) \int_{z}^{z+\Delta z} d\xi + \int_{z}^{z+\Delta z} [f(\xi) - f(z)]d\xi.$$

The past formula can be rewritten as follows:

$$\frac{F(z+\Delta z) - F(z)}{\Delta z} = f(z) + \frac{1}{\Delta z} \int_{z}^{z+\Delta z} [f(\xi) - f(z)]d\xi.$$

Let $\max |f(\xi) - f(z)|$ be a maximum difference between the functions, then the past integral can be evaluated as

$$\frac{1}{\Delta z} \int_{z}^{z+\Delta z} [f(\xi) - f(z)]d\xi \leq \frac{1}{\Delta z} \max |f(\xi) - f(z)| \Delta z$$

$$= \max |f(\xi) - f(z)|.$$

Since $f(z)$ is a continuous function, then for $\Delta z \to 0$ follows $\max |f(\xi) - f(z)| \to f(z)$, and we finally obtain

$$\lim_{\Delta z \to 0} \frac{F(z+\Delta z) - F(z)}{\Delta z} = F'(z) = f(z). \qquad (12.27)$$

The function $F(z)$ possesses a continuous derivative at all points within the region. Consequently, $F(z)$ qualifies as an analytic function, and such a function is referred to as the anti-derivative of the function $f(z)$. It is evident that adding a constant to the anti-derivative preserves its status as an anti-derivative, i.e.,

$$\int_{z_0}^{z} f(\xi)d\xi = F(z) + C,$$

where C is a constant. We can define the constant by setting the upper limit of integration as z_0, then

$$\int_{z_0}^{z_0} f(\xi)d\xi = 0 = F(z_0) + C \quad \text{or} \quad C = -F(z_0).$$

Just as in the case of a function of a real variable, the second fundamental theorem of calculus holds, i.e.,

$$\int_{z_0}^{z_1} f(\xi)d\xi = F(z_1) - F(z_0). \tag{12.28}$$

It states that the integral of a continuous analytic function $f(z)$ over a path is equal to the difference of the values of an anti-derivative of the function at the end points of the path.

12.3.3. *Cauchy integral*

Let us consider an analytic function $f(z)$ in simply connected region. Now, we create a new function

$$g(z) = \frac{f(z)}{z - a}, \tag{12.29}$$

where a is an arbitrary point in the region where $f(z)$ is analytic.

The new function $g(z)$ is analytic everywhere in the region except point a. Next, we eliminate this point a by drawing a small circle c_ε of radius ρ around a, as in Figure 12.6. Now, the function $g(z)$ is regular everywhere in the ring between contours C and c_ε. According to Cauchy's theorem (12.25), we can write

$$\int_{C^+} \frac{f(z)}{z - a} dz + \int_{c_\varepsilon^-} \frac{f(z)}{z - a} dz = 0.$$

By changing the direction of integration in the second integral, we obtain

$$\int_{C^+} \frac{f(z)}{z - a} dz = \int_{c_\varepsilon^+} \frac{f(z)}{z - a} dz.$$

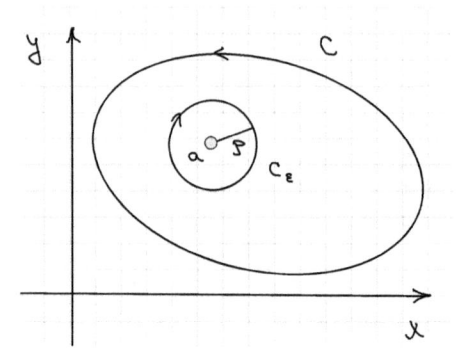

Fig. 12.6. Various paths of integration.

In the integral on the right, we set $f(z) = f(a) + f(z) - f(a)$. Then,

$$\int_{C^+} \frac{f(z)}{z-a} dz = f(a) \int_{c_\varepsilon^+} \frac{dz}{z-a} + \int_{c_\varepsilon^+} \frac{f(z) - f(a)}{z-a} dz.$$

In Example 12.5, we have already evaluated the integral

$$\int_C \frac{dz}{z-z_0} = 2\pi i.$$

Therefore, we can write

$$\int_{C^+} \frac{f(z)}{z-a} dz = 2\pi i f(a) + \int_{c_\varepsilon^+} \frac{f(z) - f(a)}{z-a} dz.$$

Next, we show that the second term on the right tends to zero when the radius $\rho \to 0$. Indeed,

$$\left| \int_{c_\varepsilon^+} \frac{f(z) - f(a)}{z-a} dz \right| \leq \frac{\max |f(z) - f(a)|}{\rho} 2\pi\rho = 2\pi \max |f(z) - f(a)|.$$

As $\rho \to 0$, z tends to a and $\max |f(z) - f(a)| \to 0$. Thus, we obtain Cauchy's integral formula

$$f(a) = \frac{1}{2\pi i} \int_C \frac{f(z)}{z-a} dz. \tag{12.30}$$

Very often this formula is written in the form

$$f(z) = \frac{1}{2\pi i} \int_C \frac{f(z')}{z'-z} dz', \tag{12.31}$$

where z' is an integration variable such that $z - z' \neq 0$.

Let the contour C be a circle of radius r centered at a. Then, using the polar coordinates $z = a + re^{i\varphi}$, we obtain

$$f(a) = \frac{1}{2\pi i} \int_C \frac{f(z)}{z-a} dz = \frac{1}{2\pi} \int_0^{2\pi} f(a + re^{i\varphi}) d\varphi = \frac{1}{2\pi r} \int_{C_r} f(z) dz.$$

This formula expresses the value of the analytic function at the center of the circle as the average of its boundary values.

Cauchy's integral (or Cauchy's formula) establishes a connection between the values of an analytic function inside a closed contour and the values of the function on the contour itself.[5] It has many applications. Later in this chapter, we study some of these.

In Cauchy's formula, the integral involves a continuous function that can be differentiated with respect to z under the integral sign as many times as need. By successively differentiating, we obtain

$$f'(z) = \frac{1}{2\pi i} \int_C \frac{f(z')}{(z'-z)^2} dz', \qquad (12.32)$$

$$f''(z) = \frac{2}{2\pi i} \int_C \frac{f(z')}{(z'-z)^3} dz', \qquad (12.33)$$

and generally

$$f^{(n)}(z) = \frac{n!}{2\pi i} \int_C \frac{f(z')}{(z'-z)^{n+1}} dz'. \qquad (12.34)$$

Here we want to note about two implications of Cauchy's formula:

(1) The modulus of an analytic function at any interior point of the region does not exceed the maximum modulus on the contour, i.e., $|f(z)| \leq M$, where M is the maximum value $|f(z')|$ on the contour.
(2) *Liouville's theorem*: If the maximum modulus of an analytic function is achieved at an interior point of its region, then the function must be constant throughout the entire region. In connection with Liouville's theorem, we note that the function $f(z) = \cos z$ is regular or analytic in the entire plane. It remains bounded for real $z = x$, but, for example, for purely imaginary values $z = iy$,

---

[5] Cauchy's formula is also directly applicable to a multiply connected domain.

it is unbounded since $z(iy) = (e^{-y} + e^y)/2$. The same behavior holds for $\sin(z)$. In this way, trigonometric functions of a complex variable differ significantly from the corresponding functions of a real variable.

12.4. Series

Let there be an infinite series with complex terms

$$S = \sum_{k=1}^{\infty} z_k = (x_1 + iy_1) + (x_2 + iy_2) + \cdots + (x_n + iy_n) + \cdots . \quad (12.35)$$

By separating real and imaginary terms, we can write

$$S = (x_1 + x_2 + \cdots + x_n + \cdots) + i(y_1 + y_2 + \cdots + y_n + \cdots).$$

The series is called convergent if the sum of its first n terms

$$S_n = (x_1 + x_2 + \cdots + x_n) + i(y_1 + y_2 + \cdots + y_n) \quad (12.36)$$

is convergent as $n \to \infty$. Then, it immediately follows that series (12.35) is convergent if and only if series with real terms $(x_1 + x_2 + \cdots + x_n + \cdots)$ and $(y_1 + y_2 + \cdots + y_n + \cdots)$ are convergent. Series (12.35) are called *absolutely convergent* series made from the modulus of complex terms

$$S_A = \sum_{k=1}^{\infty} z_k = |z_1| + |z_2| + \cdots + |z_n| + \cdots \quad (12.37)$$

is convergent. Given this, we can apply the necessary condition for convergence with respect to the absolute values of complex series $|z_n|$, along with the appropriate convergence tests (Cauchy's, d'Alembert's, and Raabe's tests) applicable to series with real terms, as discussed in Chapter 1. When needed, we apply the convergence tests separately to the series with x_n and y_n terms.

Next, we consider series where terms are functions of complex variables

$$f(z) = \sum_{k=1}^{\infty} f_k(z) = f_1(z) + f_2(z) + \cdots + f_n(z) + \cdots . \quad (12.38)$$

The Weierstrass theorem states that if the terms of the series (12.38) are analytic functions within a closed region with the contour C, and this series converges uniformly[6] on the contour C, then it converges uniformly across the entire closed region. Its sum is an analytic function inside the region, and the series can be differentiated term by term as many times as needed.

12.4.1. Power and Taylor series

Power series: An important example of a series of functions is power series

$$\sum_{k=0}^{\infty} c_k (z-a)^n, \qquad (12.39)$$

where a is a complex number. The terms of the power series are analytic functions.

Abel's theorem states that if series (12.39) converges at some point $z = b$, then it converges absolutely at any point satisfying the condition

$$|z - a| < |b - a|.$$

Moreover, it converges uniformly within a circle of radius r less than $|b - a|$. Abel's theorem has several implications for power series:

(1) If the series (12.39) is divergent at a point b, then it diverges at all points satisfying the inequality

$$|z - a| > |b - a|.$$

(2) For any power series, there exists a number R, such that within the circle $|z - a| < R$, the given power series converges, and outside this circle, it diverges.
(3) Within the circle of convergence, the power series converges to an analytic function.
(4) Inside the circle of convergence, a power series can be integrated and differentiated term by term any number of times, and the radius of convergence of the resulting series is equal to the radius

---

[6] A series is said to converge uniformly to $f(x)$ if, for any $\varepsilon > 0$, there exist an N (which may depend on $\varepsilon > 0$) such that for all $n > N$ the following holds: $|f(z) - \sum_{k=1}^{n} f_k(x)| < \varepsilon$.

of convergence of the original series. This property is a direct consequence of the theorems of Abel and Weierstrass.

(5) The radius of convergence R of the power series (12.39) is defined by

$$R = \frac{1}{\lim_{k \to \infty} \sqrt[k]{|c_k|}},$$

if this limit exists. This formula is also called Cauchy–Hadamard formula.

Taylor series: Thus, the power series within the circle of convergence defines an analytic function. We can ask a question: Can a function that is analytic within a specific circle be represented by a power series that converges to this function within that circle? Let us take an arbitrary point z inside the circle $|z - a| < R$. Then, draw a circle C_ρ with the center at a and a radius ρ such that $\rho < R$, with the point Z inside C_ρ, as shown in Figure 12.7. By applying the Cauchy integral formula (12.30), we can express $f(z)$ as integral over C_ρ:

$$f(z) = \frac{1}{2\pi i} \int_{C_\rho} \frac{f(z')}{z' - z} dz. \tag{12.40}$$

Since $|z - a| < |z' - a|$, then we can use the formula for the sum of terms of an infinitely decreasing geometric progression:

$$\frac{1}{z' - z} = \frac{1}{(z' - a)} \frac{1}{1 - \frac{z-a}{z'-a}} = \frac{1}{(z' - a)} \sum_{k=0}^{\infty} \frac{(z - a)^k}{(z' - a)^k}. \tag{12.41}$$

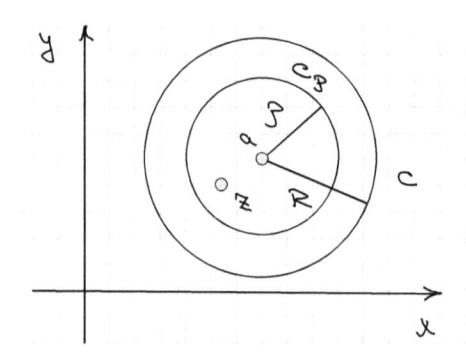

Fig. 12.7. Integration for Taylor series.

By substituting (12.41) into (12.40), and integrating term by term, we obtain

$$f(z) = \sum_{k=0}^{\infty}(z-a)^k \frac{1}{2\pi i}\int_{C_\rho}\frac{f(z')dz}{(z'-a)^{k+1}} \qquad (12.42)$$

or as

$$f(z) = \sum_{k=0}^{\infty} c_k(z-a)^k, \qquad (12.43)$$

where

$$c_k = \frac{1}{2\pi i}\int_{C_\rho}\frac{f(z')dz'}{(z'-a)^{k+1}} = \frac{f^{(k)}(a)}{k!}. \qquad (12.44)$$

The function $f(z)$, which is analytic inside the circle $|z-a| < R$, can be expanded within this circle into a convergent power series commonly known as Taylor expansion. The series (12.43) is itself referred to as the Taylor series for $f(z)$.

Example 12.6. Find Taylor series for $f(z) = e^z$.
Indeed, since $f^{(n)}(z) = e^z$, then $f^{(n)}(0) = 1$, and from (12.44) follows

$$e^z = 1 + \frac{z}{1!} + \frac{z^2}{2!} + \frac{z^3}{3!} + \cdots .$$

The function $f(z) = e^z$ is analytic in the entire z plane, therefore the Taylor expansion is valid for the entire plane. By applying the same approach, we can derive Taylor expansions for $\sin z$ and $\cos z$ as

$$\sin z = \frac{z}{1!} - \frac{z^3}{3!} + \frac{z^5}{5!} - \cdots ,$$

$$\cos z = 1 - \frac{z^2}{2!} + \frac{4^3}{4!} - \cdots . \qquad \blacksquare$$

12.4.2. The Laurent series

Now, we consider a general power series written as

$$\sum_{k=-\infty}^{\infty} c_k(z-a)^k, \qquad (12.45)$$

where a is a point in complex plane and c_k are complex numbers. Such series is called Laurent series. Let us first determine the convergence

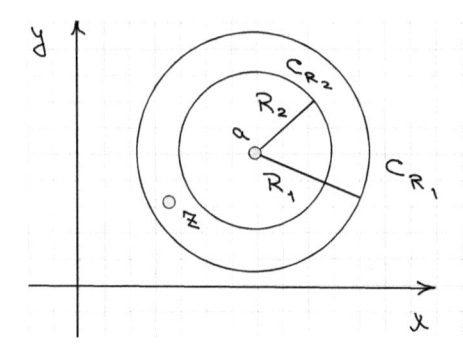

Fig. 12.8. Laurent series.

of such series. Note that the index k now extends from negative infinity to positive infinity instead of starting from zero.

The Laurent series consists of two series:

$$\sum_{k=-\infty}^{\infty} c_k(z-a)^k = \sum_{k=0}^{\infty} c_k(z-a)^k + \sum_{k=1}^{\infty} \frac{c_{-k}}{(z-a)^k}. \tag{12.46}$$

We need to determine a region, as in Figure 12.8, where both series converge. The first series is a regular power series that we considered before. The region of convergence for such series is a circle centered at a with a radius R_1. Within the circle, the series converges to some analytic function:

$$f_1(z) = \sum_{k=0}^{\infty} c_k(z-a)^k, \quad |z-a| < R_1.$$

For the second series, we introduce a new variable $z' = 1/(z-a)$. Then, this series looks like an ordinary power series:

$$c_{-1}z' + c_{-2}z'^2 + \cdots.$$

Assume that this series converges to some function, its radius of convergence is $|z'| < 1/R_2$ or $1/|z'| > R_2$. Returning to the z variable, we obtain the radius of converges as $|z - a| > R_2$, and the series converges to some function

$$f_2(z) = \sum_{k=0}^{\infty} \frac{c_k}{(z-a)^k}, \quad |z-a| > R_2.$$

Thus, the region of convergence of the entire series (12.46) is determined by two inequalities:

$$|z - a| < R_1, \quad |z - a| > R_2. \tag{12.47}$$

If $R_1 > R_2$, then there is a common region of convergence of these series, namely a circular ring $R_2 < |z - a| < R_1$ in which series (12.45) converges to the analytic function

$$f(z) = f_1(z) + f_2(z) = \sum_{k=-\infty}^{\infty} c_k(z-a)^k, \quad R_2 < |z-a| < R_1. \tag{12.48}$$

If $R_1 < R_2$, then two series do not have a common region of convergence. Thus, in this case, the series (12.45) converges nowhere to any function.

In exactly the same way as for power series, for an analytic function $f(z)$, we want to find its representation in Laurent series within a circular ring. For a point inside the ring, we can apply the Cauchy integral formula as

$$f(z) = \frac{1}{2\pi i} \int_{C_{R_1}} \frac{f(z')}{z' - z} dz + \frac{1}{2\pi i} \int_{C_{R_2}^-} \frac{f(z')}{z' - z} dz.$$

For integration along C_{R_1}, we have $|z - a|/|z' - a| < 1$, and just as in the derivation of Taylor's formula, we obtain

$$f_1(z) = \sum_{k=0}^{\infty} c_k(z - a)^k, \tag{12.49}$$

where

$$c_k = \frac{1}{2\pi i} \int_{C_{R_1}} \frac{f(z') dz'}{(z' - a)^{k+1}}, \quad k \geq 0. \tag{12.50}$$

Since for C_{R_2} we have the opposite, namely $|z' - a|/|z - a| < 1$, then after few steps (do you see how?), we obtain

$$f_2(z) = \sum_{k=0}^{\infty} \frac{c_{-k}}{(z - a)^k}, \tag{12.51}$$

where

$$c_{-k} = \frac{1}{2\pi i} \int_{C_{R_2}} \frac{f(z')}{(z'-z)^{-n+1}}, \quad n > 0. \qquad (12.52)$$

Combining both terms (12.51) and (12.52) together, we obtain, for the function $f(z)$ inside the $R_2 < |z - a| < R_1$, a representation as the Laurent series

$$f(z) = \sum_{k=-\infty}^{\infty} c_k (z-a)^k,$$

$$c_k = \frac{1}{2\pi i} \int_C \frac{f(z')}{(z'-z)^{n+1}}, \quad n = 0, \pm 1, \pm 2, \ldots, \qquad (12.53)$$

where C is an arbitrary closed contour located within the ring enclosing the point z.

Example 12.7. Expand in a Laurent series the function

$$f(z) = \frac{1}{z(z-1)(z-2)}.$$

The function has three isolated simple poles located at $z = 0$, $z = 1$, and $z = 2$. The function is analytic and single-valued in the rest of the plane. Let us consider three circular rings centered at the origin: R_1 with $0 < |z| < 1$, R_2 with $1 < |z| < 2$, and R_3 with $2 < |z| < +\infty$. In the each of the rings, we can expand the function into Laurent series. While we can proceed with employing the contour integration (12.53) for calculating the expansion coefficients c_k, in this example, we demonstrate another method based on the expansion of $f(z)$ either in powers z^k or z^{-k} or both. We start by representing the function in partial fraction form:

$$f(z) = \frac{1}{z(z-1)(z-2)} = \frac{A}{z} + \frac{B}{z-1} + \frac{C}{z-2}.$$

By multiplying both sides by the common denominator $z(z-1)(z-2)$ to clear fractions, we obtain

$$1 = A(z-1)(z-2) + Bz(z-2) + Cz(z-1).$$

Now, we substitute convenient values of z to find the constants A, B, and C. This is sometimes called the Heaviside "cover-up" method for obvious reasons! We choose $z = 0$, $z = 1$, and $z = 2$ for simplicity.

Then, we obtain the following: for $z = 0$, $A = \frac{1}{2}$, for $z = 1$, $B = -1$, and for $z = 2$, $C = \frac{1}{2}$. With these values, we can express $f(z)$ in partial fraction form:

$$f(z) = \frac{1}{z(z-1)(z-2)} = \frac{1/2}{z} - \frac{1}{z-1} + \frac{1/2}{z-2},$$

(1) **Ring R_1:** By expanding the second and the third terms into Taylor series near $z = 0$ as $-1/(z-1) = 1 + z + z^2 + z^3 + z^4 + \cdots$ and $1/(2(z-2)) = -1/4 - z/8 - z^2/16 - z^3/32 - \cdots$, we obtain

$$f(z) = \frac{1}{2z} + \frac{3}{4} + \frac{7}{8}z + \frac{15}{16}z^2 + \frac{31}{32}z^3 + \cdots.$$

This is the Laurent series which is valid for $0 < |z| < 1$.

(2) **Ring R_2:** In this ring, we expand $1/(z-1)$ in powers of $1/z$ (since this term decreases as $|z|$ increases), and we expand $1/(z-2)$ in powers of z (since this term increases with $|z|$). Then,[7]

$$\frac{1}{z-1} = \frac{1}{z}\left(\frac{1}{1-\frac{1}{z}}\right) = \sum_{k=0}^{\infty} \frac{1}{z^{k+1}}$$

and

$$\frac{1}{z-2} = -\frac{1}{2}\left(\frac{1}{1-\frac{z}{2}}\right) = -\frac{1}{2}\sum_{k=0}^{\infty} \frac{z^k}{2^k}.$$

Finally, for the ring R_2, we obtain

$$f(z) = -\frac{1}{2z} - \sum_{k=2}^{\infty} \frac{1}{z^k} - \frac{1}{4}\sum_{k=0}^{\infty} \frac{z^k}{2^k}.$$

(3) **Ring R_3:** In this ring, we expand both fractions $1/(z-1)$ and $1/(z-2)$ in powers $1/z$. (Do you see why?) Then, as we

---

[7]Here we use the geometric series representation for the function $1/(1-x)$ as

$$1/(1-x) = 1 + x + x^2 + x^3 + \cdots = \sum_{k=0}^{\infty} x^k.$$

demonstrated above,

$$\frac{1}{z-1} = \sum_{k=0}^{\infty} \frac{1}{z^{k+1}},$$

and for the other fraction, we obtain

$$\frac{1}{z-2} = \frac{1}{z}\left(\frac{1}{1-\frac{2}{z}}\right) = \sum_{k=0}^{\infty} \frac{2^k}{z^{k+1}},$$

and collecting all terms together,

$$f(z) = \sum_{k=2}^{\infty} (2^{k-1} - 1)\frac{1}{z^{k+1}}.$$

Hence, we have derived three Laurent series representing the function $f(z)$ in three regions of the plane z. ∎

12.4.3. Zeros and isolated singular points

Up to this point, our focus has been on examining functions $f(z)$ that are single-valued and analytic. However, if the function $f(z)$ is analytic everywhere except for a single point, namely $z = a$, then such a function has an *isolated singularity* at $z = a$. At this point, the function $f(z)$ may even not be defined.

Let us study the behavior of the function $f(z)$ in the vicinity of $z = a$. Since the function $f(z)$ is analytic everywhere in the ring $0 < |z - a| < R$, it can be expanded into the Laurent series. There are three types of singularities when approaching $z \to a$:

(1) The resulting Laurent series does not contain terms with negative powers of $(z - a)$. The function $f(z)$ has a finite limit.
(2) The Laurent series contains a finite number of terms with negative powers $(z - a)$. The function $f(z)$ approaches infinity $f(z) \to \infty$ as $z \to a$.
(3) Neither the first nor the second case holds, i.e., the Laurent series contains an infinite number of terms.

Removable singular point: Let us examine into the first possibility. Expansion of $f(z)$ in vicinity of a without negative powers of

$(z-a)$ is equivalent to an expansion in a Taylor series (12.43), i.e.,

$$f(z) = \sum_{k=0}^{\infty} c_k (z-a)^k, \tag{12.54}$$

$$c_k = \frac{1}{2\pi i} \int_{C_\rho} \frac{f(z')dz'}{(z'-a)^{k+1}} = \frac{f^{(k)}(a)}{k!}. \tag{12.55}$$

In such a scenario, there exists a limit $f(a) = c_0$. If the function $f(z)$ was not defined at point a, we extend its definition by setting $f(a) = c_0$. In cases where the initially specified value $f(a)$ differs from c_0, we adjust the value of the function at the point to $f(a) = c_0$. The resulting function $f(z)$ defined in this manner is analytic everywhere inside the circle $|z-a| < R$. Consequently, the discontinuity of the function $f(z)$ at point $z = a$ is resolved. Therefore, an isolated singular point $z = a$ of the function $f(z)$ – one where the expansion of $f(z)$ in a Laurent series in the neighborhood of $z = a$ lacks terms with negative powers of $(z-a)$ – is called *removable singular point*.

Now, we consider a special case when $f(a) = 0$. In order for the function to be zero at $z = a$, the first coefficient c_0 in the expansion (12.54) must be zero. If, additionally, c_1 is not equal to zero, then the point $z = a$ is referred to as a *simple zero*. In other words, the function has a root at that point, and the derivative of the function is non-zero at that point, or mathematically, $f(a) = 0$ and $f'(a) \neq 0$. A zero is said to be of *order n* if the function $f(z)$ has a root at that point, and its derivatives up to the $(n-1)$th derivative are zero at that point, but the nth derivative is non-zero. Mathematically, if $f(a) = f'(a) = \cdots = f^{(n-1)}(a) = 0$ and $f^{(n)}(a) \neq 0$, then the point $z = a$ is a zero of order n. If the function and all its derivatives are zero at that point, then such a point is called the zero of infinite order.

The behavior of a function $f(z)$ near a zero of order k at $z = a$ can be represented by its Taylor series expansion:

$$f(z) = (z-a)^k g(z),$$

where $g(z)$ is analytic and non-zero in the vicinity of a. Thus, we can evaluate the order of a zero by evaluating

$$\lim_{z \to a} \frac{f(z)}{(z-a)^k} \quad k = 1, 2, \ldots. \tag{12.56}$$

The order of the zero n is determined by the lowest value of $n = k$ for which the limit is non-zero.

Poles at isolated singularities: Let us now turn our attention to the second and the third cases. We start with the Laurent series that contains a finite number m of terms with negative powers of $(z - a)$ of the function $f(z)$ in the vicinity of the point $z = a$, i.e.,

$$f(z) = \sum_{k=-m}^{\infty} c_k (z-a)^k = \frac{c_{-m}}{(z-a)^m} + \frac{c_{-m+1}}{(z-a)^{m-1}} \\ + \cdots \frac{c_{-2}}{(z-a)^2} + \frac{c_{-1}}{z-a} + c_0 + c_1(z-a) + \cdots, \tag{12.57}$$

then the point a is called a *pole of order m* of the function $f(z)$. The coefficient c_{-1} standing at $(z-a)^{-1}$ is called the *residue* of the function $f(z)$ at the pole $z = a$, and it is denoted by $\mathrm{Res}(f, a)$.

Note that formula (12.57) can be rewritten as

$$f(z) = \frac{1}{(z-a)^m} \varphi(z), \tag{12.58}$$

where $\varphi(z)$ is an analytic function with $\varphi(a) \neq 0$. In other words, if there exists a positive integer m such that the limit

$$\lim_{z \to a} (z-a)^m f(z)$$

exists and is non-zero, while $(z-a)^{m-1} f(z)$ approaches zero as z approaches a, then a is called a pole of order m for the function $f(z)$. For example, $f(z) = 1/(z-4)^2$ has a singularity at $z = 4$, that is a pole of order 2 because $\lim_{z \to 4}(z-4)^2 f(z)$ is non-zero. If $m = 1$, the pole is called a simple pole. At a simple pole, the function $f(z)$ behaves like $\frac{1}{z-a}$ near a.

Poles are essential in the study of complex analysis, especially in the context of contour integration, residue theory, and the evaluation of complex integrals. The order of a pole influences the behavior of the function near that singularity.

Example 12.8. For the function

$$f(z) = \frac{\cos z - 1}{z^3}$$

at $z = 0$, find if it is regular, or has or a pole, and if a pole of what order it is. By expanding $\cos z - 1$ in Taylor series in the vicinity of

$z = 0$, we obtain for the function

$$f(z) \approx \frac{1 - z^2/2 + z^4/24 - \cdots - 1}{z^3} \approx -\frac{z^2}{2z^3} \approx -\frac{1}{z}.$$

Thus, the function has a simple pole at $z = 0$. ∎

Essential singularity: If the Laurent series has infinite number of terms with negative powers of $(z-a)$, then such type of singularity is called *essential singularity*. Let us consider the function $f(z) = e^{1/z}$ around its singularity at $z = 0$. The Laurent series for this function can be written as

$$e^{1/z} = 1 + \frac{1}{z} + \frac{1}{2!z^2} + \frac{1}{3!z^3} + \cdots + \frac{1}{k!z^k} + \cdots.$$

Thus, the Laurent series expansion of $e^{1/z}$ around $z = 0$ contains an infinite number of terms with negative powers of z. This characteristic property is a clear indication of an essential singularity at $z = 0$. Next, we analyze $f(z) = e^{1/z}$, when z approaches zero from different directions.

If z approaches zero along the positive x-axis, then

$$\lim_{x \to 0^+} e^{\frac{1}{x}} \to \infty.$$

If z approaches zero along the negative x-axis, then

$$\lim_{x \to 0^-} e^{\frac{1}{x}} = \lim_{x \to 0^+} e^{-\frac{1}{x}} \to 0.$$

If z approaches zero along the imaginary y-axis, then

$$\lim_{y \to 0} e^{\frac{1}{iy}} = \lim_{y \to 0} e^{-\frac{i}{y}} = \lim_{y \to 0} \left[\cos\left(\frac{1}{y}\right) - i\sin\left(\frac{1}{y}\right)\right],$$

and the function $f(z) = e^{1/z}$ oscillates without reaching any limit. This result is a consequence of *Picard's theorem*, which asserts that within any neighborhood surrounding an essential singular point, a function will take every finite value with one possible exception, an infinite number of times.

12.5. The residue theorem and its applications

Let the point a be an isolated singular point of an analytic function $f(z)$. Based on previous considerations, in the vicinity of this point, the function $f(z)$ can be expanded in a Laurent series (12.53). Now, let's integrate the function $f(z)$ over a small closed contour C around the point a, where the Laurent series is uniformly convergent:

$$\int_C f(z)dz = \int_C \sum_{k=-\infty}^{\infty} c_k(z-a)^k dz = \sum_{k=-\infty}^{\infty} c_k \int_C (z-a)^k dz. \tag{12.59}$$

We are interested in evaluating the integral

$$\int_C (z-a)^k dz$$

for integer values of k with $-\infty \leq k \leq \infty$. We set the path of integration C as circle of radius ρ centered at point $z = a$, and the direction of integration is counterclockwise. By using the polar form as $z = a + \rho e^{i\varphi}$ with $dz = i\rho e^{i\varphi} d\varphi$ and $0 \leq \varphi \leq 2\pi$, we obtain

$$\int_C (z-a)^k dz = i\rho^{k+1} \int_0^{2\pi} e^{i(k+1)\varphi} d\varphi.$$

By direct integration, we can see that the integral is equal to zero for all k but $k = -1$ (do you see why?). For $k = -1$, we obtain

$$i \int_0^{2\pi} d\varphi = 2\pi i.$$

Thus, for the integral (12.59), we obtain

$$\int_C f(z)dz = c_{-1} 2\pi i. \tag{12.60}$$

The complex coefficient c_{-1} is called the *residue* of $f(z)$ at a and denoted as

$$\text{Res}[f(z), z = a] \equiv \text{Res}(f, a) = c_{-1} = \frac{1}{2\pi i} \int_C f(z)dz. \tag{12.61}$$

It is evident that if the point a is a regular point, or a removable singular point of the function $f(z)$, then the residue of the function at this point is equal to zero.

Functions of Complex Variables

The *residue theorem* states that if a function $f(z)$ is analytic everywhere in a closed domain except for a finite number of isolated singular points $a_k (k = 1, \ldots, N)$, then the integral along a contour located inside the domain and surrounding all $z_k = a_k$ is equal to the sum of residues at the indicated singular points multiplied by $2\pi i$ as follows:

$$\int_{C^+} f(z)dz = 2\pi i \sum_{k=1}^{N} \text{Res}(f, a_k), \qquad (12.62)$$

where the contour C^+ is traversed in the positive direction. The residue theorem is a powerful tool to evaluate integrals of complex functions around closed contours when we know how to evaluate the residue $\text{Res}(f, a) = c_{-1}$.

12.5.1. *Evaluating the residue*

While the residue at isolated points can be calculated by direct integration (12.61), there are simpler methods for evaluating the residue for different types of singularities:

(1) Simple poles (Poles of order one)
Let point a be a first-order pole of the function $f(z)$. Then, in the vicinity of this point, from the Laurent expansion, we have

$$f(z) = c_{-1}(z-a)^{-1} + c_0 + c_1(z-a) + c_2(z-a)^2 + \cdots.$$

Multiplying both sides of the above equation by $(z-a)$ and taking the limit $z \to a$, we obtain

$$\text{Res}(f, a) = c_{-1} = \lim_{z \to a} (z-a) f(z). \qquad (12.63)$$

We can note that in this case, the function $f(z)$ can also be represented as a ratio of two analytic functions:

$$f(z) = \frac{g(z)}{h(z)}, \qquad (12.64)$$

where $g(z) \neq 0$ and $h(z)$ has a pole of the first order (or simple zero) at $z = a$, i.e., we can write

$$h(z) = (z-a)h'(a) + \frac{1}{2}(z-a)^2 h''(a) + \cdots,$$

with $h'(a) \neq 0$. Then, from (12.63), we obtain

$$\text{Res}(f, a) = c_{-1} = \frac{g(a)}{h'(a)}. \tag{12.65}$$

(2) Higher-order poles

Let the point a be a pole of order n of the function $f(z)$. Then, in the vicinity of this point, the Laurent expansion takes the form

$$f(z) = c_{-n}(z-a)^{-n} + \cdots + c_{-1}(z-a)^{-1} + c_0 + c_1(z-a) + \cdots.$$

By multiplying both sides of the above expansion on $(z-a)^n$, we obtain

$$(z-a)^n f(z) = c_{-n} + c_{-n+1}(z-a) + \cdots c_{-1}(z-a)^{n-1} + \cdots.$$

By differentiating both sides $(n-1)$ times, and taking the limit $z \to a$, we obtain

$$\text{Res}(f, a) = c_{-1} = \frac{1}{(n-1)!} \lim_{z \to a} \frac{d^{n-1}}{dz^{n-1}}[(z-a)^n f(z)]. \tag{12.66}$$

(3) Removable singularities

If at $z = a$ the function $f(z)$ has a removable singularity, then the residue is zero. For example, $f(z) = \sin(z)/z$ has removable singularity at $z = 0$, and the residue at this point is zero.

(4) Essential singularities

If a is an essential singularity, evaluating the residue might involve more complex techniques, and in some cases, the residue may not be easy to calculate at all. However, if a function can easily be represented by Laurent series, then the answer is straightforward as the coefficient at $(z-a)^{-1}$ term. For example, $f(z) = e^{1/z}$ has $\text{Res}(e^{1/z}, z = 0) = 1$. (Do you see why?)

Example 12.9. Find the residues of the function

$$f(z) = \frac{1}{z^3 + 2z^2 - 3z}$$

at the indicated points: $z = 0$, $z = -1$, and $z = 1$. The original function can be rewritten as

$$f(z) = \frac{1}{z(z-1)(z+3)}$$

to make it easier to find the residues. Then, the residue at $z = 0$

$$\text{Res}\left(\frac{1}{z(z-1)(z+3)}, z = 0\right) = \lim_{z \to 0} \frac{z}{z(z-1)(z+3)} = -\frac{1}{3}.$$

In the same way, we can evaluate residues at the other two points: The residue at $z = 1$ is $1/4$, and the residue at $z = -3$ is $1/12$. ∎

Example 12.10. Consider the function

$$f(z) = \frac{\sin(z)}{z^2 + 4z + 4}$$

and find the residues at the poles of the function. First, we need to find the poles and their order. We can factorize the denominator as $z^2 + 4z + 4 = (z+2)^2$. It means that the function $f(z)$ has a second-order pole at $z = -2$. By using (12.66), we obtain for $n = 2$

$$\text{Res}\left(\frac{1}{(z+2)^2}, z = -2\right) = \lim_{z \to -2} \frac{d}{dz}\left((z+2)^2 \frac{\sin(z)}{(z+2)^2}\right)$$

$$= \lim_{z \to -2} \frac{d}{dz} \sin(z) = \cos(-2). \blacksquare$$

12.5.2. Evaluating definite integrals by using residues

Evaluating definite integrals by using residues is a powerful technique, especially when dealing with real integrals that can be transformed into complex integrals. But first we need to discuss an analytic extension of a function $f(x)$ from the real axis x into a complex domain z. The process of analytic continuation involves several steps:

(1) Start with a function $f(x)$ defined on a the real axis.
(2) Write $f(x)$ in terms of complex variables, for example, by replacing x on z.

(3) Check for analyticity, i.e., ensure that the complex function $f(z)$ is differentiable in a neighborhood of the real axis.
(4) Extend the function to a larger domain in the complex plane while maintaining differentiability.

Example 12.11. Let's consider the analytic extension of the real function $f(x) = \sqrt{x}$, which is defined for $x \geq 0$. We'll extend it to a larger domain in the complex plane. We can express $f(x)$ as a complex function $f(z) = \sqrt{z}$ with z as a complex number. The function $\sqrt{z}$ is differentiable except at $z = 0$ (do you see why?) ∎

Now, we are ready to outline the method for evaluating definite integrals by applying the residue theorem:

(1) Choose a closed contour in the complex plane that includes the segment of the real axis corresponding to the definite integral that you want to evaluate. The contour should enclose all singularities of the integrand.
(2) Identify all singularities (poles) of the integrand within the contour, and classify them as simple poles, higher-order poles, or essential singularities.
(3) Calculate the residues at each of the singularities inside the contour.
(4) Apply the residue theorem (12.62).
(5) If the contour contains parts where the integrand is not defined or has different behavior, evaluate those contributions separately. Often, these contributions will go to zero as the radius of the contour becomes large.
(6) Combine the contributions from the residues and the contour to obtain the value of the integral.

Here we consider a couple of types of definite integrals than can be evaluate by using the residue theorem:

(1) Integrals of $\int_0^{2\pi} F(\sin\theta, \cos\theta) d\varphi$ type: In this integral, the integrand is a function of $\sin\varphi$ and $\cos\theta$. First, we change the variable of integration θ by introducing a complex variable z as

$$z = e^{i\theta} \quad \text{with} \quad d\theta = \frac{1}{i}\frac{dz}{z}. \tag{12.67}$$

Then,

$$\sin\theta = \frac{1}{2i}(e^{i\theta} - e^{-i\theta}) = \frac{1}{2i}\left(z - \frac{1}{z}\right),$$
$$\cos\theta = \frac{1}{2}(e^{i\theta} + e^{-i\theta}) = \frac{1}{2}\left(z + \frac{1}{z}\right).$$
(12.68)

When θ changes from 0 to 2π, the complex variable z runs through a closed contour, as a circle of radius one, in the positive direction. Thus, the integral of $\int_0^{2\pi} F(\sin\theta, \cos\theta)d\theta$ type turns into an integral of a function of a complex variable:

$$\frac{1}{i}\int_{|z|=1} F\left(z - \frac{1}{z}, z + \frac{1}{z}\right)\frac{dz}{z}.$$
(12.69)

Example 12.12. Evaluate the following definite integral:

$$I = \int_0^{2\pi} \frac{d\theta}{5 + 3\cos\theta}.$$

By using (12.67) and (12.68), we obtain

$$I = \frac{1}{i}\int_{|z|=1} \frac{1}{5 + \frac{3}{2}\left(z + \frac{1}{z}\right)}\frac{dz}{z} = \frac{1}{i}\int_{|z|=1} \frac{dz}{\frac{3}{2}z^2 + 5z + \frac{3}{2}}$$
$$= \frac{1}{i}\int_{|z|=1} \frac{dz}{\frac{1}{2}(3z + 1)(z + 3)}.$$

This integral has two simple poles: one at $z = -\frac{1}{3}$ and one at $z = -3$. Only the first pole is located inside the contour on integration. The residue of

$$\frac{1}{\frac{1}{2}(3z + 1)(z + 3)}$$

at $z = -\frac{1}{3}$ is

$$\operatorname{Res}\left(f, -\frac{1}{3}\right) = \lim_{z \to -\frac{1}{3}}\left(z + \frac{1}{3}\right)\frac{1}{\frac{1}{2}(3z + 1)(z + 3)} = \frac{1}{4}.$$

Then, from the residue theorem follows

$$I = 2\pi i \frac{1}{i}\operatorname{Res}\left(f, -\frac{1}{3}\right) = \frac{\pi}{2}. \qquad\blacksquare$$

(2) Integrals of $\int_{-\infty}^{\infty} f(x)dx$ type: If the function $f(x)$ is defined everywhere on the real axis $-\infty < x < \infty$, it has no singularities on the real line $y = 0$, it can be analytically continued to $\text{Im}(z) \geq 0$, and $f(z) \to 0$ at least as fast as $1/|z|^2$ as $|z| \to \infty$, then the improper integral of the first kind can be expressed as

$$\int_{-\infty}^{\infty} f(x)dx = 2\pi i \sum_{k=1}^{N} \text{Res}(f, a_k), \qquad (12.70)$$

where a_k are isolated singularity points in the *upper* half-plane. If the function $f(x)$ can be analytically continued to the lower half-plane $\text{Im}(z) \leq 0$, then

$$\int_{-\infty}^{\infty} f(x)dx = -2\pi i \sum_{k=1}^{N} \text{Res}(f, a_k), \qquad (12.71)$$

where a_k are isolated singularity points in the *lower* half-plane.

Example 12.13. Evaluate the improper integral of the first kind

$$I = \int_{-\infty}^{\infty} \frac{1}{5x^2 + 4x + 1} dx.$$

By extending the function in the complex plane, we obtain

$$\frac{1}{5z^2 + 4z + 1}.$$

The denominator of this function has two roots:

$$z_1 = \frac{1}{5}(-2 + i) \quad \text{and} \quad z_2 = \frac{1}{5}(-2 - i).$$

Only the first root is located in the upper plane, i.e., $\text{Im}(z) \geq 0$. The residue at this point is

$$\text{Rez}\left(\frac{1}{5z^2 + 4z + 1}, \frac{1}{5}(-2 + i)\right) = -\frac{i}{2}.$$

Then, by applying (12.70), we obtain

$$I = 2\pi i \left(-\frac{i}{2}\right) = \pi. \qquad \blacksquare$$

(3) Integrals of $\int_{-\infty}^{\infty} e^{i\beta x} f(x) dx$ type: If the function $f(x)$ does not have isolated singularities on $-\infty < x < \infty$, and it can be analytically extended in the upper plane $\text{Im}(z) \geq 0$, and it converges uniformly to zero when $|z| \to \infty$, then for $\beta > 0$,

$$\int_{-\infty}^{\infty} e^{i\beta x} f(x) dx = 2\pi i \sum_{k=1}^{N} \text{Res}(e^{i\beta z} f, a_k), \qquad (12.72)$$

where a_k are isolated singularity points in the upper half-plane $\text{Im}(z) \geq 0$. If the function $f(x)$ can be analytically extended to the lower plane $\text{Im}(z) \leq 0$, then

$$\int_{-\infty}^{\infty} e^{i\beta x} f(x) dx = -2\pi i \sum_{k=1}^{N} \text{Res}(e^{i\beta z} f, a_k), \qquad (12.73)$$

where a_k are isolated singularity points in the lower half-plane $\text{Im}(z) \leq 0$.

Example 12.14. Evaluate the following integral:

$$I = \int_{-\infty}^{\infty} \frac{\cos(ax)}{x^2 + b^2} dx, \quad a > 0, \quad b > 0.$$

We can rewrite this integral as

$$I = \text{Re} \int_{-\infty}^{\infty} \frac{e^{iax}}{x^2 + b^2} dx.$$

By the extension in the complex plane, we obtain the function $f(z)$ as

$$f(z) = \frac{e^{iaz}}{z^2 + b^2}.$$

This function has one only isolated singular point in the upper half-plane at $z_1 = ib$. The residue at this point is

$$\text{Rez}\left(\frac{e^{iaz}}{z^2 + b^2}, ib\right) = \frac{e^{-ab}}{i2b},$$

and finally, we obtain

$$\int_{-\infty}^{\infty} \frac{\cos(ax)}{x^2 + b^2} dx = \frac{\pi}{2b} e^{-ab}. \qquad \blacksquare$$

So far, we have assumed that the function $f(x)$ does not have isolated singularities on the real axis x. However, with minor additional considerations, we can extend the above considerations on functions $f(z)$ with singularities on the real axis. As an example, we consider the following integral.

Example 12.15. Evaluate integral

$$I = \int_0^\infty \frac{\sin ax}{x} dx, \quad a > 0.$$

The integral can be rewritten as

$$I = \frac{1}{2} \text{Im} \int_{-\infty}^\infty \frac{e^{iax}}{x} dx.$$

The integral above exists only as a so-called *Principal Value* (PV) integral, namely

$$\text{PV} \int_{-\infty}^\infty \frac{e^{iax}}{x} dx = \lim_{\substack{\varepsilon \to 0 \\ R \to \infty}} \left[\int_{-R}^{-\varepsilon} \frac{e^{iax}}{x} dx + \int_{\varepsilon}^{R} \frac{e^{iax}}{x} dx \right].$$

Let us consider a contour C consistent of four parts: from $-R$ to $-\varepsilon$, a semicircle C_ε, then from ε to R, and finally a semicircle C_R, as illustrated in Figure 12.9. Since the function e^{iaz}/z is analytic inside the contour, then according to the Cauchy's integral theorem (12.23),

$$\int_C f(z) dz = \int_{-R}^{-\varepsilon} \frac{e^{iax}}{x} dx + \int_{\varepsilon}^{R} \frac{e^{iax}}{x} dx + \int_{C_\varepsilon^-} \frac{e^{iaz}}{z} dz + \int_{C_R^+} \frac{e^{iaz}}{z} dz = 0.$$

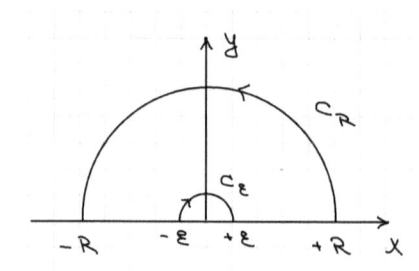

Fig. 12.9. Integration around an isolated singularity at $x = 0$.

We start with the last integral. Let us change the variable as $z = Re^{i\varphi}$. Since also $|z| = R$, we obtain

$$\left| \int_{C_R^+} \frac{e^{iaz}}{z} dz \right| \leq \frac{1}{R} R \int_0^\pi \left| e^{iaRe^{i\varphi}} \right| d\varphi = 2 \int_0^{\pi/2} e^{-aR\sin\varphi} d\varphi.$$

Next, we use the result[8]

$$\frac{\sin\varphi}{\varphi} \geq \frac{2}{\pi}.$$

Then,

$$2 \int_0^{\pi/2} e^{-aR\sin\varphi} d\varphi < 2 \int_0^{\pi/2} e^{-2aR\varphi/\pi} d\varphi = 2 \frac{\pi}{2aR} (1 - e^{-aR}) \to 0$$

as $R \to \infty$. Thus, the last integral tends to zero as $R \to \infty$. This is a special case of *Jordan's lemma* which states the following: Let $f(z)$ be an analytic function in the upper half-plane $\text{Im}\, z > 0$ except for a finite number of isolated singular points, and suppose it tends uniformly to zero as $|z| \to \infty$, then for $a > 0$,

$$\lim_{R \to \infty} \int_{C_R} e^{iaz} f(z)\, dz = 0,$$

where C_R is a semicircular contour in the upper half-plane defined by the arc of the circle $|z| = R$ from $-\pi$ to π.

For the third integral, we change the variable to $z = \varepsilon e^{i\varphi}$, and we integrate clockwise. So, the integral becomes

$$\int_{C_\varepsilon^-} \frac{e^{iaz}}{z} dz = i \int_\pi^0 e^{ia\varepsilon(\cos\varphi + i\sin\varphi)} d\varphi \to -i\pi \quad \text{as } \varepsilon \to 0.$$

Then, we obtain

$$\int_{-R}^{-\varepsilon} \frac{e^{iax}}{x} dx + \int_\varepsilon^R \frac{e^{iax}}{x} dx = -\int_{C_\varepsilon^+} \frac{e^{iaz}}{z} dz = i\pi.$$

Hence, for $R \to \infty$ and $\varepsilon \to 0$, the value of the integral is

$$I = \int_0^\infty \frac{\sin ax}{x} dx = \frac{1}{2} \text{Im} \int_{-\infty}^\infty \frac{e^{iax}}{x} dx = \frac{\pi}{2}. \qquad \blacksquare$$

---

[8]It is a good exercise to prove it.

Integrals from physics

The Fresnel integrals: These are useful in the study of optics and diffraction. They are defined as

$$C(x) = \int_0^x \cos t^2 dt; \quad S(x) = \int_0^x \sin t^2 dt.$$

But here we wish to evaluate the integrals

$$S(\infty) = \int_0^\infty \sin x^2 dx \quad \text{and} \quad C(\infty) = \int_0^\infty \cos x^2 dx$$

using wedge-shaped contour C with angle $\pi/4$ shown in the Figure 12.10. We choose the obvious integrand

$$f(z) = e^{iz^2},$$

and by Cauchy's theorem,

$$\int_C e^{iz^2} dz = 0,$$

or, because the arc AB is represented by $z = Re^{i\theta}$,

$$\int_0^R e^{ix^2} dx + \int_{AB} e^{iz^2} dz + \int_R^0 e^{iz^2} dz$$

$$= \int_0^R e^{ix^2} dx + \int_0^{\pi/4} e^{iR^2 e^{2i\theta}} iRe^{i\theta} d\theta + \int_R^0 e^{iz^2} dz = 0.$$

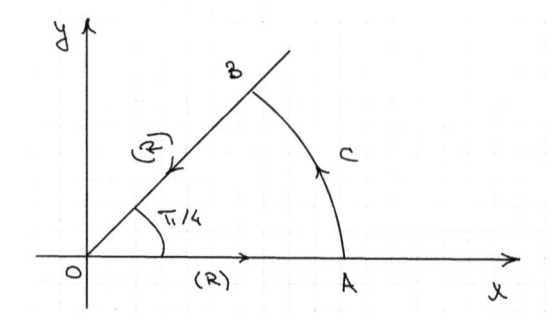

Fig. 12.10. Contour of integration.

How should we represent the third integral? The equation of the radius BO is

$$z = Re^{i\pi/4} = \frac{r}{\sqrt{2}}(1+i), \quad R \geq r \geq 0,$$

so

$$e^{iz^2} = e^{-r^2},$$

and therefore

$$\int_0^R e^{ix^2}\,dx \equiv \int_0^R \left(\cos x^2 + i\sin x^2\right)dx$$

$$= e^{i\pi/4}\int_0^R e^{-r^2}\,dr - \int_0^{\pi/4} e^{iR^2\cos 2\theta - R^2\sin 2\theta}\,iRe^{i\theta}\,d\theta.$$

Now, consider the limit of this equation as $R \to \infty$:

$$\int_0^\infty \left(\cos x^2 + i\sin x^2\right)dx \equiv [C(\infty) + iS(\infty)]$$

$$= e^{i\pi/4}\int_0^\infty e^{-r^2}\,dr - \lim_{R\to\infty}\left[\int_0^{\pi/4} e^{iR^2\cos 2\theta - R^2\sin 2\theta}\,iRe^{i\theta}\,d\theta\right].$$

Now,

$$\int_0^\infty e^{-r^2}\,dr = \frac{\sqrt{\pi}}{2},$$

and for the last integral on the right,

$$I_{AB} \equiv \int_0^{\pi/4} e^{iR^2\cos 2\theta - R^2\sin 2\theta}\,iRe^{i\theta}\,d\theta$$

$$\leq \left|\int_0^{\pi/4} e^{iR^2\cos 2\theta - R^2\sin 2\theta}\,iRe^{i\theta}\,d\theta\right|$$

$$\leq \int_0^{\pi/4} Re^{-R^2\sin 2\theta}\,d\theta = \frac{R}{2}\int_0^{\pi/2} e^{-R^2\sin\phi}\,d\phi,$$

where $\phi = 2\theta$. Using the easily-established inequality

$$\sin\phi \geq \frac{2\phi}{\pi}, \quad 0 \leq \phi \leq \frac{\pi}{2},$$

this last integral is

$$\frac{R}{2}\int_0^{\pi/2} e^{-R^2 \sin\phi}d\phi \le \frac{R}{2}\int_0^{\pi/2} e^{-2R^2\phi/\pi}d\phi = \frac{\pi}{4R}\left(1 - e^{-R^2}\right).$$

Clearly,

$$\lim_{R\to\infty} \frac{\pi}{4R}\left(1 - e^{-R^2}\right) = 0,$$

so

$$C(\infty) + iS(\infty) = \frac{\sqrt{\pi}}{2}\left(\frac{1+i}{\sqrt{2}}\right),$$

and the final results are

$$\int_0^\infty \cos x^2 dx = \frac{1}{2}\sqrt{\frac{\pi}{2}} = \int_0^\infty \sin x^2 dx.$$

Figure 12.11 illustrates the behavior of both functions, $C(x)$ and $S(x)$. Recall from the fundamental theorem of calculus that

$$\frac{d}{dx}\int_c^x f(t)dt = f(x).$$

Therefore, $C'(x) = \cos x^2$ and so $C'(0) = 1$; similarly, $S'(x) = \sin x^2$ and $S'(0) = 0$. So *now* we know which graph is which!

The Airy integral: Evaluate the integral

$$\int_0^\infty \cos\left(\frac{\omega^3}{3}\right) d\omega$$

by considering

$$\oint_C e^{i\omega^3/3} d\omega$$

over the closed wedge-shaped contour shown in the Figure 12.12 using the result

$$\int_0^\infty e^{-r^3/3} dr = 3^{-2/3}\Gamma(1/3).$$

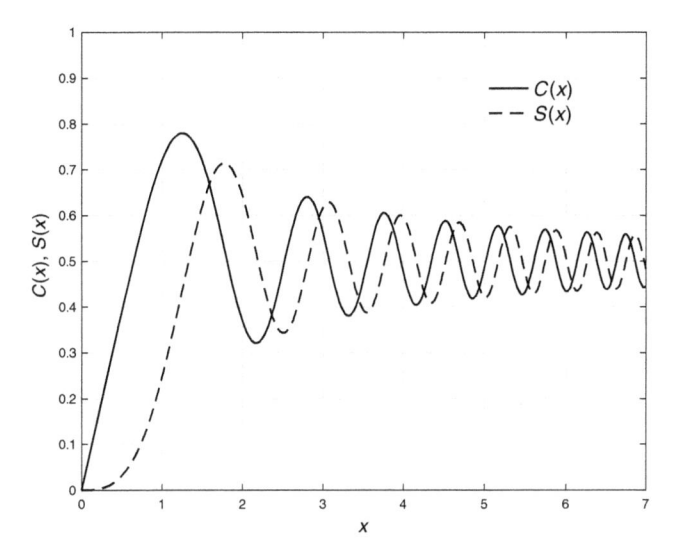

Fig. 12.11. $C(x)$ and $S(x)$ functions.

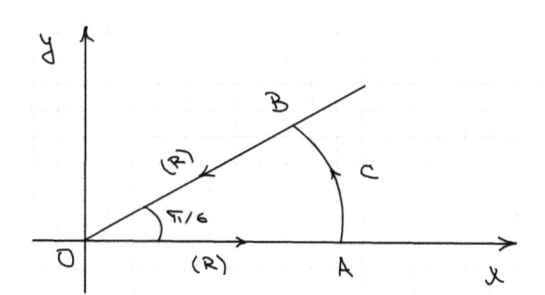

Fig. 12.12. Contour of integration.

By Cauchy's theorem,

$$\oint_C e^{iz^3/3}dz = 0 = \int_{OA} e^{iz^3/3}dz + \int_{AB} e^{iz^3/3}dz + \int_{BO} e^{iz^3/3}dz.$$

On OA, $z = x$, $0 \leq x \leq R$; on AB, $z = Re^{i\theta}$, $0 \leq \theta \leq \pi/6$; on BO, $z = re^{i\pi/6}$, $R \geq r \geq 0$. Therefore,

$$\int_0^R e^{ix^3/3}dx + \int_0^{\pi/6} e^{iR^3 e^{3i\theta}/3}\left(iRe^{i\theta}\right)d\theta + \int_R^0 e^{ir^3 e^{i\pi/2}/3}\left(e^{i\pi/6}\right)dr = 0,$$

i.e.,

$$\int_0^R \left(\cos x^3/3 + i \sin x^3/3\right) dx$$

$$= e^{i\pi/6} \int_0^R e^{-r^3/3} dr - \int_0^{\pi/6} e^{iR^3(\cos 3\theta + i \sin 3\theta)/3} \left(iRe^{i\theta}\right) d\theta.$$

Consider the limit as $R \to \infty$. The first integral on the right is equal to

$$3^{-2/3}\Gamma(1/3) \left(\sqrt{3}+i\right)/2.$$

For the second integral on the right,

$$\left|\int_0^{\pi/6} e^{iR^3(\cos 3\theta + i \sin 3\theta)/3} \left(iRe^{i\theta}\right) d\theta\right| \leq \int_0^{\pi/6} e^{-(R^3 \sin 3\theta)/3} R d\theta,$$

so in terms of the quantity $\phi = 3\theta$ and the inequality $\sin \phi \geq 2\phi/\pi$, we are able to write that

$$\int_0^{\pi/6} e^{-(R^3 \sin 3\theta)/3} R d\theta \leq \frac{R}{3} \int_0^{\pi/2} e^{-2R^3 \phi/3\pi} d\phi = \frac{\pi}{2R^2}\left[1 - e^{-(R^3/3)}\right],$$

and this tends to zero as $R \to \infty$. Then, we have found that

$$\int_0^\infty \left(\cos x^3/3 + i \sin x^3/3\right) dx = 3^{-2/3}\Gamma(1/3)\left(\frac{\sqrt{3}+i}{2}\right),$$

so on equating the real and imaginary parts of this expression, it follows that

$$\int_0^\infty \cos(x^3/3) dx = \frac{3^{-1/6}}{2}\Gamma\left(\frac{1}{3}\right)$$

and

$$\int_0^\infty \sin(x^3/3) dx = \frac{3^{-2/3}}{2}\Gamma\left(\frac{1}{3}\right).$$

12.6. Exercises and problems

(1) Write the following $f(z)$ functions in the form $f(z) = u(x,y) + iv(x,y)$, and using the Cauchy–Riemann conditions to find if they are analytic:

(a) $f(z) = e^{iz}$
(b) $f(z) = z^3$
(c) $f(z) = ze^z$
(d) $f(z) = ze^{z^*}$
(e) $f(z) = e^z \sin(z)$.

(2) Show that $u(x,y)$ is harmonic and find corresponding $v(x,y)$ function, such that $f(x,y) = u(x,y) + iv(x,y)$ is analytic:

(a) $u(x,y) = 2y - 2xy$
(b) $u(x,y) = \sinh(x)\cos(y)$
(c) $u(x,y) = y^3 - 3x^2 y - 6y$
(d) $u(x,y) = \dfrac{y}{x^2 + y^2}$.

(3) Evaluate $\int_L z^2 dz$ along the indicated paths Figure 12.13:

(a) path (a)
(b) path (b)
(c) path (c)
(d) from $x = -1$ to $x = 1$ along the real axis.

(4) For each of the following functions, find the first few terms of the Laurent series:

(a) $f(z) = \dfrac{1}{1+z}$ at (i) $0 \le |z| < 1$ and (ii) $1 < |z| < \infty$

(b) $f(z) = \dfrac{1}{z^2(2-z)}$ at (i) $0 < |z| < 2$, (ii) $z = 2$, and (iii) $2 < |z| < \infty$

(c) $f(z) = z^2 \sin\left(\dfrac{1}{z}\right)$ in $0 < |z| < \infty$

(d) $f(z) = \dfrac{1}{z(z^2+2)}$ for (i) $0 < |z| < \sqrt{2}$ and (ii) $\sqrt{2} < |z| < \infty$.

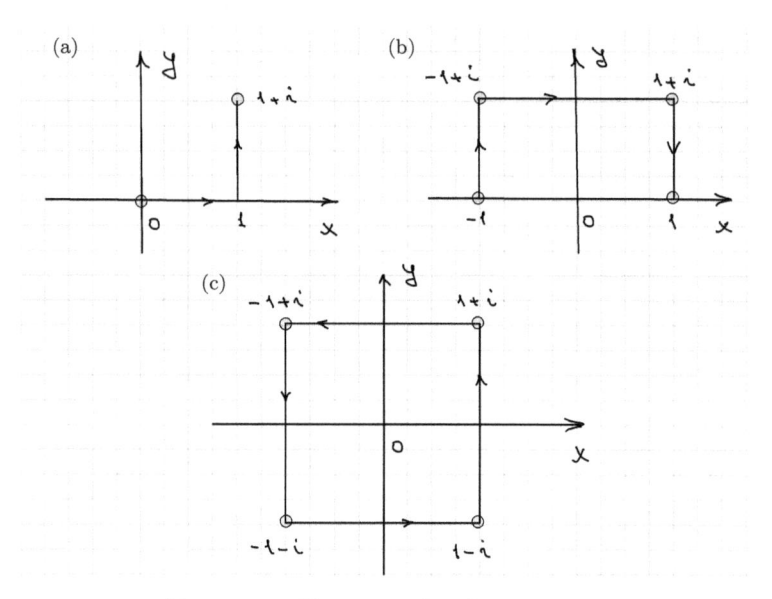

Fig. 12.13. Various paths of integration.

(5) Find the residue of the functions:

(a) $f(z) = \dfrac{1}{z(1-z)}$

(b) $f(z) = \dfrac{1 - \cos(z)}{z^n}$ for $n = 3, 4, 5$

(c) $f(z) = \sinh(z)/z^4$

(d) $f(z) = \dfrac{\sin(z)}{(z^2 + 1)^2}$.

(6) By using one of methods discussed in Section 12.5.2, evaluate the following definite integrals:

(a) $\displaystyle\int_0^{2\pi} \dfrac{1}{3 - 2\sin\theta}\, d\theta$

(b) $\displaystyle\int_{-\infty}^{\infty} \dfrac{x^2}{x^4 + 1}\, dx$

(c) $\displaystyle\int_{-\infty}^{\infty} \dfrac{x^2}{(x^2 + 1)^2(x^2 + 4)}\, dx$

(d) $\displaystyle\int_{-\infty}^{\infty} \dfrac{\cos(2x)}{(x^2 + 1)^2}\, dx$.

Answers and solutions

(1) (a) $u(x,y) = \cos(x)\cosh(y) + \sin(x)\cosh(y)$,
$v(x,y) = \cos(x)\sinh(y) - \sin(x)\sinh(y)$, analytic.

(b) $u(x,y) = x^3 - 3xy^2$, $v(x,y) = 3x^2y - y^3$, analytic.

(c) $u(x,y) = xe^x \cos(y) - ye^x \sin(y)$,
$v(x,y) = xe^x \sin(y) + ye^x \cos(y)$, analytic.

(d) $u(x,y) = xe^x \cos(y) - ye^x \sin(y)$,
$v(x,y) = xe^x \sin(y) + ye^x \cos(y)$, not analytic.

(e) $u(x,y) = e^x \cdot \sin(x) \cdot \cosh(y)$, $v(x,y) = e^x \cdot \cos(x) \cdot \sinh(y)$, analytic.

(2) (a) harmonic, $v(x,y) = x^2 - 2x - y^2$.

(b) harmonic, $v(x,y) = \cosh(x)\sin(y)$.

(c) harmonic, $x^3 - 3xy^2 - 6x$.

(d) harmonic, $v(x,y) = \dfrac{x}{x^2 + y^2}$.

(3) (a) $-\dfrac{2}{3} + i\dfrac{2}{3}$.

(b) $\dfrac{2}{3}$.

(c) 0.

(d) $\dfrac{2}{3}$.

(4) (a) (i) $1 - z + z^2 - z^3 + z^4 - z^5 + \cdots$.

(ii) $\dfrac{1}{z} - \dfrac{1}{z^2} + \dfrac{1}{z^3} - \dfrac{1}{z^4} + \dfrac{1}{z^5} - \cdots$.

(b) (i) $\dfrac{1}{2z^2} + \dfrac{1}{4z} + \dfrac{1}{8} + \dfrac{z}{16} + \dfrac{z^2}{32} + \dfrac{z^3}{64} + \cdots$.

(ii) $\dfrac{1}{4(z-2)} + \dfrac{1}{4} - \dfrac{3(z-2)}{16} + \dfrac{1}{8}(z-2)^2 - \dfrac{5}{64}(z-2)^3 + \dfrac{3}{64}(z-2)^4 - \cdots$.

(iii) $-\dfrac{1}{z^3} - \dfrac{2}{z^4} - \dfrac{4}{z^4} - \dfrac{8}{z^6} - \dfrac{16}{z^7} - \cdots$.

(c) $z - \dfrac{1}{6z} + \dfrac{1}{120z^3} - \dfrac{1}{5040z^5} + \cdots$.

(d) (i) $\dfrac{1}{2z} - \dfrac{z}{4} + \dfrac{z^3}{8} - \dfrac{z^5}{16} + \cdots$.

(ii) $\dfrac{1}{z^3} - \dfrac{2}{z^5} + \dfrac{4}{z^7} - \dfrac{8}{z^9} + \dfrac{16}{z^{11}} + \cdots$.

(5) (a) $\mathrm{Res}\left(\dfrac{1}{z(1-z)}, z=0\right) = 1$, $\mathrm{Res}\left(\dfrac{1}{z(1-z)}, z=1\right) = -1$.

(b) $\mathrm{Res}\left(\dfrac{1-\cos(z)}{z^n}, z=0\right) : \dfrac{1}{2}(n=3), 0(n=4), -\dfrac{1}{24}(n=5)$.

(c) $\mathrm{Res}\left(\dfrac{\sinh(z)}{z^4}, z=0\right) = \dfrac{1}{6}$.

(d) $\mathrm{Res}\left(\dfrac{\sin(z)}{(z^2+1)^2}, z=\pm i\right) = -\dfrac{1}{4e}$.

(6) (a) $2\pi/\sqrt{5}$.
(b) $\pi/\sqrt{2}$.
(c) $\pi/18$.
(d) $3\pi/(2e^2)$.

Index

A

Abel's theorem, 435
absolutely convergent series, 434
affine coordinate system, 65
Airy integral, 458
alternating series, 14
analytic function, 418

B

Bessel identity, 227
Bessel inequality, 227
brachistochrone problem, 382

C

Cartesian coordinate system, 65
Cauchy's integral formula, 432
Cauchy's integral test, 10
Cauchy's integral theorem, 427
Cauchy's root test, 7
Cauchy–Hadamard theorem, 20
Cauchy–Hadamard formula, 436
Cauchy–Riemann conditions, 418
chain rule for partial differentiation, 142
cofactor, 110
complex conjugate, 39
complex conjugate matrix, 107
contour integral, 424

convexity, 415
Cramer's rule, 118
curl, 92

D

d'Alembert's ration test, 8
de Moivre's formula, 45
Dirac's delta function, 332
Dirichlet conditions, 233
divergence, 92

E

essential singularity, 445
Euler's Basel problem, 13
Euler's formula, 40
Euler's method, undetermined coefficients, 293
Euler–Lagrange equation, 390
Euler–Poisson equation, 395

F

Fermat's principle, 407
Fredholm Alternative, 372
Fresnel integrals, 456
Frobenius series, 302
functional, 383
fundamental lemma of the calculus of variations, 388

G

geometric series, 2
gradient, 86
Green's function, 331
Green's theorem, 202

H

harmonic series, 5
Heaviside cover-up method, 440
Hermitian matrix, 107
homogeneous system, 117
hyperbolic functions, 51

I

identity matrix, 105
infinite sequence, 1
infinite series, 2
inverse matrix, 106
isolated singularity, 442
isoperimetric problem, 400

J

Jacobian, 209

K

Kummer's transformation, 15

L

Lagrange multipliers, 157
Lagrange reminder, 23
Lagrange's method variation of parameters, 285
Laurent series, 437
Legendre polynomials, 224
line integral, first kind, 174
line integral, second kind, 179
linear independence, 62
Liouville's formula, 282
Liouville's theorem, 433

M

Maclaurin series, 24
main comparison test, 7
matrix transpose, 107
minor, 110

N

necessary condition for a local extremum, 388
non-homogeneous system, 117

O

orthogonal matrix, 107
orthogonality condition, 72
orthonormal set, 222

P

Parseval's identity, 228
partial derivative, 138
Picard's theorem, 445
pole of order n, 444
positive series, 6
power series, 19
Principal Value integral, 454
principle of least action, 405

R

Raabe's test, 9
removable singular point, 443
residue, 444
residue theorem, 447
Riemann zeta function, 13

S

series convergent, 2
series reminder, 3
simple zero, 443
Sturm–Liouville problem, 373
symmetric matrix, 107

T

Taylor series, 23, 149, 437
total differential, 141
trace, 107

W

Weierstrass's theorem, 228
WKB approximation, 310
Wronskian, 281

Z

zero of order n, 443
zero vector, 59

www.ingramcontent.com/pod-product-compliance
Lightning Source LLC
LaVergne TN
LVHW011603060925
820435LV00022B/195